GENES AND PROTEINS
IN ONCOGENESIS

P & S BIOMEDICAL SCIENCES SYMPOSIA Series

HENRY J. VOGEL, Editor
College of Physicians and Surgeons
Columbia University
New York, New York

Henry J. Vogel (Editor). *Nucleic Acid–Protein Recognition*, 1977

Arthur Karlin, Virginia M. Tennyson, and Henry J. Vogel (Editors). *Neuronal Information Transfer*, 1978

Benvenuto Pernis and Henry J. Vogel (Editors). *Cells of Immunoglobulin Synthesis*, 1979

Benvenuto Pernis and Henry J. Vogel (Editors). *Regulatory T Lymphocytes*, 1980

Georgiana Jagiello and Henry J. Vogel (Editors). *Bioregulators of Reproduction*, 1981

Hymie L. Nossel and Henry J. Vogel (Editors). *Pathobiology of the Endothelial Cell*, 1982

I. Bernard Weinstein and Henry J. Vogel (Editors). *Genes and Proteins in Oncogenesis*, 1983

GENES AND PROTEINS IN ONCOGENESIS

Edited by

I. BERNARD WEINSTEIN
HENRY J. VOGEL

College of Physicians and Surgeons
Columbia University
New York, New York

1983

ACADEMIC PRESS

A Subsidiary of Harcourt Brace Jovanovich, Publishers

New York   London
Paris   San Diego   San Francisco   São Paulo   Sydney   Tokyo   Toronto

COPYRIGHT © 1983 BY ACADEMIC PRESS, INC.
ALL RIGHTS RESERVED.
NO PART OF THIS PUBLICATION MAY BE REPRODUCED OR
TRANSMITTED IN ANY FORM OR BY ANY MEANS, ELECTRONIC
OR MECHANICAL, INCLUDING PHOTOCOPY, RECORDING, OR ANY
INFORMATION STORAGE AND RETRIEVAL SYSTEM, WITHOUT
PERMISSION IN WRITING FROM THE PUBLISHER.

ACADEMIC PRESS, INC.
111 Fifth Avenue, New York, New York 10003

United Kingdom Edition published by
ACADEMIC PRESS, INC. (LONDON) LTD.
24/28 Oval Road, London NW1 7DX

Library of Congress Cataloging in Publication Data
Main entry under title:

Genes and proteins in oncogenesis.

(P & S biomedical sciences symposia series)
Includes index.
1. Carcinogenesis--Congresses. 2. Gene
expression--Congresses. 3. Tumor proteins--
Congresses. 4. Mutagenesis--Congresses. 5. Co-
carcinogens--Congresses. I. Weinstein, I. Bernard,
Date. II. Vogel, Henry James, Date.
III. Series.
RC268.5.G43 1983   616.99'2042   83-7046
ISBN 0-12-742420-2

PRINTED IN THE UNITED STATES OF AMERICA

83 84 85 86   9 8 7 6 5 4 3 2 1

Contents

List of Participants	xiii
Preface	xxi

OPENING ADDRESS

Opening Address: Genes and Proteins in Oncogenesis
HOWARD M. TEMIN

Introduction	3
Evolution of Highly Oncogenic Retroviruses	5
Multistage Carcinogenesis	10
Other Carcinogenic Agents	10
Conclusions	11
References	12

PART I. DNA MODIFICATION OF CARCINOGENS

Conformational Changes in DNA Induced by Chemical Carcinogens
DEZIDER GRUNBERGER AND REGINA M. SANTELLA

Introduction	13
Conformation of AAF- and AF-Modified DNA	14
Z-DNA Conformation in Alternating Purine–Pyrimidine Polymers	23
Conclusions	36
References	38

Studies on DNA Damage and Mutagenesis by Ultraviolet Light
KWOK MING LO, DOUGLAS E. BRASH, WILLIAM A. FRANKLIN, JUDITH A. LIPPKE, AND WILLIAM A. HASELTINE

Introduction	41
Discussion	43
Summary	53
References	53

Membrane-Mediated Chromosomal Damage

P. A. CERUTTI, I. EMERIT, AND P. AMSTAD

Introduction	55
The Clastogenic Action in PMA in Human Lymphocytes Is Membrane-Mediated	59
Concluding Remarks	64
References	65

Molecular Mechanism of Mutagenesis by the Ultimate Carcinogen N-Acetoxy-N-2-Acetylaminofluorene

ROBERT P. P. FUCHS, NICOLE KOFFEL-SCHWARTZ, AND MICHAEL P. DAUNE

Introduction	69
Materials and Methods	70
Results	72
Discussion	76
Summary	80
References	81

Nitrated Polycyclic Aromatic Hydrocarbons: A Newly Recognized Group of Ubiquitous Environmental Agents

HERBERT S. ROSENKRANZ, ELENA C. McCOY, AND ROBERT MERMELSTEIN

Text	83
References	93

PART II. MULTISTAGE EVENTS

Cellular Events in Multistage Carcinogenesis

I. BERNARD WEINSTEIN, ANN HOROWITZ, ALAN JEFFREY, AND VESNA IVANOVIC

Multistage Carcinogenesis	99
Mechanism of Action of Tumor Promoters	101
Promoter-like Activities of Polycyclic Aromatic Hydrocarbons	106
Chromosomal Effects and Activated Oxygen	107
Summary	108
References	109

Potent New Mutagens from Pyrolysates of Proteins and New Naturally Occurring Potent Tumor Promoters

HIROTA FUJIKI AND TAKASHI SUGIMURA

Introduction	111
New Mutagens from Pyrolysates of Amino Acids, Proteins, and Proteinaceous Foods	112
New Naturally Occurring Potent Tumor Promoters	114
Indole Alkaloids: Dihydroteleocidin B, Teleocidin, and Lyngbyatoxin A	116
In Vivo Carcinogenicity Test with Dihydroteleocidin B, Teleocidin, and Lyngbyatoxin A	118
Polyacetates: Aplysiatoxin and Debromoaplysiatoxin	119
References	120

Multistage Skin Tumor Promotion in Mouse Skin: Critical Protein Changes During Tumor Promotion

T. J. SLAGA AND K. G. NELSON

Introduction	125
Tumor Promotion	126
Inhibitors of Tumor Promotion	128
Critical Protein Changes During Tumor Promotion	130
Multistage Promotion	133
Protein Changes in Papillomas and Squamous Cell Carcinomas	138
Conclusion	140
References	141

Tumor Promotion in the Skin of Hairless Mice by Halogenated Aromatic Hydrocarbons

ALAN POLAND, JOYCE KNUTSON, EDWARD GLOVER, AND ANDREW KENDE

Introduction	143
Epidermal Changes Produced by Halogenated Aromatic Hydrocarbons	146
References	160

PART III. GROWTH FACTORS AND RECEPTORS

Transforming Growth Factors Produced by Viral-Transformed and Human Tumor Cells

GEORGE J. TODARO, HANS MARQUARDT, DANIEL R. TWARDZIK, FRED H. REYNOLDS, JR., AND JOHN R. STEPHENSON

Introduction	165
EGF Binding by Retrovirus-Transformed and Human Tumor Cells	166

Transforming Growth Factor Production 168
Purification of Rat, Mouse, and Human TGFs 170
Rat and Human TGF Amino Acid Sequences 175
Tyrosine Phosphorylation of the EGF Receptor 176
Conclusions 180
References 181

Receptor-Mediated Endocytosis of Peptide Hormones and Macromolecules in Cultured Cells

MARK C. WILLINGHAM AND IRA H. PASTAN

Introduction 183
The Morphologic Structures of Endocytosis on the Cell Surface 183
Intracellular Fate of the Receptosome 191
Compartmentalization within the Golgi System 192
Pathologic Processes: Toxins and Viruses 193
The Physiological Significance of Coated Pit Endocytosis 193
Summary 194
References 194

Thyroid Dependence of Neoplastic Transformation

DUANE L. GUERNSEY, CARMIA BOREK, PAUL B. FISHER, AND ISIDORE S. EDELMAN

Introduction 197
Historical Background 198
Thyroid Hormone Modulation of Neoplastic Transformation in Cell Culture 199
Thyroid Hormone Modulation of Adenovirus Transformation of Cells in Culture 204
T_3 Nuclear Receptors in Normal and Transformed Cells 206
Conclusion 208
References 208

PART IV. VIRAL ONCOGENES

A Novel Retrovirus, Adult T-Cell Leukemia Virus (ATLV): Characterization and Relation to Malignancy

YORIO HINUMA

Introduction 213
Properties of ATLV 214
Relation of ATLV to ATL 219
Concluding Remarks 221
References 222

The v-*myc* and c-*myc* Genes: Roles in Oncogenesis

WILLIAM S. HAYWARD

Introduction	225
The c-*myc* Gene	227
Activation of c-*myc* by ALV	228
Conclusions	229
References	231

Homologous to the *mos* Transforming Gene of Moloney Sarcoma Virus

ALESIA WOODWORTH, MARIANNE OSKARSSON, DONALD G. BLAIR, MARY LOU McGEADY, MICHAEL TAINSKY, AND GEORGE F. VANDE WOUDE

Introduction	233
Results	234
References	240

Enhanced Expression and Amplification of a Cellular Oncogene in Human Tumors

U. ROVIGATTI AND S. M. ASTRIN

Introduction	241
Discussion	248
References	250

Cellular DNA Sequences Involved in Chemical Carcinogenesis

PAUL T. KIRSCHMEIER, SEBASTIANO GATTONI-CELLI, ARCHIBALD S. PERKINS, AND I. BERNARD WEINSTEIN

Introduction: Genes Involved in Carcinogenesis	253
Studies of Cellular onc Genes in Normal and Transformed Murine Cells	254
Studies on the Expression of Poly(A)$^+$ RNA-Containing LTR-Like Sequences	255
Development of an Expression and Transduction Vector	260
Transduction with MULV Helper Virus	262
Conclusion	264
References	265

PART V. GENE AMPLIFICATION AND TRANSFECTION

Gene Amplification and Methotrexate Resistance in Cultured Animal Cells

ROBERT T. SCHIMKE, PETER C. BROWN, RANDAL N. JOHNSTON, BRIAN MARIANI, AND THEA TLSTY

Introduction	269
The Mechanism of Gene Amplification	270
Single-Step Selection for Methotrexate Resistance	272
Pretreatment Regimens and Enhanced Frequencies of MTX Resistance	276
DHFR Gene Amplification under Nonselecting Conditions	277
Discussion	279
References	282

Identification, Isolation, and Characterization of Three Distinct Human Transforming Genes

M. WIGLER, M. GOLDFARB, K. SHIMIZU, M. PERUCHO, Y. SUARD, O. FASANO, E. TAPAROWSKY, AND J. FOGH

Detecting Human Transforming Genes	285
Isolating Transforming Genes from Human Sources	287
The Relationship of Human Transforming Genes to Known Retroviral Oncogenes	288
Activation of Normal Cellular Genes into Transforming Genes	290
Conclusion	290
References	291

PART VI. TRANSFORMATION-RELATED PROTEINS AND TRANSCRIPTS

Do Variant SV40 Sequences Have a Role in the Maintenance of the Oncogenic Transformed Phenotype?

ROBERT POLLACK, CAROL PRIVES, AND JAMES MANLEY

Introduction	295
SV40 Viral Genes and Transformation	297
Initial Hypothesis	299
References	302

Protein Kinases Encoded by Avian Sarcoma Viruses and Some Related Normal Cellular Proteins

R. L. ERIKSON, TONA M. GILMER, ELEANOR ERIKSON, AND J. G. FOULKES

Introduction	305
Results and Discussion	306
References	320

The Control of Cellular Transcripts in Transformed Cells

ARNOLD J. LEVINE, TED SCHUTZBANK, AND ROBIN ROBINSON

Introduction	323
Results	324
Discussion	333
References	335

PART VII. NUCLEAR MATRIX, CYTOSKELETON, AND MITOCHONDRIA IN CARCINOGENESIS

The Concept of DNA Rearrangement in Carcinogenesis: The Nuclear Matrix

DONALD S. COFFEY

Introduction	339
DNA Rearrangement in Cancer	339
DNA Rearrangement in Normal Development and Expression of the Oncogene	342
Nuclear Structure and Function	344
References	347

A Point Mutation in β-Actin Gene and Neoplastic Transformation

TAKEO KAKUNAGA, JOHN LEAVITT, HIROSHI HAMADA, AND TADASHI HIRAKAWA

Introduction	351
A Variant Form of Actin Is Present in a Chemically Transformed Human Cell Line	352
A Variant Form of Actin Encoded by the β-Actin Gene That Has a Point Mutation in Its Structural Gene	355
Additional Change in the β-Actin Molecule in a Variant of the Transformed Cells	359
Incremental Increases in the Malignancy Associated with the Increased Abnormality of Mutated β-Actin Molecules	363

A Hypothesis on the Role of Actin Mutation in the Neoplastic Transformation 364
References 366

Studies of Mitochondria in Carcinoma Cells with Rhodamine-123

LAN BO CHEN, THEODORE J. LAMPIDIS,
SAMUEL D. BERNAL, KAREN K. NADAKAVUKAREN, AND
IAN C. SUMMERHAYES

Introduction 369
Prolonged Retention of Rhodamine-123 by Carcinogen-Transformed
    Epithelial Cells and Carcinoma-Derived Cells 370
Selective Toxicity of Rhodamine-123 in Carcinoma Cells in Vitro 375
Rhodamine-123 Selectivity Reduces Clonal Growth in Carcinoma Cells in
    Vitro 377
Anticarcinoma Activity in Vivo of Rhodamine-123 and the Combination
    Therapy with 2-Deoxyglucose 380
Measurement of Rhodamine-123 Uptake by Carcinoma Cells 382
Basis for Rhodamine-123 Toxicity 384
Summary 386
References 386

Index 389

List of Participants

AL SEDAIRY, S. T., Department of Pathology, College of Physicians and Surgeons, Columbia University, New York, New York 10032

AMSTAD, P., Swiss Institute for Experimental Cancer Research, CH-1066 Epalinges s/Lausanne, Switzerland

ANDERSON, DEBORAH, Dana-Farber Cancer Institute, 44 Binney Street, Boston, Massachusetts 02115

ASTRIN, S. M., Institute for Cancer Research, Fox Chase Cancer Center, Philadelphia, Pennsylvania 19111

BACKER, JOSEPH M., Institute for Cancer Research, College of Physicians and Surgeons, Columbia University, New York, New York 10032

BERNAL, SAMUEL D., Dana-Farber Cancer Institute, 44 Binney Street, Boston, Massachusetts 02115

BLAIR, DONALD G, Laboratory of Molecular Oncology, National Cancer Institute, National Institutes of Health, Bethesda, Maryland 20205

BOREK, CARMIA, Department of Pathology, College of Physicians and Surgeons, Columbia University, New York, New York 10032

BRASH, DOUGLAS E., Dana-Farber Cancer Institute, 44 Binney Street, Boston, Massachusetts 02115

BRAUNHUT, SUSAN J., Department of Physiology, Harvard Medical School, Boston, Massachusetts 02115

BROWN, PETER C., Department of Biological Sciences, Stanford University, Stanford, California 94305

BUCHHAGEN, DOROTHY, Department of Microbiology and Immunology, Downstate Medical Center, State University of New York, New York, New York 11203

CARLINO, JOSEPH A., Department of Pathology, College of Physicians and Surgeons, Columbia University, New York, New York 10032

CERUTTI, PETER A., Swiss Institute for Experimental Cancer Research, CH-1066 Epalinges s/Lausanne, Switzerland

CHAGANTI, R. S. K., Memorial Sloan-Kettering Cancer Center, 1275 York Avenue, New York, New York 10021

CHEN, LAN BO, Dana-Farber Cancer Institute, 44 Binney Street, Boston Massachusetts 02115

COFFEY, DONALD S., Johns Hopkins Hospital, Baltimore, Maryland 21205

CONJALKA, MICHAEL, Memorial Sloan-Kettering Cancer Center, 1275 York Avenue, New York, New York 10021

DAUNE, MICHAEL P., Institut de Biologie Moléculaire et Cellulaire, 15, Rue Descartes, 67084 Strasbourg, France

DEMPSEY, ROY, Department of Pathology, Tufts University School of Medicine, Boston, Massachusetts 02111
DOLAN, KEVIN P., Institute for Cancer Research, College of Physicians and Surgeons, Columbia University, New York, New York 10032
DOWNES, BARBARA A., Department of Pathology, College of Physicians and Surgeons, Columbia University, New York, New York 10032
EDELMAN, ISIDORE S., Department of Biochemistry, College of Physicians and Surgeons, Columbia University, New York, New York 10032
EISENBERG, MAX, Department of Biochemistry, College of Physicians and Surgeons, Columbia University, New York, New York 10032
ELLIS, LELAND, Department of Anatomy and Cell Biology, College of Physicians and Surgeons, Columbia University, New York, New York 10032
EMERIT, I., Laboratory of Experimental Cytogenetics, Institut Biomédical des Cordeliers, Université Pierre et Marie Curie, Paris, France
ERIKSON, ELEANOR, Department of Pathology, University of Colorado Medical School, 4200 East 9th Avenue, Denver, Colorado 80262
ERIKSON, RAYMOND L., Department of Pathology, University of Colorado Medical School, 4200 East 9th Avenue, Denver, Colorado 80262
FAIRBANKS, K., Department of Medicine, College of Physicians and Surgeons, Columbia University, New York, New York 10032
FARINAS, ELIZABETH M., Department of Pathology, College of Physicians and Surgeons, Columbia University, New York, New York 10032
FARRELL, MICHAEL, Dana-Farber Cancer Institute, 44 Binney Street, Boston, Massachusetts 02115
FASANO, O., Cold Spring Harbor Laboratory, Cold Spring Harbor, New York 11724
FISHER, PAUL B., Department of Microbiology, College of Physicians and Surgeons, Columbia University, New York, New York 10032
FOGH, J., Sloan-Kettering Institute for Cancer Research, Rye, New York 10580
FOULKES, J. G., Center for Cancer Research, Massachusetts Institute of Technology, Cambridge, Massachusetts 02139
FRANKLIN, WILLIAM A., Dana-Farber Cancer Institute, 44 Binney Street, Boston, Massachusetts 02115
FRIEND, CHARLOTTE, Center for Experimental Biology, Mt. Sinai School of Medicine, New York, New York 10029
FUCHS, ROBERT P. P., Institut de Biologie Moléculaire et Cellulaire, 15, Rue Descartes, 67084 Strasbourg, France
FUJIKI, HIROTA, National Cancer Center Research Institute, Tsukiji, Chuo-ku, Tokyo 104, Japan
GATTONI-CELLI, SEBASTIANO, Institute for Cancer Research, College of Physicians and Surgeons, Columbia University, New York, New York 10032
GEARD, CHARLES R., Department of Radiology, College of Physicians and Surgeons, Columbia University, New York, New York 10032
GILMER, TONA M., Department of Pathology, University of Colorado Medical School, 4200 East 9th Avenue, Denver, Colorado 80262

LIST OF PARTICIPANTS

GINSBERG, HAROLD S., Department of Microbiology, College of Physicians and Surgeons, Columbia University, New York, New York 10032
GLOVER, EDWARD, McArdle Laboratory for Cancer Research, University of Wisconsin, Madison, Wisconsin 53706
GODFREY, MAURICE, Department of Pathology, College of Physicians and Surgeons, Columbia University, New York, New York 10032
GODMAN, GABRIEL C., Department of Pathology, College of Physicians and Surgeons, Columbia University, New York, New York 10032
GOLDBERGER, ROBERT F., Office of the Provost, Columbia University, New York, New York 10027
GOLDFARB, M., Cold Spring Harbor Laboratory, Cold Spring Harbor, New York 11724
GREATON, CYNTHIA J., Department of Pathology, College of Physicians and Surgeons, Columbia University, New York, New York 10032
GREGORY, TANIA, Department of Pathology, College of Physicians and Surgeons, Columbia University, New York, New York 10032
GRUNBERGER, DEZIDER, Institute for Cancer Research, College of Physicians and Surgeons, Columbia University, New York, New York 10032
GUERNSEY, DUANE L., Department of Physiology and Biophysics, University of Iowa, Iowa City, Iowa 52242
HAMADA, HIROSHI, Cell Genetics Section, National Cancer Institute, National Institutes of Health, Bethesda, Maryland 20205
HANAFUSA, HIDESABURO, The Rockefeller University, 1230 York Avenue, New York, New York 10021
HASELTINE, WILLIAM A., Dana-Farber Cancer Institute, 44 Binney Street, Boston, Massachusetts 02115
HAYWARD, WILLIAM, Sloan-Kettering Institute, 1275 York Avenue, New York, New York 10021
HINUMA, YORIO, Institute for Virus Research, Kyoto University, Kyoto 606, Japan
HIRAKAWA, TADASHI, Cell Genetics Section, National Cancer Institute, National Institutes of Health, Bethesda, Maryland 20205
HOCHSTADT, JOY, Catholic Medical Center, Woodhaven, New York 11421
HOROWITZ, ANN, Institute of Cancer Research, College of Physicians and Surgeons, Columbia University, New York, New York 10032
HSIAO, WENDY, Institute for Cancer Research, College of Physicians and Surgeons, Columbia University, New York, New York 10032
IVANOVIC, VESNA, Institute for Cancer Research, College of Physicians and Surgeons, Columbia University, New York, New York 10032
JEFFREY, ALAN, Institute for Cancer Research, College of Physicians and Surgeons, Columbia University, New York, New York 10032
JHANWAR, SURESH, Memorial Sloan-Kettering Cancer Center, 1275 York Avenue, New York, New York 10021
JIANG, S., Department of Pathology, Downstate Medical Center, State University of New York, New York, New York 11203

JOHNSTON, RANDAL N., Department of Biological Sciences, Stanford University, Stanford, California 94305
KAKUNAGA, MARIKO, Cell Genetics Section, National Cancer Institute, National Institutes of Health, Bethesda, Maryland 20205
KAKUNAGA, TAKEO, Cell Genetics Section, National Cancer Institute, National Institutes of Health, Bethesda, Maryland 20205
KENDE, ANDREW, Department of Chemistry, University of Rochester, Rochester, New York 14627
KIRSCHMEIER, PAUL T., Institute for Cancer Research, College of Physicians and Surgeons, Columbia University, New York, New York 10032
KNUTSON, JOYCE, McArdle Laboratory for Cancer Research, University of Wisconsin, Madison, Wisconsin 53706
KOFFEL-SCHWARTZ, NICOLE, Institut de Biologie Moléculaire et Cellulaire, 15, Rue Descartes, 67084 Strasbourg, France
KORNBLUTH, RICHARD S., Department of Pathology, College of Physicians and Surgeons, Columbia University, New York, New York 10032
KOWALIK, SHARON, Department of Pathology, College of Physicians and Surgeons, Columbia University, New York, New York 10032
LAMBERT, MICHAEL, Institute for Cancer Research, College of Physicians and Surgeons, Columbia University, New York, New York 10032
LAMPIDIS, THEODORE J., Dana-Farber Cancer Institute, 44 Binney Street, Boston, Massachusetts 02115
LANDRETH, KENNETH, Sloan-Kettering Institute for Cancer Research, Rye, New York 10580
LAZZARINO, DEBORAH A., Department of Pathology, College of Physicians and Surgeons, Columbia University, New York, New York 10032
LEAVITT, JOHN, Linus Pauling Institute, Palo Alto, California 94306
LEVINE, ARNOLD J., Department of Microbiology, State University of New York at Stony Brook, School of Medicine, Stony Brook, New York 11794
LI, HENG-CHUN, Department of Biochemistry, Mt. Sinai School of Medicine, New York, New York 10029
LIPPKE, JUDITH A., Dana-Farber Cancer Institute, 44 Binney Street, Boston, Massachusetts 02115
LO, KWOK MING, Dana-Farber Cancer Institute, 44 Binney Street, Boston, Massachusetts 02115
LOPEZ, CECILIA A., Department of Pathology, College of Physicians and Surgeons, Columbia University, New York, New York 10032
MCCOY, ELENA C., Center for the Environmental Health Sciences, School of Medicine, Case Western Reserve University, Cleveland, Ohio 44106
MCGEADY, MARY LOU, Laboratory of Molecular Oncology, National Cancer Institute, National Institutes of Health, Bethesda, Maryland 20205
MANLEY, JAMES, Department of Biological Sciences, Columbia University, New York, New York 10027
MARIANI, BRIAN, Department of Biological Sciences, Stanford University, Stanford, California 94305
MARQUARDT, HANS, Laboratory of Viral Carcinogenesis, National Cancer In-

stitute, Frederick, Maryland 21701

MATSUI, M., Institute for Cancer Research, College of Physicians and Surgeons, Columbia University, New York, New York 10032

MELERA, PETER W., Sloan-Kettering Institute for Cancer Research, Rye, New York 10580

MERMELSTEIN, ROBERT, Joseph C. Wilson Center for Technology, Xerox Corporation, Webster, New York 14580

MILLER, DAVID L., New York State Institute for Basic Research, 1050 Forest Hill Road, Staten Island, New York 10314

MILLER, DOROTHY A., Department of Human Genetics and Development, College of Physicians and Surgeons, Columbia University, New York, New York 10032

MILLER, ORLANDO J., Department of Human Genetics and Human Development, College of Physicians and Surgeons, Columbia University, New York, New York 10032

MUNK, GARY B., Department of Pathology, College of Physicians and Surgeons, Columbia University, New York, New York 10032

NADAKAVUKAREN, KAREN K., Dana-Farber Cancer Institute, 44 Binney Street, Boston, Massachusetts 02115

NELSON, K. G., Department of Pathology, University of North Carolina, Chapel Hill, North Carolina 27514

NICOLAIDES, MARIA N., Department of Pathology, College of Physicians and Surgeons, Columbia University, New York, New York 10032

OSKARSSON, MARIANNE, Laboratory of Molecular Oncology, National Cancer Institute, National Institutes of Health, Bethesda, Maryland 20205

OSMAN, MOHAMED M., Department of Pathology, College of Physicians and Surgeons, Columbia University, New York, New York 10032

OZZELLO, LUCIANO, Department of Pathology, College of Physicians and Surgeons, Columbia University, New York, New York 10032

PASTAN, IRA H., Laboratory of Molecular Biology, National Cancer Institute, National Institutes of Health, Bethesda, Maryland 20205

PENMAN, SHELDON, Department of Biology, Massachusetts Institute of Technology, Cambridge, Massachusetts 02139

PERKINS, ARCHIBALD S., Institute for Cancer Research, College of Physicians and Surgeons, Columbia University, New York, New York 10032

PERNIS, BENVENUTO, Department of Microbiology, College of Physicians and Surgeons, Columbia University, New York, New York 10032

PERUCHO, M., Cold Spring Harbor Laboratory, Cold Spring Harbor, New York 11724

PFENNINGER, KARL H., Department of Anatomy and Cell Biology, College of Physicians and Surgeons, Columbia University, New York, New York 10032

PICCININI, LINDA A., Department of Pathology, College of Physicians and Surgeons, Columbia University, New York, New York 10032

PIROLLO, KATHLEEN, Wistar Institute, Philadelphia, Pennsylvania 19104

POLAND, ALAN, McArdle Laboratory for Cancer Research, University of Wisconsin, Madison, Wisconsin 53706

POLLACK, ROBERT E., Office of the Dean, Columbia College, Columbia University, New York, New York 10027

PRINCIPATO, MARY ANN, Department of Pathology, College of Physicians and Surgeons, Columbia University, New York, New York 10032

PRIVES, CAROL, Department of Biological Sciences, Columbia University, New York, New York 10027

REYNOLDS, FRED H., Jr., Laboratory of Viral Carcinogenesis, National Cancer Institute, Frederick, Maryland 21701

RIZZOLO, PHILIP V., Department of Medicine, College of Physicians and Surgeons, Columbia University, New York, New York 10032

ROBINSON, ROBIN, Department of Microbiology, State University of New York at Stony Brook, School of Medicine, Stony Brook, New York 11794

RODI, DIANE J., Institute for Cancer Research, College of Physicians and Surgeons, Columbia University, New York, New York 10032

ROSENKRANZ, HERBERT S., Center for the Environmental Sciences, School of Medicine, Case Western Reserve University, Cleveland, Ohio 44106

ROVIGATTI, UGO, Division of Virology, St. Jude Children's Research Hospital, P.O. Box 318, Memphis, Tennessee 38101

RUDDLE, NANCY H., Department of Epidemiology, Yale University Medical School, New Haven, Connecticut 06510

SAMUELS, R., Department of Pathology, College of Physicians and Surgeons, Columbia University, New York, New York 10032

SANTELLA, REGINA M., Institute for Cancer Research, College of Physicians and Surgeons, Columbia University, New York, New York 10032

SCHIMKE, ROBERT T., Department of Biological Sciences, Stanford University, Stanford, California 94305

SCHUTZBANK, TED, Department of Microbiology, State University of New York at Stony Brook, School of Medicine, Stony Brook, New York 11794

SHIMIZU, K., Cold Spring Harbor Laboratory, Cold Spring Harbor, New York 11724

SIMMONS, TOBIANNE, Department of Pathology, Saint Anthony's Hospital, Woodhaven, New York 11421

SLAGA, THOMAS J., University of Texas System Cancer Center, Science Park, Research Division, P.O. Drawer 389, Smithville, Texas 78957

SLATE, DORIS, Pfizer Central Research, Groton, Connecticut 06340

SLOVIN, SUSAN F., Department of Pathology, Tufts University School of Medicine, Boston, Massachusetts 02111

STEPHENSON, JOHN R., Laboratory of Viral Carcinogenesis, National Cancer Institute, Frederick, Maryland 21701

SUARD, Y., Cold Spring Harbor Laboratory, Cold Spring Harbor, New York 11724

SUGIMURA, TAKASHI, National Cancer Center Research Institute, Tsukiji, Chuo-ku, Tokyo 104, Japan

SUMMERHAYES, IAN C., Dana-Farber Cancer Institute, 44 Binney Street, Boston, Massachusetts 02115

TAINSKY, MICHAEL, Laboratory of Molecular Oncology, National Cancer Institute, National Institutes of Health, Bethesda, Maryland 20205

TAPAROWSKY, E., Cold Spring Harbor Laboratory, Cold Spring Harbor, New York 11724

TAPLEY, DONALD F., Office of the Dean, College of Physicians and Surgeons, Columbia University, New York, New York 10032

TEMIN, HOWARD M., McArdle Laboratory, University of Wisconsin, Madison, Wisconsin 53706

TLSTY, THEA, Department of Biological Sciences, Stanford University, Stanford, California 94305

TODARO, GEORGE J., Laboratory of Viral Carcinogenesis, National Cancer Institute, Frederick, Maryland 21701

TOLIDJIAN, BETTY, Department of Pathology, College of Physicians and Surgeons, Columbia University, New York, New York 10032

TOWLE, LAURIE R., Department of Radiology, College of Physicians and Surgeons, Columbia University, New York, New York 10032

TSE-ENG, DORIS, Department of Pathology, College of Physicians and Surgeons, Columbia University, New York, New York 10032

TWARDZIK, DANIEL, Laboratory of Viral Carcinogenesis, National Cancer Institute, Frederick, Maryland 21701

VANDE WOUDE, GEORGE F., Laboratory of Molecular Oncology, National Cancer Institute, National Institutes of Health, Bethesda, Maryland 20205

VOGEL, HENRY J., Department of Pathology, College of Physicians and Surgeons, Columbia University, New York, New York 10032

VOGEL, RUTH H., Department of Pathology, College of Physicians and Surgeons, Columbia University, New York, New York 10032

WEINSTEIN, I. BERNARD, Institute for Cancer Research, College of Physicians and Surgeons, Columbia University, New York, New York 10032

WIGLER, MICHAEL, Cold Spring Harbor Laboratory, Cold Spring Harbor, New York 11724

WILLINGHAM, MARK C., Laboratory of Molecular Biology, National Cancer Institute, National Institutes of Health, Bethesda, Maryland 20205

WOODWORTH, ALESIA, Laboratory of Molecular Oncology, National Cancer Institute, National Institutes of Health, Bethesda, Maryland 20205

XUE, BIN, Department of Pathology, New York University Medical Center, New York, New York 10016

YAKURA, HIDETAKA, Memorial Sloan-Kettering Cancer Center, 1275 York Avenue, New York, New York 10021

YEN, ANDREW, Department of Physiology and Biophysics, University of Iowa, Iowa City, Iowa 52242

Preface

Environmental factors in carcinogenesis have been the subject of biomedical studies for over two hundred years. Sir Percival Pott gave his classical description of cancer of the scrotum in chimney sweeps in 1775, when the medical faculty of King's College, one of the roots of the College of Physicians and Surgeons, was eight years old. Genetic factors have been considered since at least 1914, when Boveri proposed that carcinogens produce chromosomal alterations. Such environmental and genetic aspects, in recent years, have witnessed a remarkable flowering in the multistep theory of carcinogenesis, on the one hand, and, on the other, in the recognition of the role of specific virus-borne and cellular genes in the oncogenic process. The amalgamation of the environmental and genetic aspects, in molecular terms, is now in sight, if not at hand. In the immediate past, this field has generated a strikingly high level of excitement and promise.

A symposium on "Genes and Proteins in Oncogenesis" was held at Arden House, on the Harriman Campus of Columbia University, from June 4 through June 6, 1982. The meeting was the seventh of the P & S Biomedical Sciences Symposia. The proceedings are contained in this volume.

Dr. Donald F. Tapley, Dean of the College of Physicians and Surgeons (P & S), which sponsors the symposia, welcomed the participants.

To Dr. Howard M. Temin, we express our sincere thanks for his delivery of the opening address. We are grateful to Dr. Dezider Grunberger, Dr. Benvenuto Pernis, Dr. Hidesaburo Hanafusa, Dr. Charlotte Friend, Dr. Harold S. Ginsberg, and Dr. Karl H. Pfenninger, who acted as session chairmen.

Dr. Ruth H. Vogel's contributions to the organization of the symposium and the preparation of this volume are much appreciated.

I. Bernard Weinstein
Henry J. Vogel

OPENING ADDRESS

Opening Address: Genes and Proteins in Oncogenesis

HOWARD M. TEMIN

McArdle Laboratory
University of Wisconsin
Madison, Wisconsin

INTRODUCTION

The vertebrate genome contains many different types of DNA sequences. These include coding sequences, controlling sequences, and movable genetic elements.

Retroviruses are vertebrate viruses that appear to have evolved from movable genetic elements (1,2). Viral oncogenes have evolved from proto-oncogenes (3). The evolution of some parts of the vertebrate genome, especially cDNA genes, may involve processes similar to those involved in the evolution of retroviruses, and the formation of active transforming genes in nonviral cancers may also involve processes similar to those involved in the formation of highly oncogenic retroviruses (Fig. 1) (4).

Movable genetic elements are defined as parts of a cell genome that are not essential, that are present in different locations in different cells or individuals, and that are capable of moving to another position in the cell genome (5). Those movable genetic elements that have been sequenced have been found to have a special sequence organization—direct repeat of cell DNA, inverted repeat of element DNA, unique sequences, inverted repeat of element DNA, and direct repeat of cell DNA. In addition, larger movable genetic elements often have a large terminal direct repeat. Remarkably, this same organization is found in retrovirus proviruses. Even more remarkably, the same terminal sequences (TG. . . . CA) are found in all retroviruses and in movable genetic elements of *Drosophila* and yeast (2).

The evolution of retroviruses from cellular movable genetic elements might have followed steps similar to these involved in the evo-

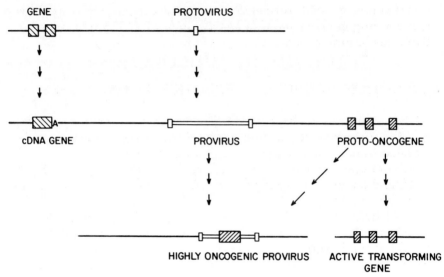

Fig. 1. Evolution of retroviruses, cDNA genes, and oncogenes (4).

lution of bacterial transposons, that is, transposition, deletion resulting in recombination, transposition, etc. (6,7).

Parts of cell DNA, not in retroviruses, might be capable of transposition through an RNA intermediate as in the formation of cDNA genes. cDNA genes are dispersed genes without intervening sequences, with a 3'-poly(A) track, and surrounded by a small direct repeat. cDNA genes or pseudogenes have been described for α-globin, Alu sequences, small nuclear RNA sequences, tubulin, and immunoglobulin light chains (8–15).

Highly oncogenic retroviruses differ from other retroviruses by the presence in their coding sequences of viral oncogenes. Thus, the genome of highly oncogenic retroviruses can be described as controlling sequences, oncogene coding sequences, and controlling sequences.

When a highly oncogenic retrovirus infects a sensitive cell in the appropriate state of differentiation, it transforms it. How restrictive this requirement is differs among retroviruses. Thus, Rous sarcoma virus (RSV) transforms most sensitive chicken cells that it infects, except hemopoetic cells, whereas reticuloendotheliosis virus strain T transforms only a very small fraction of spleen or bone marrow cells that it infects.

Three requirements are, therefore, necessary for neoplastic transformation by highly oncogenic retroviruses—formation of active transforming gene(s), presence of target(s) for product(s) of active transforming gene(s), and absence of inhibitor(s) of product(s) of active transforming gene(s).

EVOLUTION OF HIGHLY ONCOGENIC RETROVIRUSES

The sequences in highly oncogenic retroviruses coding for the transforming product(s) are known as the viral oncogene(s) and are descended from cellular sequences known as proto-oncogenes.

For example, we have been studying the viral oncogene of REV-T, *v-rel*, and its cellular counterpart, *c-rel*. *c-rel* consists of 5 or 6 exons spread over 25 kbp (kilobase pairs) of cell DNA and coding for a 4 kb poly(A) RNA (16). *v-rel* consists of 1.4 kbp with no intervening sequences and produces no separate poly(A) RNA, that is, its mRNA has 5' and 3' viral sequences and is 2.5 kb. In addition, there are some base pair changes between homologous sequences in *c-rel* and *v-rel*.

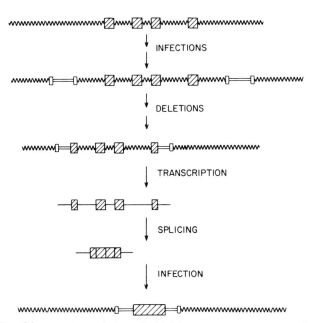

Fig. 2. Possible steps in evolution of a highly oncogenic retrovirus. The insertion and deletion of the 3' retrovirus could also happen at a later time or as a result of recombination of RNA molecules from a 5' recombinant and a retrovirus.

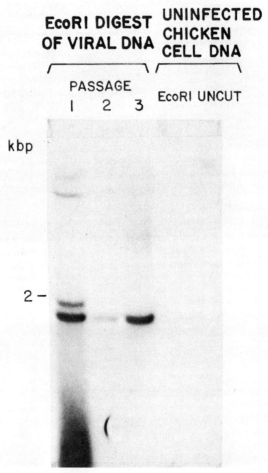

Fig. 3. Loss of intervening sequences after passage of virus recovered from pSNV-TKα-globinΔter(R1). [See maps in Fig. 4 and experimental details in (17).] Passages were at 3-day intervals, and Hirt supernatant DNA was isolated at the time of passage. 1.9-kpb band is parental insert; smaller insert is progeny virus insert with spliced globin (K. Shimotohno, personal communication).

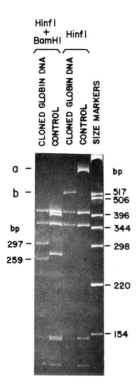
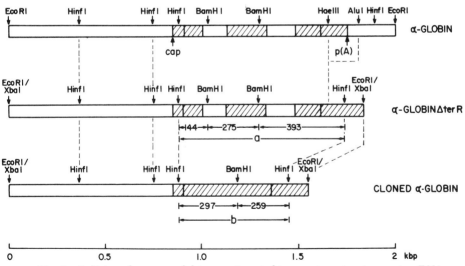

Fig. 4. Splicing of mouse α-globin gene inserted in an avian retrovirus vector. DNA clones were made from inserts in progeny virus as described in (17) and mapped by restriction enzyme digestion. Control is α-globinΔterR DNA (K. Shimotohno, personal communication).

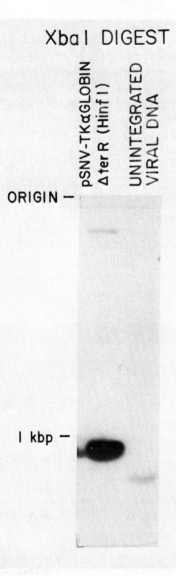

Fig. 5. Loss of intervening sequences after growth of virus containing globin DNA without its promoter. [See maps in Fig. 4 and experimental details in (17).] Control plasmid is in left lane. Hirt supernatant DNA isolated 3 days after infection with virus recovered 5 days after transfection is in the right lane (K. Shimotohno, personal communication).

Finally, c-rel RNA is present in most normal cells, but expression of v-rel mRNA in spleen cells results in leukemogenesis.

Thus, the evolution of c-rel to v-rel involved at least the following steps: recombination of c-rel with a nondefective reticuloendotheliosis virus with concomitant deletion of 5' and 3' cellular c-rel controlling sequences and some 3' c-rel sequences expressed in stable RNA, loss of c-rel intervening sequences, and base pair mutations in c-rel coding sequences (see Fig. 2). In addition, c-rel sequences have fused in phase with viral coding sequences.

Loss of intervening sequences may be a spontaneous process after recombination between retrovirus and cell sequences. For example, we inserted DNA of a mouse α-globin gene with two intervening sequences in a retrovirus vector, recovered progeny virus and showed that the two intervening sequences were precisely removed at a rate of about 10% per replication cycle (Figs. 3 and 4) (17). This splicing took place in the absence of the globin promoter and, therefore, took place in viral genomic RNA (Fig. 5).

However, even these steps from c-onc to v-onc are not sufficient to give a highly oncogenic retrovirus. We made a recombinant virus by

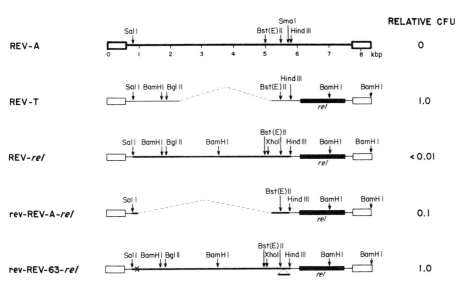

Fig. 6. Recombinants between REV-T and REV-A and revertants of recombinants between REV-T and REV-A. Maps of viral DNA and relative colony forming activity are shown. (See ref. 18.)

inserting *v-rel* into REV-A (Fig. 6). However, this recombinant virus is essentially nontransforming (18).

A highly transforming virus is found as a spontaneous variant of REV-A *rel* (0.1% frequency). The variants have deleted some REV-A coding sequences (Fig. 6). [Interestingly, in the case of a recombinant originally made with a mutant REV lacking the proper initiation codon for the *gag* gene (clone 63), an insertion or a mutation not detectable by restriction enzyme mapping allows recovery of transforming activity.]

Thus, we see that spontaneous mutational processes followed by selection drive the evolution of highly oncogenic retroviruses. Similar processes probably were involved in the original evolution of retroviruses themselves.

MULTISTAGE CARCINOGENESIS

All carcinogenesis appears to involve several steps. In chemical carcinogenesis, these steps are called initiation, promotion, and carcinogenesis or, alternatively, morphological transformation, anchorage-independent multiplication, and tumorigenesis (19).

The evolution of highly oncogenic retroviruses and oncogenesis by weakly oncogenic retroviruses also involves several steps. Can we make direct analogies between these processes? I think not yet. The virus work can indicate hypotheses to be tested in other systems and techniques to test them. However, each system will have to be worked out separately.

OTHER CARCINOGENIC AGENTS

The end result of the evolution of highly oncogenic retroviruses is insertional activation of possibly mutated cellular coding sequences. This activation comes about as a result of integration of virus DNA, deletion of virus and cell DNA, recombination of virus and cell DNA, processing of RNA transcripts, reverse transcription, and mutations. Is it likely that either this same result or these same processes or both are involved in carcinogenesis by other carcinogenic agents?

Alternatively, since there are other cellular requirements for transformation by highly oncogenic retroviruses (presence of targets and absence of inhibitors), insertional inactivation of a gene for an inhibitor could also be involved. In addition, gene amplification or other genetic processes might be important.

There are a wide variety of carcinogenic agents known in addition to highly oncogenic retroviruses: weakly oncogenic retroviruses, integrating DNA tumor viruses, nonintegrating DNA tumor viruses, irradiation, electrophilic chemicals, other chemicals, and solid bodies. For each of these we can consider what hypotheses the retrovirus model indicates to describe the carcinogenesis.

Some weakly oncogenic retroviruses cause tumors by formation by integration and deletion of a structure resembling that of a highly oncogenic retrovirus, that is, viral controlling sequences near cellular coding sequences, so-called downstream promotion or promoter insertion (20,21). However, additional steps appear necessary for tumor formation by these viruses (22).

Integrating DNA tumor viruses either code directly for an oncogenic product and/or cause some type of insertional mutation. Both mechanisms are related to the action of retroviruses.

Nonintegrating DNA tumor viruses cannot cause insertional mutations, but they might code for an oncogenic product. However, no such product has yet been described.

Irradiation and electrophilic chemicals might act directly on cell DNA to inactivate sequences coding for inhibitors of product(s) of active transforming genes or to activate proto-oncogenes by mutation. Alternatively, they might act indirectly to stimulate cellular mutational processes that would have these effects.

Other chemicals and solid bodies might also act indirectly to stimulate mutational processes or to alter the state of differentiation of cells and, thus, lead to production of targets of product(s) of active transforming genes or to the loss of inhibitors of such products.

CONCLUSIONS

It is now possible to describe possible steps in the evolution of highly oncogenic retroviruses and to duplicate many of these steps in the laboratory. Two major genetic mechanisms—insertional activation and base pair mutation—seem most important for oncogenesis by retroviruses. These mechanisms lead to formation of active transforming genes. These genes cause cancer when present in cells containing targets for their product and containing no inhibitors of the action of their product. The purification of some of these products enables direct study to be made of these products, their targets, and inhibitors of their action. Retroviruses, thus, provide an analogy and possible model for all carcinogenesis.

ACKNOWLEDGMENTS

The work in my laboratory is supported by PHS Grants CA-07175 and 22443. The author is an American Cancer Society Research Professor.

REFERENCES

1. Temin, H. M. (1970) *Perspect. Biol. Med.* **14**, 11–26.
2. Temin, H. M. (1980) *Cell* **21**, 599–60.
3. Bishop, J. M. (1981) *Cell* **23**, 5–6.
4. Temin, H. M. (1982) *J. Cell. Biochem.* **19**, 105–118.
5. Calos, M. P., and Miller, J. H. (1980) *Cell* **20**, 579–595.
6. Kleckner, N. (1977) *Cell* **11**, 11–23.
7. Shimotohno, K., and Temin, H. M. (1981) *Cold Spring Harbor Symp. Quant. Biol.* **45**, 719–730.
8. VanArsdell, S. W., Denison, R. A., Bernstein, L. B., Weiner, A. M., Manser, T., and Gesteland, R. F. (1981) *Cell* **26**, 11–17.
9. Jagadeeswaran, P., Forget, B. G., and Weissman, S. M. (1981) *Cell* **26**, 141–142.
10. Nishioka, Y., Leder, A., and Leder, P. (1980) *Proc. Natl. Acad. Sci. U.S.A.* **77**, 2806–2809.
11. Vanin, E. F., Goldberg, G. I., Tucker, P. W., and Smithies, O. (1980) *Nature (London)* **286**, 222–226.
12. Hollis, G. F., Hieter, P. A., McBride, O. W., Swan, D., and Leder, P. (1982) *Nature (London)* **296**, 321–325.
13. Lemischka, L., and Sharp, P. A. (1982) *Nature (London)* **300**, 330–335.
14. Wilde, C. D., Crowther, C. E., Cripe, T. P., Lee, M. G. S., and Cowan, N. J. (1982) *Science* **217**, 549–552.
15. Grimaldi, G., and Singer, M. F. (1982) *Proc. Natl. Acad. Sci. U.S.A.* **79**, 1497–1500.
16. Chen, I. S. Y., Wilhelmsen, K., and Temin, H. M. (1983) *J. Virol.* **45**, 105–118.
17. Shimotohno, K., and Temin, H. M. (1982) *Nature (London)* **299**, 265–268.
18. Chen, I. S. Y., and Temin, H. M. (1982) *Cell* **31**, 111–120.
19. Thomassen, D. G., and DeMars, (1982) *Cancer Res.* **42**, 4054–4063.
20. Hayward, W. S., Neel, B. G., and Astrin, S. M. (1981) *Nature (London)* **290**, 475–480.
21. Payne, G. S., Courtneidge, S. A., Crittenden, L. B., Fadly, A. M., Bishop, J. M., and Varmus, H. E. (1981) *Cell* **23**, 311–322.
22. Cooper, G. M., and Neiman, P. E. (1981) *Nature (London)* **292**, 857–858.

PART I

DNA MODIFICATION BY CARCINOGENS

Conformational Changes in DNA Induced by Chemical Carcinogens

DEZIDER GRUNBERGER AND REGINA M. SANTELLA

Cancer Center/Institute of Cancer Research, and
Department of Biochemistry and
Division of Environmental Sciences
Columbia University
College of Physicians and Surgeons
New York, New York

INTRODUCTION

In the past 20 years, it has become evident that most chemical carcinogens require metabolic activation *in vivo* to "ultimate carcinogens," i.e., the derivatives that are actually involved in the induction of tumors (1). Ultimate carcinogens are strong electrophiles and react nonenzymatically to form covalent bonds with nucleophilic groups in the nucleic acids and proteins of target cells (3). It appears that these reactions are critical to the initiation of the carcinogenic process.

Since there are many nucleophilic sites in cellular macromolecules, multiple DNA, RNA, and protein-bound derivatives of carcinogens are formed (4). DNA, however, has a unique role in the storage of genetic information and in mutation as a consequence of modification. Thus, the importance of DNA modification in the initiation of carcinogenesis becomes evident. To understand carcinogenesis at a molecular level, it is necessary to determine the complete chemical structure of the carcinogen–macromolecular adducts and the associated conformational changes in the target macromolecules and finally to relate these chemical and physical alterations to possible aberrations in the functional properties of the chemically modified macromolecules.

A complicating feature in attempting to relate chemical structure to functional effects is the fact that with most of these agents more than one type of nucleoside adduct in DNA is formed. This is also true with the simpler alkylating agents in which almost every nitrogen and oxygen residue of all the nucleic acid bases can be modified (5).

One of the extensively studied interactions of a metabolically activated carcinogen is the covalent attachment of the reactive metabolite of N-2-acetylaminofluorene (AAF) to nucleic acids (6). When N-hydroxy-AAF, the metabolically activated derivative, is administered to rats *in vivo* or to hepatocytes in culture, at least three types of DNA adducts are detected (7–9). The major product obtained from hydrolysates of the modified DNA is N-(deoxyguanosin-8-yl)AAF and a minor component is 3-(deoxyguanosin-N^2-yl)AAF (Fig. 1) (10). These two adducts are also formed after *in vitro* reaction of native DNA with the N-acetoxy derivative of AAF (11). Furthermore, *in vivo* a third product, the deacetylated C-8-guanosine adduct (G-AF), can be detected (9,12,13). Modification of guanosine residues at the C-8 position with AAF or AF, as well as at the N^2 position with AAF in DNA, could be associated with different conformational changes. The functional properties and the repair of the modified DNA could also be differently affected (8,9,14). Therefore, exploration of the conformation of all three types of adducts at the site of modification became necessary.

CONFORMATION OF AAF- AND AF-MODIFIED DNA

GUANOSINE MODIFIED AT C-8 WITH AAF

The major modified nucleoside formed after reaction of N-acetoxy-AAF with DNA *in vitro*, as well as *in vivo*, is the result of AAF binding at the C-8 position of guanosine (Fig. 1b). This site is not directly involved in base pairing with the complementary C residues during replication, transcription, or in codon–anticodon interaction in translation. However, we have found that the modified G residues cannot

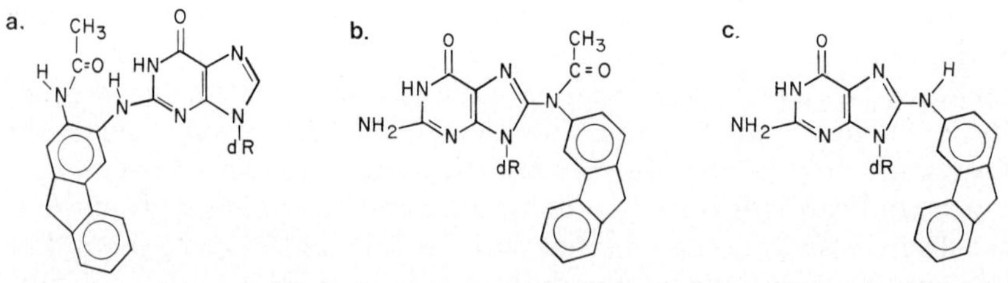

Fig. 1. Structures of DNA adducts: (a) 3-(deoxyguanosine-N^2-yl)-N-acetyl-2-aminofluorene; (b) N-(deoxyguanosine-8-yl)-N-acetyl-2-aminofluorene; (c) N-(deoxyguanosin-8-yl)-2-aminofluorene.

function normally in systems in which base pairing takes place (15–18). These results suggested that attachment of the bulky AAF to G residues is associated with gross conformational alterations of the nucleic acids.

Molecular model building of AAF-modified oligonucleotides, circular dichroism (CD), and nuclear magnetic resonance (NMR) spectra in combination with enzymatic studies provided considerable evidence that AAF modification of the C-8 position of G resulted in a dramatic conformational distortion at the site of modification (17,19,20). On the basis of these results, we proposed a specific, three-dimensional conformation called the "base displacement" model (21–23).

The first feature of this model is that the attachment of the AAF residue to the C-8-position of G is associated with a change in the glycosidic N-9—C-1' conformation from the anti form of nucleosides in nucleic acids with Watson–Crick geometry to the syn form (Fig. 2). Evidence for this is restricted to a study of molecular models of AAF-G, which indicates severe steric hindrance between AAF and the ribose or deoxyribose of the nucleoside unless the guanine base is rotated about the glycoside bond from the anti to syn conformation.

The second major feature of this model is that there is a stacking interaction between AAF and a base adjacent to the substituted G residue. These changes are best illustrated in a computer-generated stereoscopic display of a double stranded DNA fragment (Fig. 3). The computer display allows one to readily perform rotations around appropriate bond angles while obtaining a three-dimensional image of the molecular structure on a video screen. In the display, the modified

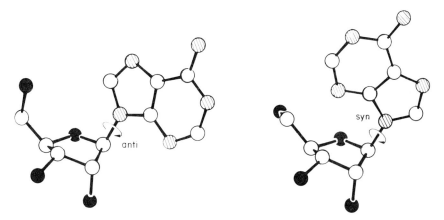

Fig. 2. Anti and syn conformations of nucleosides.

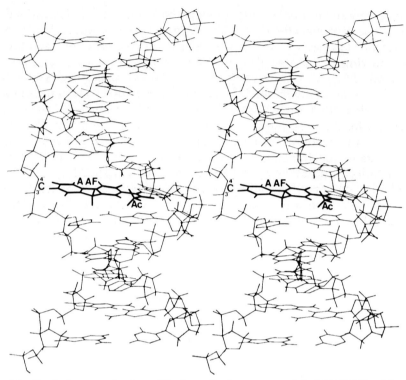

Fig. 3. A stereoscopic view of the base displacement model of AAF-DNA adducts. The guanine to which the AAF is attached has been rotated out of the helix, and the AAF moiety is inserted into the helix and stacked with neighboring bases. AC designates acetyl group of AAF. The cytosine (marked C) residue on the opposite strand would overlap with the AAF residue, therefore it has been removed and the 3′ and 4′ carbon atoms of the corresponding deoxyribose in the DNA backbone are indicated. In reality, this C probably rotates out from the helix to accommodate the AAF; the exact conformation is not known, although there is evidence that this region of DNA is "single-stranded." To view this image in stereo, two lenses of about 20 cm focal length should be mounted about 14 cm above the images and 6.5 cm apart.

base has been rotated around the glycosidic bond from anti to the syn conformation to avoid any steric hindrance. In addition, the planar fluorene ring system is inserted into the helix occupying the former position of the displaced guanine residue. It is also evident that the G residue displaced by AAF in the double helix cannot base pair with the C residue on the complementary strand and that, during the process of replication or transcription, no base pairing at this position could occur.

A similar model, called the "insertion–denaturation" model has been proposed by Fuchs and Daune (24,25). They have determined by electric dichroism the orientation of the covalently bound fluorene ring to the long axis of DNA. The results clearly indicated that the fluorene ring lies almost perpendicular to the helix axis, the angle being 80° (26).

These conformational changes cause a marked distortion of the double stranded DNA helix at sites of AAF modification. The best evidence for localized regions of denaturation comes from the studies on the susceptibility of AAF-modified DNA to digestion by S_1 nuclease, a single-strand specific endonuclease from *Aspergillus oryzae* (27,28). The results indicate that modification of native DNA by covalent attachment of AAF residues leads to localized regions of denaturation, because the modified regions are excised by S_1 nuclease (Fig. 4). The estimated number of base pairs destabilized by a single AAF modifi-

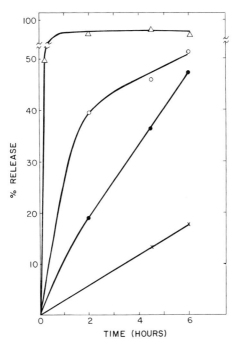

Fig. 4. Digestion of [^{14}C]AAF-modified DNA with S_1 nuclease. Native, heat-denatured, and [^{14}C]AAF-modified DNA were incubated with S_1 nuclease for the indicated periods of time. △, A_{260} released from heat-denatured DNA; X, A_{260} released from native DNA; ●, A_{260} released from modified DNA; ○, [^{14}C]AAF residues released from modified DNA. For details see ref. 28.

cation is in the range of 5 to 50, depending on the extent of modification of the DNA, the length of the nuclease digestion period, and the NaCl concentration during the digestion (29). Becasue attachment of AAF to G residues requires rotation of the base about the glycosidic bond and there is less hindrance to the rotation of bases in single-stranded than in double-stranded regions, it follows that single-stranded regions of nucleic acids are more susceptible to AAF modification than the double-stranded ones (20,30).

The base displacement model was confirmed by Broyde and Hingerty (31) using minimized potential energy calculations for dCpdG modified with AAF. For all computed conformers with energies less than 5 kcal/mol, the fluorene ring is stacked approximately co-planar with the neighboring cytidine; the G residue is syn, and the plane G is twisted perpendicular to the plane of the fluorene (Fig. 5). This is in agreement with our experimental findings that led to the base displacement model.

GUANOSINE MODIFIED AT N^2 WITH AAF

Administration of N-OH-AAF in rats results in about 15% of total AAF residues bound to the N^2 position of G (8,9) (Fig. 1a). It is inter-

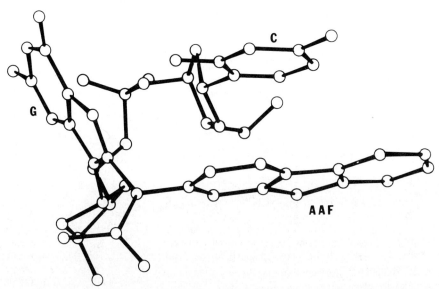

Fig. 5. The base-displaced lowest energy conformation of dCpdG-AAF. Deoxyriboses are both C-2′-endo and the guanine is in syn conformation. Based on ref. 31.

esting that this adduct is formed only when N-acetoxy AAF reacts with native DNA but not with single-stranded DNA. Nor is it formed in RNA *in vivo* or *in vitro*. Furthermore, it is not possible to modify DNA only in N^2 position of G, since a mixture of C-8 and N^2-G adducts are always obtained.

Because there are differences in the steric aspects connected with the modification of C-8 and N^2 positions of G, differences might also exist between the conformational distortions in the DNA helix associated with these two types of adducts. Following incubation of the modified DNA with S_1 nuclease, the undigested fraction was precipitated with cold ethanol (28). This material and the released oligonucleotides present in the supernatant fraction were then separately hydrolyzed to nucleosides and analyzed by Sephadex LH-20 column chromatography (Fig. 6). The profile of the undigested fraction of the DNA demonstrates a decrease in the ratio of the C-8 to N^2-guanine AAF adducts. In addition, the C-8, but not the N^2 adduct was detected in the fraction released by S_1 nuclease. Thus, the enzyme recognized

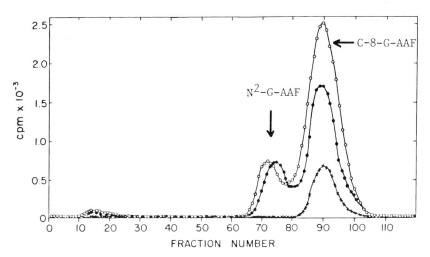

Fig. 6. Sephadex LH-20 column chromatography of nucleoside hydrolysates of the total, the *Neurospora crassa* nuclease-digested fraction, and the *N. crassa* nuclease-resistant fraction of [^{14}C]AAF-modified DNA. The [^{14}C]AAF-modified DNA was completely digested to nucleosides after heat denaturation. *N. crassa* nuclease digestion was performed in 10 mM Tris–HCl (pH 7.9), 100 mM NaCl, and 1 mM MgCl$_2$ for 3 hr at 37°C. The resistant fraction was precipitated with 2.5 vol of ethanol, collected by centrifugation, and hydrolyzed to nucleosides. (○), total [^{14}C]AAF-modified DNA; (●) *N. crassa* nuclease-resistant fraction; (X), *N. crassa* nuclease-digested fraction. Based on ref. 28.

only the C-8-G-AAF sites as single-stranded regions on the modified DNA.

In contrast to the C-8 adduct, substitution of AAF on the N^2 position of guanine does not produce a major change in conformation of the DNA helix. Although the precise conformation of the helix at the latter sites has not been determined, model-building studies indicate that the fluorene residue on the N^2-G could simply occupy the minor groove of the DNA without any distortion of the helix. Thus, the base displacement model may apply only to the C-8 and not to the N^2-G adduct of AAF.

GUANOSINE MODIFIED AT C-8 WITH AF

After *in vivo* administration of N-OH-AAF, a major fraction of the DNA-bound carcinogen in rat liver is deacetylated (7,9,13). This lesion has been identified as N-(deoxyguanosin -8-yl)-2-AF, and it is accumulated in the target tissues (Fig. 1c).

There are differences in biological properties between C-8-G-AFF and C-8-G-AF adducts in DNA, which may be related to different conformational alterations in DNA. The deacetylated adduct may cause less distortion in the molecule than the acetylated one. Scribner *et al.* (32) proposed that the AF residue can be accommodated more easily in the helix than in the acetylated residue. Evans *et al.* (33) investigated the conformation G-AF by proton magnetic resonance spectra. They found that the deacetylated adduct preferred the anti form in contrast to the acetylated form which existed preferentially in the syn conformation. Using space filling models, a possible structure of AF-modified DNA was proposed that had less distortion of the helix than that of the acetylated adduct.

Results from our laboratory have also suggested differences in conformation between AF and AAF modified DNAs. We have studied the conformation of the modified dimer, dApdG, as a model compound for DNA (34). The CD spectra (Fig. 7) show that attachment of AAF to the C-8 position of the deoxyguanosine residue induces large CD bands compared to the unmodified dimer. These results indicate a strong intramolecular interaction between fluorene and the adjacent base. With AF binding, the induced CD bands are only about one-third those seen in dApdG-AAF and suggests that there is some interaction between the AF residue and adjacent base, but less stacking than with AAF. The spectra in methanol are also shown in Fig. 7. With both samples, a large decrease in $[\Theta]$ values indicate diminished inter-

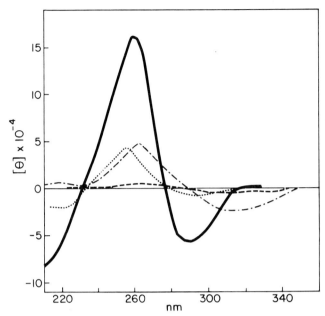

Fig. 7. Circular dichroism spectra of dApdG-AAF in 0.005 M phosphate buffer (pH 7) (–) and methanol (····) and of dApdG-AF in buffer (_._._.) and methanol (---). From ref. 34.

action between the fluorene residue and adjacent bases. Similar results are seen with increasing temperature which like methanol is known to break up stacking interactions (Fig. 8).

Proton magnetic resonance spectra of dApdG-AAF and dApdG-AF also indicated different interactions between adenine and AAF or AF residues (34). In the AF modified dimer, less stacking interaction occurs than in the AAF-modified sample.

These differences in the conformation of AF- and AAF-modified monomers and dimers could be indicative of differences in the conformation of these adducts in DNA. Kriek and Spelt (35) using S_1 nuclease digestion of DNA modified either by AAF or AF showed that the type of lesion induced by AF substitution is cleaved much more slowly than are those induced by AAF. These results suggest that the local regions of denaturation induced by AF substitution are smaller than those associated with AAF modification.

Some additional information has been obtained using antibodies to

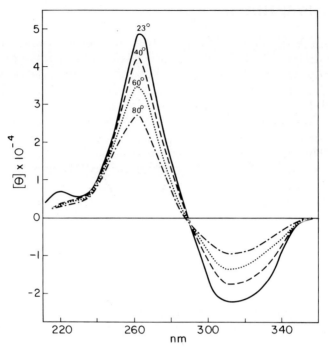

Fig. 8. Temperature dependence of the CD spectra of dApdG-AF in 0.005 M phosphate buffer. From ref. 34.

cytidine, which react with denatured but not native DNA (36). Sage *et al.* (37) found that these antibodies react weakly with AF-modified DNA, indicating little denaturation of the molecule. In contrast, these antibodies react well with AAF-modified DNA.

Taken together, the results with modified oligomers as well as with DNA suggest that the AAF and AF adducts induce different types of local conformational changes in DNA. Although modification of G residues with AAF causes large distortions of the DNA as a result of rotation of the base about the glycosidic bond from anti to syn conformation, such large conformational alterations may not be necessary when the bulky acetyl group is not present. It is possible that the different biological properties, especially the different rates of excision repair of AF and AAF residues in DNA, are related to the different types of local conformational changes induced in DNA by these residues.

Z-DNA CONFORMATION IN ALTERNATING PURINE-PYRIMIDINE POLYMERS

A recent X-ray crystallographic study of the self-complementary hexamer d(CpGpCpGpCpG) by Wang *et al.* (38) led to the discovery of a family of new DNA double helixes, termed Z-DNA (39,40). Z-DNA differs markedly from B-DNA in that it forms a left-handed helix with 12 base pairs per turn (Fig. 9). Further, the sugar-phosphate backbone follows a zig-zag course, thus the term Z-DNA. The repeat unit in Z-DNA is a dinucleotide, rather than a mononucleotide as in B-DNA. The guanosine bases are rotated about the glycosidic bond to the syn conformation, in contrast to the anti conformation in B-DNA (Fig. 2). The sugar conformations are also not the same, dG is the C-3'-endo and dC is C-2'-endo. A left-handed DNA conformation was also demonstrated by X-ray diffraction of crystals of d(CpGpCpG) and fibers of alternating G-C polymers (41–43).

It had been known for some time that spectroscopic properties of poly(dG-dC)·poly(dG-dC) in solution show major changes with increasing ionic strength. In low salt solutions (<2.5 M NaCl), circular dichroism (CD) (44,45), and Raman spectra (46) are characteristic of a B-type DNA. Increased concentrations of salts or ethanol induce an inversion of the CD spectrum and changes of the Raman spectrum indicative of a cooperative intramolecular transition (47,48). [^{31}P]NMR studies demonstrated that the (dG-dC)$_8$ duplex in high salt solutions has two different types of nucleotide and P conformations, including a set of glycosidic torsion angles and phosphodiester linkages that are different from B-DNA (49,50). All the spectroscopic data suggest that the high salt form of poly(dG-dC)·poly(dG-dC) in solution corresponds to the Z-type conformation elucidated by crystallographic analysis.

Construction of recombinant pBR322 derivatives containing stretches of up to 40 base pairs of dG-dC made it possible to show that (dG-dC) regions can exist in left-handed conformation when joined to a natural DNA segment (51). CD, [^{31}P] NMR, and Raman spectral studies have shown that a portion of this DNA exists in a Z-type conformation in 4.5–5.0 M NaCl (51,52). Moreover, a distorted junction occurs between a DNA region in a Z-type conformation and an adjacent region that previously assumed a B-type conformation (53).

Indirect evidence for the existence of Z-DNA *in vivo* comes from studies using antibodies to Z-DNA (54,55). These antibodies were shown to bind in reproducible patterns exclusively in interband re-

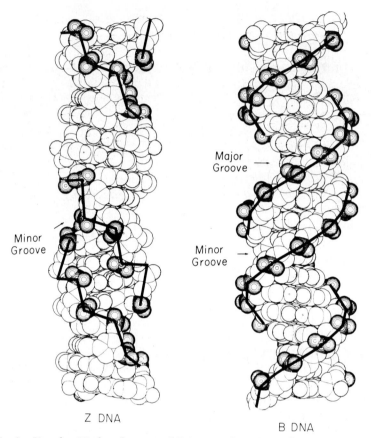

Fig. 9. Van der Waals sideviews of Z-DNA and B-DNA. The irregularity of the Z-DNA backbone is illustrated by the heavy lines that go from phosphate to phosphate residues along the chain. The minor groove in Z-DNA is quite deep extending to the axis of the double helix. In contrast, B-DNA has a smooth line connecting the phosphate groups and the two grooves, neither one of which extends into the helix axis of the molecule. (From ref. 40.)

gions of *Drosophila melanogaster* polytene chromosomes (56), which are associated with transcription of certain genes. Antibodies that react with Z-DNA were also found in the sera of mice with an autoimmune disease similar to human systemic lupus erythematosus. They occur spontaneously, but the immunogen is unknown.

As a direct consequence of the syn conformation of the deoxyguanosine residues in Z-DNA, the C-8 position is exposed on the outer surface of the molecule (Fig. 10). Since AAF modification of deoxyguano-

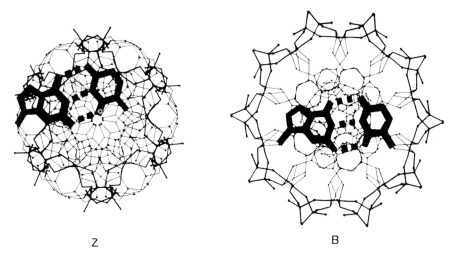

Fig. 10. End views of the regular idealized helical forms of Z- and B-DNA. Heavier lines are used for the phosphate-ribose backbone. A guanine-cytosine base pair is shown by shading. The difference in the positions of the base pairs is quite striking; they are near the center of B-DNA but at the periphery of Z-DNA. From ref. 40.

sine in DNA results in the rotation of the guanine from anti to the syn conformation, a similar change in conformation would be expected in poly(dG-dC)·poly(dG-dC), and it might stabilize the conformation of the polymer in the Z form. Modification of this position by AAF may, therefore, not require such drastic distortion of the conformation as is seen in B-DNA where the C-8 position is crowded inside the helix. In such a case, Z-DNA may be more susceptible to carcinogen modification than B-DNA. To explore these possibilities, we have investigated the conformational changes induced in poly(dG-dC)·poly(dG-dC) with AAF modification (57).

INDUCTION OF Z-TYPE CONFORMATION IN
Poly(dG-dC)·Poly(dG-dC) MODIFIED WITH AAF

Since CD spectra are particularly sensitive to changes in base orientations in polymers, it would seem well suited for studying DNA conformational alterations. We have used this method to investigate the effect of AFF modification on the conformation of poly(dG-dC)·poly(dG-dC). Modification was performed by the reaction of the polymer with N-acetoxy-AAF. In the polymer, the AFF residue binds only to the C-8 of guanosine. Figure 11 shows the CD spectra of

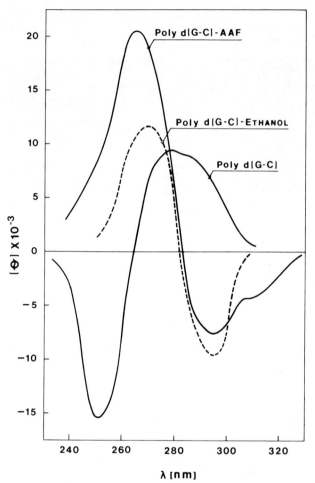

Fig. 11. CD spectra of poly(dG-dC)·poly(dG-dC). In 1 mM phosphate buffer in 60% ethanol and modified by AAF to an extent of 28% in 1 mM phosphate buffer. From ref. 57.

poly(dG-dC)·poly(dG-dC) in buffer and 60% ethanol. In aqueous solution, a B-DNA spectrum is seen, whereas in ethanol the CD is inverted and characteristic of Z-DNA. Also shown is a sample modified to the extent of 28% with AAF. With this high level of modification, the CD spectrum resembles that of Z-DNA. With lower levels of modification (3%), the sample still shows a CD characteristic of B-DNA, but is converted to that of Z-DNA at lower ethanol concentrations than for unmodified poly(dG-dC)·poly(dG-dC) (Fig. 12). Sage and Leng

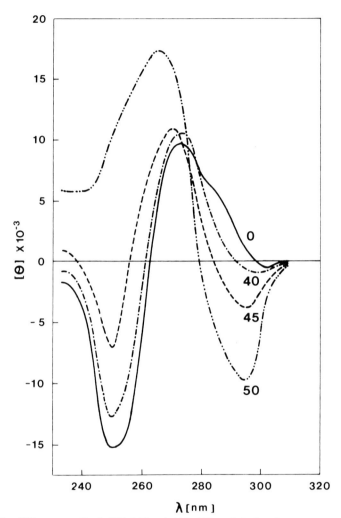

Fig. 12. CD spectra of poly(dG-dC)·poly(dG-dC) modified with AAF to an extent of 3% at various ethanol concentrations. (—), in 1 mM phosphate buffer; (—··—··), in 40% ethanol; (-----), in 45% ethanol; (—·—·), in 50% ethanol. From ref. 57.

(58) have also shown that, with low levels of AAF modification, the polymer undergoes the B to Z transition at lower ethanol concentrations. On the other hand, poly(dG)·poly(dC), a homopolymer that cannot undergo the B to Z transition (57), did not show any changes in CD spectra with high levels of AAF modification (Fig. 13).

To obtain additional information about the conformation of the

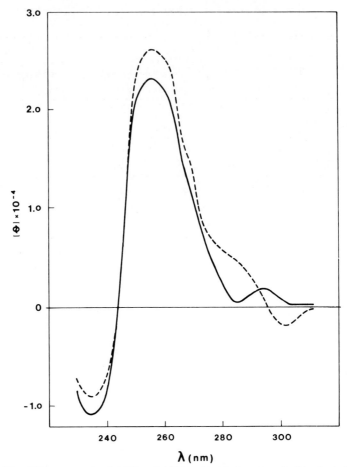

Fig. 13. CD spectra of poly(dG)·poly(dC) in 1 mM phosphate buffer. (—), Unmodified; (-----), modified to 17% with AAF. From ref. 57.

modified polymer, its susceptibility to S_1 nuclease digestion was determined. Table I shows that AAF-modified DNA and AAF-modified poly(dG)·poly(dC) were digested and, therefore, contain significant single-stranded regions. In contrast, poly(dG-dC)·poly(dG-dC) modified heavily with AAF was essentially resistant to S_1 nuclease digestion and must be double-stranded, proving that AAF modification did not induce localized regions of denaturation in this alternating purine-pyrimidine polymer.

Denatured sites in a double-stranded polymer can also be detected

TABLE I
Nuclease S_1 Digestion[a]

	Modification (%)	Digestion (%) Based on	
		(A_{260})	(cpm)
Native DNA	0	5	
Denatured DNA	0	92	
AAF-modified DNA	20	75	83
AAF-modified poly(dG-dC)·poly-(dG-dC)	28	11	3
AAF-modified poly-(dG)·poly(dC)	19	59	54

[a] Adapted from ref. 57.

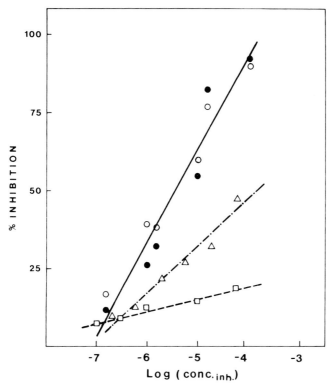

Fig. 14. Radioimmunoassay at nonequilibrium conditions in which the binding of purified anti-C antibodies to [^{3}H]dDNA was measured in the presence of various concentrations of dDNA (●), DNA-AAF 11% modified (○), poly(dG-dC)·poly(dG-dC)-AAF 21% modified (□), and poly(dG)·poly(dC)-AAF 5% modified (△). (Based on ref. 59.)

by radioimmunoassay using anticytidine antibodies (35). These antibodies react specifically with cytidine residues, which are accessible in the single-stranded regions of a polymer. They precipitate a tracer of ^{3}H-denatured DNA (dDNA), but not native DNA (59). Addition of a competitor for reaction with the antibodies, such as nonlabeled dDNA, inhibits the precipitation of the radioactive tracer. This is shown in Fig. 14 where addition of dDNA or DNA-AAF inhibit the precipitation of the [^{3}H]dDNA by competing for reaction with the antibodies. Poly(dG-dC)·poly(dG-dC) with a 21% modification level does not inhibit the binding of the antibodies to the tracer indicating that this polymer does not react with the antibodies and therefore has no significant single-stranded regions. In contrast, a sample of AAF modified poly(dG)·poly(dC) does react with the antibodies and thus contains some denatured regions.

In order to propose a model for the modified polymer potential energy calculations were performed on the dCpdG model system in collaboration with Drs. Suse Broyde and Brian Hingerty (59). A low energy conformation whose DNA backbone is very similar to that of the dCpdG segment of Z_1-DNA is shown (Fig. 15). The guanine is syn and approximately co-planar with cytidine, and the AAF residue is twisted nearly perpendicular to the G; the deoxyriboses are alternately C-3'-endo in guanosine and C-2'-endo in cytidine. This energy minimized

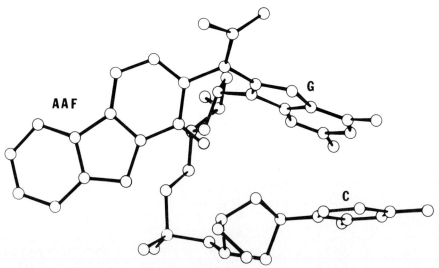

Fig. 15. Minimum energy conformation of dCpdG-AAF with DNA backbone torsion angles similar to dCpdG segment of Z-DNA. Based on ref. 59.

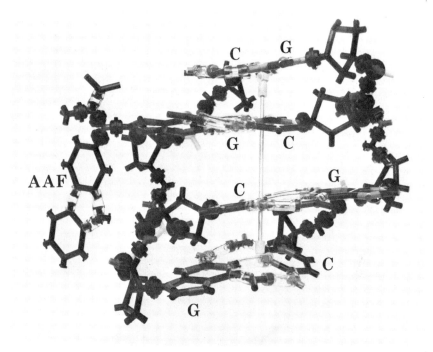

Fig. 16. Minimum energy conformation of dCpdG-AAF shown in Fig. 7 inserted into a Z_1-DNA tetrameric duplex of dGpdCpdGpdC. The AAF is linked externally to the Z-DNA helix and is mobile, with rotational flexibility about its long axis. Details in ref. 59.

structure readily fits into the Z-DNA helix as shown in Fig. 16 where the computed adduct of Fig. 15 is inserted into a model of the Z_1-DNA tetrameric duplex, dGpdCpdGpdC. The long axis of fluorene makes an angle of approximately 35° with the helix axis.

Taken together, the data with S_1 nuclease digestion and anticytidine antibodies seem to be in accord with the suggestion that the inverted CD spectra is indicative of a Z-DNA type conformation.

AAF MODIFIED Poly(dG-m^{5}dC)·Poly(dG-m^{5}dC) ADOPTS A
Z-TYPE CONFORMATION

The dinucleotide sequence m^{5}dC-dG occurs frequently in eukaryotic DNA and in many organisms it composes more than one-half of all dCpdG sequences (60). Furthermore, the presence of methylated sites within a structural gene has been implicated in the inhibition of transcription of certain eukaryotic genes (61). Recently, it has been shown that the methylated polynucleotide poly(dG-m^{5}C)·poly(dG-m^{5}C) undergoes a transition from B to Z form at much lower salt concentrations than required to convert the nonmethylated form (62). The Z form of this polymer has thus been shown to be stable under typical physiological conditions.

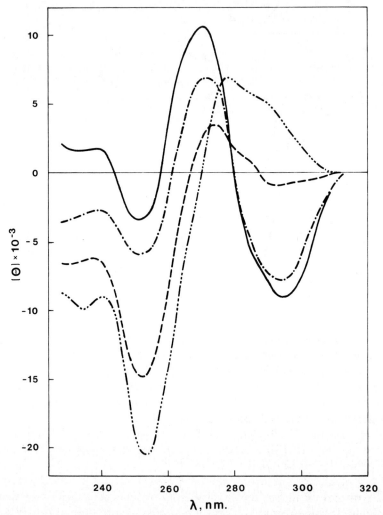

Fig. 17. CD spectra of poly(dG-m^{5}C)·poly(dG-m^{5}C) in 50 mM NaCl, 5 mM Tris pH 8.0, and with various levels of AAF modification. (–···–), control; (----), 6.5% modification; (–··–··–), 8.8 % modification; (–), 10.4% modification. From ref. 63.

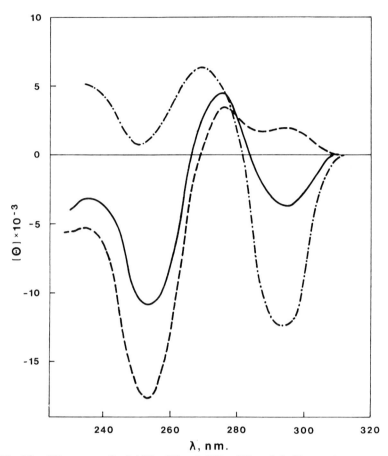

Fig. 18. CD spectra of poly(dG-m⁵C)·poly(dG-m⁵C) modified by AAF to an extent of 3.5% in various concentrations of Mg^{2+}. (---), No Mg^{2+}; (—), 0.2 mM Mg^{2+}; (-·-·), 0.4 mM of Mg^{2+}. All samples contained 50 mM NaCl, 5 mM Tris, pH 8.0. (From ref. 63.)

We have investigated the conformational changes of poly(dG-m⁵C)·poly(dG-m⁵C) with AAF modification (63). The CD spectra of the polymer bound with various levels of AAF is shown in Fig. 17. At modification levels above 6%, there begins to appear a negative band at 295 nm characteristic of Z-DNA. With 10.4% modification, the polymer is completely in the Z form. Poly(dG-m⁵C)·poly(dG-m⁵C) undergoes a B to Z transition with increasing concentrations of Mg^{2+} with a midpoint at 0.6 mM Mg^{2+}, compared to 0.7 M for the nonmethylated polymer. With low levels of AAF modification, it is possible to further decrease the amount of Mg^{2+} needed to induce the B to Z transition. A sample of poly(dG-m^{5}C)·poly(dG-m^{5}C) with 3.5% AAF modification has a midpoint of transition at 0.3 mM Mg^{2+} (Fig. 18). Thus, both methylation at the C-5 of cytosine and AAF modification at C-8 of guanosine favor induction of the Z conformation.

The susceptibility of this modified polymer to S_1 nuclease was also investigated. Figure 19 shows the level of digestion of the various polymers with time. Heat-denatured DNA is completely hydrolyzed by S_1 nuclease, whereas native DNA is quite resistant. A DNA sample modified to a level of 8.5% showed about 50% digestion after 3 hours. In contrast, a sample of poly(dG-m^{5}C)·poly(dG-m^{5}C) with 17.5% modification shows only 14% digestion after a 3-hour incubation with S_1 nuclease. These results are similar to the resistance of modified poly(dG-dC)·poly(dG-dC) to S_1 nuclease digestion and indicate that base pairing also remains intact when the methylated polymer is modified.

To investigate the possible preferential reactivity of the Z-type conformation the binding level of the various polymers with increasing levels of AAF was determined. The results shown in Fig. 20, indicate that the order of reactivity follows the ease of formation of the Z conformation: poly(dG-m^{5}C)·poly(dG-m^{5}C) > poly(dG-dC)·poly(dG-

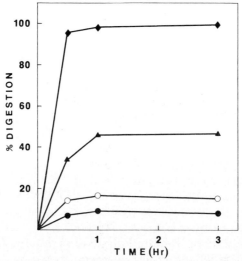

Fig. 19. Nuclease S_1 digestion of native calf thymus DNA (●), denatured calf thymus DNA (◆), poly(dG-m^{5}C)·poly(dG-m^{5}C) modified with AAF to an extent of 17.5% (○), and calf thymus DNA modified with AAF to an extent of 8.5% (▲). From ref. 63.

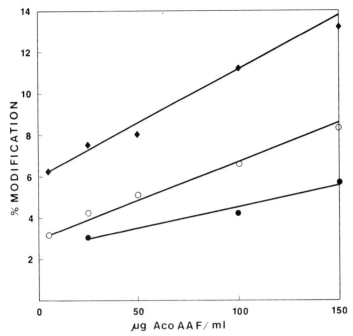

Fig. 20. The extent of binding of AAF to synthetic polydeoxyribonucleotides in 50 mM NaCl, 5 mM Tris, pH 8.0. Poly(dG)·poly(dC) (●); poly(dG-dC)·poly(dG-dC) (○); and poly(dG-m⁵C)·poly(dG-m⁵C) (◆). From ref. 63.

dC) > poly(dG)·poly(dC). The highest reactivity of the methylated polymer could be related to its ease of formation of Z-DNA. Nonmethylated poly(dG-dC)·poly(dG-dC) adopts the Z conformation next most readily and has the second highest level of reactivity, whereas poly(dG)·poly(dC) cannot form Z-DNA and has the lowest reactivity. A previous study has also shown that the alternating polymer poly(dG-dC)·poly(dG-dC) is 1.5 to 3 times more reactive than the homopolymer poly(dG)·poly(dC) under the same conditions (64).

It is not unexpected to find that AAF reacts more rapidly with the Z form of DNA than with the B form. In Z-DNA, the C-8 position of the guanine ring is located on the outside of the molecule where there are no structural constraints that would impede the access of a large molecule such as AAF. On the other hand, in B-DNA, the C-8 H atom of guanine is in Van der Waals contact with the ribose-phosphate chain and reaction can occur only if there is considerable distortion in the backbone. From a steric point of view, one would thus anticipate that the Z-DNA would have a higher rate of reactivity for AAF than B-DNA.

On the other hand, Pullman and his colleagues (65) have carried out studies of the electrostatic potential reactivity of the various components of the guanine ring and, in particular, have compared the potential at the C-8 position in both the Z and the B conformations. Their calculations indicate that a higher degree of reactivity is likely at the C-8 position of guanine in the B conformation than in the Z conformation. There are thus two opposing tendencies, one of which facilitates a higher reactivity for B-DNA due to electronegativity and the other a greater accessibility in Z-DNA. Our results suggest that the Z form of the molecule is more reactive to AAF and may indicate that the stereochemical factors are more important in this case than the reactivity factors.

CONCLUSIONS

Taken together, our data indicate two different conformations for AAF modified DNA: the base displacement and the Z-type, depending on whether the modified residues are in random or alternating purine-pyrimidine sequences. Table II summarizes the properties of these two types of polymers. In random-sequence DNA, the conformation with AAF modification is best represented by the base displacement model, which involves disruption of base pairing and covalent intercalation of the AAF residue. This partial denaturation of the DNA is detected by heat denaturation, by increased susceptibility to digestion by S_1 nuclease, and by increased reactivity with anti cytidine antibodies. The CD spectrum of AAF-modified DNA is essen-

TABLE II
Comparison of Properties of AAF-Modified DNA and Poly(dG-dC)·Poly(dG-dC)

Properties	Type of AAF-modified polymers	
	DNA	Poly(dG-dC)·poly(dG-dC)
Conformation	Base displacement	Z-DNA
CD spectra	B-DNA type	Inverted B-DNA
Base pairing	Disrupted	Intact
Nuclease S_1 susceptibility	Sensitive	Resistant
Anticytidine antibodies	Reactive	Nonreactive
Biological effect	Frame-shift mutation	Gene regulation?

tially that of B-form DNA. In contrast, alternating purine-pyrimidine sequences, when modified by AAF, adopt the Z conformation. Thus, modified poly(dG-dC)·poly(dG-dC) shows a CD spectrum characteristic of Z-form DNA (51). Although the deoxyguanosine residues adopt the syn conformation, as in the base displacement model, the base pairing remains intact in Z-DNA. This is indicated by the resistance of modified poly(dG-dC)·poly(dG-dC) to digestion with S_1 nuclease (51) and the lack of reactivity with anticytidine antibodies (54).

What is the biological significance of the conformational changes induced in DNA by AAF? One possibility is that, as a consequence of the altered conformation, the modified dG cannot base-pair with the complementary dC resulting in frame-shift mutation (Fuchs, this volume).

Another possibility is the altered expression of genetic information coded by the DNA. Proper expression requires the interaction of various proteins with specific regions of the DNA recognized by their base sequence and conformation (66). The binding of bulky carcinogens, such as AAF, distorts the conformation of DNA at specific sites and may block the stable interaction between a protein and the modified DNA.

A special case is the Z-type conformation. Methylation of the CpG sequence has been implicated in the inhibition of transcription of some eukaryotic genes (61). It is possible to assume that methylation induces the formation of Z-DNA segments, which then act as a conformational switch in the regulation of gene expression. Since AAF can also stabilize the Z-DNA conformation under physiological conditions, the proposal could be extended to chemical carcinogens. This carcinogen could inhibit gene expression by causing a conformational switch in a region that, under normal conditions, would be transcribed.

It has also been suggested that if there are regions within the eukaryotic genome containing Z-DNA, the regular structure of chromatin would be disrupted (67). If this is the case, then AAF would prevent formation of normal chromatin structure. All these suggestions, however, require further investigations.

ACKNOWLEDGMENTS

We thank Drs. I. B. Weinstein and A. Rich for helpful suggestions; Dr. A. Nordheim for a gift of poly(dG-m^{5}C)·poly(dG-m^{5}C), Dr. B. Erlanger for anti cytidine antibodies, and Ms. S. Allen for preparation of the manuscript.

This investigation was supported by PHS Grants CA 21111 and CA 31696 from the National Institutes of Health, DHHS.

REFERENCES

1. Miller, E. C. (1978) *Cancer Res.* **38**, 1479–1496.
2. Miller, E. C., and Miller, A. J. (1981) *Cancer Res.* **47**, 1055–1064.
3. Miller, E. C., and Miller, A. J. (1981) *Cancer Res.* **47**, 2327–2345.
4. Miller, A. J., and Miller, E. C. (1979) *In* "Environmental Carcinogenesis" (P. Emmelot and E. Kriek, eds.), pp. 25–50. Elsevier/North-Holland Biomedical Press, Amsterdam.
5. Singer, B. (1976) *Nature (London)* **264**, 333–339.
6. Kriek, E., and Westra, J. G. (1978) *In* "Chemical Carcinogens and DNA" (P. L. Grover, ed.), Vol. II, pp. 1–28. CRC Press, Boca Raton, Florida.
7. Irving, C. C., and Veazey, R. A. (1969) *Cancer Res.* **29**, 1799–1804.
8. Kriek, E. (1972) *Cancer Res.* **32**, 2042–2048.
9. Howard, P. C., Casciano, D. A., Beland, F. A., and Shaddock, J. G., Jr. (1981) *Carcinogenesis* **2**, 97–102.
10. Westra, J. G., Kriek, E., and Hittenhausen, H. (1976) *Chem.-Biol. Interact.* **15**, 149–164.
11. Kriek, E. (1965) *Biochim. Biophys. Acta* **20**, 793–799.
12. Westra, J. G., and Visser, A. (1979) *Cancer Lett.* **8**, 155–162.
13. Poirier, M. C., Williams, G. M., and Yuspa, S. H. (1980) *Mol. Pharmacol.* **18**, 581–587.
14. Beland, F. A., Dooley, K. L., and Casciano, D. A. (1979) *J. Chromatogr.* **174**, 177–186.
15. Grunberger, D., Nelson, J. H., Cantor, C. R., and Weinstein, I. B. (1970) *Proc. Natl. Acad. Sci. U.S.A.* **66**, 488–496.
16. Grunberger, D., and Weinstein, I. B. (1971) *J. Biol. Chem.* **246**, 1123–1128.
17. Grunberger, D., Blobstein, S. H., and Weinstein, I. B. (1974) *J. Mol. Biol.* **82**, 459–468.
18. Millette, R. L., and Fink, L. M. (1975) *Biochemistry* **14**, 1426–1431.
19. Nelson, J. H., Grunberger, D., Cantor, C. R., and Weinstein, I. B. (1971) *J. Mol. Biol.* **62**, 331–346.
20. Levine, A. F., Fink, L. M., Weinstein, I. B., and Grunberger, D. (1974) *Cancer Res.* **34**, 319–327.
21. Weinstein, I. B., and Grunberger, D. (1974) *In* "Chemical Carcinogenesis" (P. O. P. Ts'o and Y. DiPaolo, eds.), Part A, pp. 217–235. Dekker, New York.
22. Grunberger, D., and Weinstein, I. B. (1976) *In* "Biology of Radiation Carcinogenesis" (J. M. Yuhas, R. W. Tennant, and J. D. Regan, eds.), pp. 175–187. Raven Press, New York.
23. Grunberger, D., and Weinstein, I. B. (1978) *In* "Chemical Carcinogens and DNA" (P. L. Grover, eds.), Vol. II, pp. 59–93. CRC Press, Boca Raton, Florida.
24. Fuchs, R., and Daune, M. (1971) *FEBS Lett.* **14**, 206–208.
25. Fuchs, R., and Daune, M. (1972) *Biochemistry* **11**, 2659–2666.
26. Fuchs, R. P. P., Lefevre, J.-F., Pouyet, J., and Daune, M. P. (1976) *Biochemistry* **15**, 3347–3351.
27. Fuchs, R. P. P. (1975) *Nature (London)* **257**, 151–152.
28. Yamasaki, H., Pulkrabek, P., Grunberger, D., and Weinstein, I. B. (1977) *Cancer Res.* **37**, 3756–3760.
29. Yamasaki, H., Leffler, S., and Weinstein, I. B. (1977) *Cancer Res.* **37**, 684–691.
30. Fujimura, S., Grunberger, D., Carvajal, G., and Weinstein, I. B. (1972) *Biochemistry* **11**, 3629–3635.

31. Broyde, S., and Hingerty, B. E. (1982) *Chem.-Biol. Interact.* **40**, 113-119.
32. Scribner, J. D., Fisk, S. R., and Scribner, N. K. (1979) *Chem.-Biol. Interact.* **26**, 11-25.
33. Evans, F. E., Miller, D. W., and Beland, F. A. (1980) *Carcinogenesis* **1**, 955-959.
34. Santella, R. M., Kriek, E., and Grunberger, D. (1980) *Carcinogenesis* **1**, 897-902.
35. Kriek, E., and Spelt, C. E. (1979) *Cancer Lett.* **1**, 147-154.
36. Erlanger, B. R., and Beiser, S. M. (1964) *Proc. Natl. Acad. Sci. U.S.A.* **52**, 68-74.
37. Sage, E., Spodheim-Maurizot, M., Leng, M., Pascale, R. I. O., and Fuchs, R. P. P. (1979) *FEBS Lett.* **108**, 66-68.
38. Wang, A. H., Quigley, G. J., Kolpak, F. J., Crawford, J. L., van Boom, J. H., Van der Marel, G., and Rich, A. (1979) *Nature (London)* **282**, 680-686.
39. Crawford, J. L., Kolpack, F. J., Wang, A. H. J., Quigley, G. J., van Boom, J. H., Van der Marel, G., and Rich, A. (1980) *Proc. Natl. Acad. Sci. U.S.A.* **77**, 4016-4020.
40. Wang, A. H. J., Quigley, G. J., Kolpak, F. J., Van der Marel, G., Van Boom, J. A., and Rich, A. (1981) *Science* **211**, 171-176.
41. Drew, H., Takano, T., Tanaka, S., Itakura, K., and Dickerson, R. E. (1980) *Nature (London)* **286**, 567-573.
42. Arnott, S, Chandrasekaran, R., Birdsell, D. L., Leslie, A. G. W., and Ratliff, R. L. (1980) *Nature (London)* **283**, 743-745.
43. Dickerson, R. E., Drew, H. R., Conner, B. N., Wing, R. M., Fratini, A. V., and Kopta, M. L. (1982) *Science* **216**, 475-484.
44. Wells, R. D., Larson, J. E., Grant, R. C., Shortle, B. E., and Cantor, C. R. (1970) *J. Mol. Biol.* **54**, 465-497.
45. Pohl, F. M., and Jovin, T. M. (1972) *Mol. Biol.* **67**, 375-396.
46. Pohl, F. M., Ramade, A., and Stockburger, M. (1973) *Biochim. Biophys. Acta* **335**, 85-92.
47. Pohl, F. M. (1976) *Nature (London)* **260**, 365-366.
48. Thamann, T. J., Lord, R. C., Wang, A.H.-J., and Rich, A. (1981) *Nucleic Acids Res.* **9**, 5443-5457.
49. Patel, D. J., Canuel, L. L., and Pohl, F. M. (1979) *Proc. Natl. Acad. Sci. U.S.A.* **76**, 2508-2511.
50. Simpson, R. T., and Shindo, H. (1980) *Nucleic Acids Res.* **8**, 2093-2103.
51. Klysik, J., Stirdivant, S. M., Larson, J. E., and Hart, P. A. (1981) *Nature (London)* **290**, 671-677.
52. Zacharias, W., Larson, J. E., Klysik, J., Stirdivant, S. M., and Wells, R. M. (1982) *J. Biol. Chem.* **257**, 2775-2782.
53. Wartell, R. M., Klysik, J., Hiller, W., and Wells, R. D. (1982) *Proc. Natl. Acad. Sci. U.S.A.* **79**, 2549-2553.
54. Lafer, E. M., Moller, A., Nordheim, A., Stollar, B. D., and Rich, A. (1981) *Proc. Natl. Acad. Sci. U.S.A.* **78**, 3546-3550.
55. Malfoy, B., Hartmann, B., and Leng, M. (1981) *Nucleic Acids Res.* **9**, 5659-5669.
56. Nordheim, A., Pardue, M. L., Lafer, E. M., Moller, A., Stollar, B. D., and Rich, A. (1981) *Nature (London)* **294**, 417-422.
57. Santella, R. M., Grunberger, D., Weinstein, I. B., and Rich, A. (1981) *Proc. Natl. Acad. Sci. U.S.A.* **78**, 1451-1455.
58. Sage, E., and Leng, M. (1980) *Proc. Natl. Acad. Sci. U.S.A.* **77**, 4597-4601.
59. Santella, R. M., Grunberger, D., Broyde, S., and Hingerty, B. E. (1981) *Nucleic Acids Res.* **9**, 5459-5467.
60. Ehrlich, M., and Wang, R. Y. H. (1981) *Science* **212**, 1350-1357.
61. Razin, A., and Riggs, A. D. (1980) *Science* **210**, 604-610.

62. Behe, M., and Felsenfeld, G. (1981) *Proc. Natl. Acad. Sci. U.S.A.* **78**, 1619–1623.
63. Santella, R. M., Grunberger, D., Nordheim, A., and Rich, A. (1982) *Biochem. Biophys. Res. Commun.* **106**, 1226–1232.
64. Harvan, D. J., Hass, J. R., and Lieberman, M. W. (1977) *Chem.-Biol. Interact.* **17**, 203–210.
65. Zakrzewska, K., Lavery, R., Pullman, A., and Pullman, B. (1981) *Nucleic Acids Res.* **8**, 3717–3932.
66. Wells, R. D., Goodman, T. C., Hillen, W., Horn, G. T., Klein, R. D., Larson, J. E., Muller, U. R., Neuendorf, S. K., Panayotatos, N., and Stirdivant, S. M. (1980) *Prog. Nucleic Acids Res. Mol. Biol.* **24**, 167–267.
67. Nichol, J., Behe, M., and Felsenfeld, G. (1982) *Proc. Natl. Acad. Sci. U.S.A.* **79**, 1771–1775.

Studies on DNA Damage and Mutagenesis by Ultraviolet Light

KWOK MING LO
DOUGLAS E. BRASH
WILLIAM A. FRANKLIN
JUDITH A. LIPPKE
WILLIAM A. HASELTINE

Dana-Farber Cancer Institute
Harvard Medical School
Boston, Massachusetts

INTRODUCTION

Damage to DNA may have several biological consequences. Extensive DNA damage sometimes results in cell death. Some lethal damage may be repaired with no obvious consequence or may result in permanent phenotypic or genotypic alterations.

Most potent mutagens are agents that damage DNA as are many of the powerful carcinogens. One suspects that such carcinogens initiate the cancer process via permanent alterations in DNA structure. However, in no case has this supposition been proved. It is known that the tumor cell phenotype is heritable; the cancer cells give rise to cancer cells upon division. This elementary observation establishes only the property of phenotype stability, a stability that may be achieved via alterations in regulatory networks as well as by rearrangements or changes in the structure of DNA.

The uncertainty as to the fundamental character of the initiated tumor cell reflects ignorance of the events that follow DNA insult. Such events may be traced provided one has knowledge of (1) the relevant DNA alterations, (2) the enzymes that process such lesions, (3) regulatory pathways that respond to DNA damage, and (4) the genetic mechanisms that regulate cellular response to DNA damage. The challenging task that lies ahead is to map this difficult terrain and to provide detail to areas that are now blank.

Toward this end, we have initiated a series of experiments designed to trace the consequences of UV light irradiation. Ultraviolet light was selected for these studies as we believed that the chemistry of the biologically relevant lesion was known. DNA efficiently absorbs ultraviolet light in the range of 240–300 nm. Although solar radiation below 300 nanometers drops sharply due to absorption by atmospheric constituents, enough radiation reaches the surface to produce substantial damage to the DNA of human skin. Short-term effects—sunburn—as well as long-term effects—skin cancer and premature aging phenomena—can be attributed, at least in part, to such damage.

The chemical consequences of absorption of ultraviolet light by DNA have been explored by a number of workers. Progress to 1976 is nicely summarized in S. Y. Wang's book (1). Most of the ultraviolet light work has been done at a wavelength of 254 nm, very near the absorption maximum of DNA. However, recent evidence suggests that wavelengths between 350 and 400 nm may also result in significant biological effects (2).

Chemical analysis of irradiated DNA nucleosides and free bases have revealed a plethora of products. The major products observed in the DNA are the cyclobutane pyrimidine dimers T<>T, T<>C, C<>T, C<>C formed between adjacent pyrimidines (1). Other significant photoproducts include photohydrates of the pyrimidines and a product formed between thymine and cytosine called, by Wang, the Thy(6-4)Pyo (1).

In recent years, attention has been centered on the role of cyclobutane pyrimidine dimers. Physical and genetic evidence suggests that these lesions do contribute substantially to the biological consequences of UV exposure (2,3). At UV fluences, cyclobutane dimers, particularly T<>T, predominate as the major chemical lesion. Prokaryotic and eukaryotic cells deficient in the ability to remove pyrimidine dimers from DNA are much more sensitive to the lethal and mutagenic effects of UV light than are their normal counterparts.

The lethal effects of UV light can be largely overcome by exposure of cells to high doses of visible light. This photoreactivation phenomenon, which has been characterized in some bacteria and in yeast, is mediated by enzymes that directly reverse the formation of pyrimidine dimers. Ultraviolet light-induced mutations can, for the most part, also be reversed by photoreactivation activities although certain exceptions have been reported (4–6).

These arguments seemed sufficiently persuasive for us to begin the study of the physiology of pyrimidine dimer lesions in human DNA. Our intent was to apply recent advances in recombinant DNA technology and sequencing to repair of such lesions. Our previous studies using defined fragments of DNA as substrates for the repair activities of the pyrimidine dimer specific endonucleases of *Micrococcus luteus* and phage T4 had convinced us of the value of this approach (7,8). Biochemical events at individual nucleotide sequences can be monitored using this approach. The surprising nature of the conclusions from this work, that both the *M. luteus* and T4 enzymes possess pyrimidine dimer N-glycosylase and not simple endonuclease activity, was initially deduced from analysis of individual scission products using such methods (7,8).

DISCUSSION

The initial question we addressed was: Is the distribution of pyrimidine dimer damage within a defined sequence the same upon irradiation of naked DNA as it is upon irradiation of intact cells? A positive answer would permit a broad extrapolation of DNA damage obtained by *in vitro* radiation of DNA samples.

The spectrum of DNA damage in the alphoid sequence of DNA was analyzed. The alpha sequence is a tandemly repeated 342 base pair-long sequence that comprises about 1% of the human DNA (9). It can be purified from the bulk of cellular DNA by cleavage by the restriction enzyme *Eco*RI. The strategy of our experiments was to purify total cellular DNA from both untreated and irradiated human cell lines in culture. Experiments were done using DNA cleaved with *Eco*RI. A 342 base-long DNA fragment was separated by centrifugation from the bulk of cellular DNA. The DNA was labeled with radioactive phosphorous at the termini and subsequently cleaved with the enzyme *Eco*RI*. Such cleavage produced a 90 base pair-long fragment labeled at one end. This fragment was purified by gel electrophoresis (9).

Application of the Maxam-Gilbert DNA sequencing technique (10) to alpha fragments prepared in this manner yielded homogeneous sequences. The location of the pyrimidine dimers within the alpha sequence was determined by cleavage of the DNA with an excess of the *M. luteus* UV-specific endonuclease. Under the conditions used, cleavage is quantitative at all four dimer sites. Treatment of irradiated, but not unirradiated, DNA resulted in cleavage at sites of adjacent pyrimidines. No such breaks were observed in unirradiated DNA samples so treated (Fig. 1). The extent of dimer formation was measured at each site by determination of the amount of radioactivity in

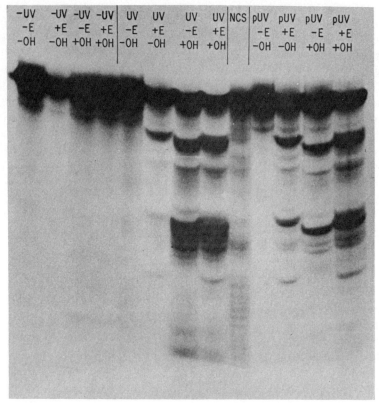

Fig. 1. Comparison of UV light induced damage to the alpha sequence irradiated as naked DNA or as cellular DNA. The 342 base pair-long alpha DNA fragment of human DNA was prepared from HeLa cells before (lanes 1–9) or after (lanes 10–13) irradiation with 5000 J/m² of UV light. The DNA was labelled at the 3′ termini in reactions that included [α-^{32}P]dATP and [α-^{32}P]TTP and the Klenow fragment of *E. coli* DNA polymerase I. The DNA was digested with the restriction endonuclease *Eco*RI* and a 92 base pair-fragment was separated from other labeled DNA fragments by electrophoresis on a nondenaturing polyacrylamide gel. The DNA was treated as described below prior to layering on a urea containing 8% polyacrylamide gel. The sequence of the unirradiated DNA fragment determined by the chemical DNA sequencing reactions is indicated. Lanes 1–4: DNA prepared from unirradiated cells untreated (lane 1), treated with *M. luteus* pyrimidine dimer endonuclease (lane 2), treated with 1 *M* piperidine at 90°C for 20 min (lane 3), treated with *M. luteus* pyrimidine dimer endonuclease followed by treatment with 1 *M* piperidine at 90°C for 20 min (lane 4). Lane 5–8: DNA extracted from unirradiated cells, exposed to 5000 J/m², and then subjected to the same four treatments in the same order prior to layering as described for lanes 1–4. Lane 10–13: DNA purified from cells exposed to 5000 J/m² treated as described for DNA of lanes 1–4 prior to layering. Lane 9: DNA purified from unirradiated cells treated with noecarzinostatin.

each scission product. There was no significant difference in the relative distribution of dimers formed under the two conditions. However, the effective dose of UV irradiation received by cellular DNA was about one-half that received by naked DNA, most likely due to shielding of the nuclear DNA by cellular constituents.

The experiment in Fig. 1 presented a surprising result. To ensure quantitative cleavage at dimer sites (the AP endonuclease activity of some enzyme preparations is low), DNA treated with the *M. luteus* enzyme was also treated with hot alkali (0.1 N NaOH, 90°C, 15 min.). Such treatment of irradiated DNA resulted in the formation of new scission products. The alkali-induced scission products were also evident in DNA samples that had not been treated with the *M. luteus* enzyme. Moreover, the scission events did not occur at the *M. luteus* enzyme-induced breaks. The relative distribution of alkali-labile sites was similar in DNA irradiated before or after extraction, again with a twofold reduction in apparent dose for irradiation of cells.

These experiments answered our initial question. Within a defined sequence, the distribution of pyrimidine dimer damage is much the same *in vivo* as it is in the cellular environment. The results also revealed a new type of UV lesion detected at sites of alkaline lability. The unusual dose response of this lesion suggested that the chemistry was substantially different from that of cyclobutane dimers, as a photosteady state was not achieved for such lesions at high UV doses.

Determination of the quantitative yield of alkali-labile sites in UV-irradiated DNA revealed striking differences in the rate of formation of such lesions, dependent upon sequence. At some sequences, the alkali-labile sites were produced at a frequency greater than that of nearby T<>T dimers, even at UV fluences of 15 to 50 J/m^2. At other sequences the rate of formation as a function of dose was 20- to 40-fold lower. Large differences in the rate of formation of T<>T dimers were not observed (Fig. 2) (11).

The high frequency of formation at some sites suggested the alkali-labile lesions might account for some of the observed effects of UV light. In particular, we speculated that the sites of frequent formation of alkali-induced lesions might correspond to UV induced mutagenic hotspots.

To test this hypothesis, we measured the distribution of both pyrimidine dimer and alkali-labile lesions in a segment of the *lacI* gene of *E. coli*. Miller and his colleagues had previously measured the relative distribution of UV-induced missense and chain terminating mutations in this gene (12,13). The precise nucleotide changes for a collection of about 700 independently isolated chain-terminating mutations was obtained by elegant genetic methods. A dose of about 100 J/m^2

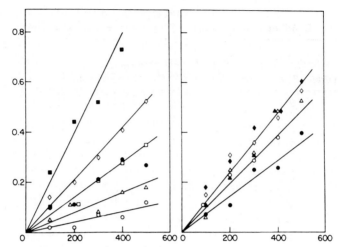

Fig. 2. Dose-response of UV-induced (6-4) lesions and cyclobutane pyrimidine dimers at several sites in the *lacI* gene. Plasmid pMC1 containing the *lacI* gene was 3' end-labeled at the bp 565 BstEII or bp 770 Sau3AI site. DNA was irradiated with 500 J/m^{-2} of UV light (254 nm), incubated with piperidine to cause strand scission at (6-4) sites or with *M. luteus* UV endonuclease to incise at cyclobutane-type pyrimidine dimers, and analysed on 8% urea-containing polyacrylamide gels. Bands were cut from the polyacrylamide gels and analysed by Cerenkov counting, the data were corrected for multiple cuts within the same DNA fragment, and the percentage of initial molecules carrying scissions at a particular site was computed. Left: (6-4) lesions at CC dinucleotides 627 (○), 671 (△) and 676 (□), and at TC dinucleotides 665 (●), 684 (◇) and 689 (■). Right: Cyclobutane pyrimidine dimers at CT dinucleotides 661 (●) and 691 (△), and at TT dinucleotides 637 (○), 655 (◇), 709 (▲) and 673 (◆). Abscissa: dose (J/m^2). Ordinate: % scission.

was used by Miller *et al* to induce the mutants. Our measurements of relative damage frequency were done at a dose of 500 J/m^2. However, the dose response for both pyrimidine dimers and alkali-labile lesions was linear within this range (11).

The extent of pyrimidine dimer damage and alkali-labile lesions formed at mutation sites within a segment of the *lacI* DNA sequence is summarized in Table I.

Can one conclude from these data that the alkali-labile lesions are premutagenic? In favor of this hypothesis is the observation that hotspots for the formation of alkali-labile lesions are also hotspots for UV-induced mutations. Figure 3 illustrates the observed proportionality between the relative rate of formation of alkali-labile lesions and mutation frequency. Such a correlation cannot be made with cyclobutane

TABLE I
Distribution of UV-Induced Base Damage in the *lacI* Gene

Dinucleotide	Site[a]	Sequence	(6-4) Product (%)	Dimer (%)	Mutants	Substitution	Nonsense site number
TC	652	*TC*	0.86	0.82	43	C → T	O24
	658	T*TC*	0.49	0.36	80	C → T	A23
	672	T*TC*C	0.60	0.86	1	C → A	O25
	678	T*TC*C	0.83	—	4	C → A	O26
	689	C*TC*C	0.57	0.48	39	C → T	U6
	706	TTT*TC*	0.98	0.91	60	C → T	O27
	732	CCC*TC*	1.08	—	0	C → A	A25
CC	418	*CC*[b]	0.01	0.35	7	C → T	A15
	484	*CC*	0.16	0.25	20	C → T	A16
	688	C*TC*C	0.15	0.53	9	C → T	A24
TT	636	T*TT*	0.08	2.08	1	T → A	Y4

[a] First base of nucleotide. [b] 5-methylcytosine

pyrimidine dimers. In particular, no chain terminating hotspots occur at TT sequences, although these comprise a major set of the most reactive sites for UV-induced DNA damage.

On the other hand, the selection for nonsense mutations probably

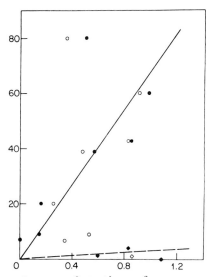

Fig. 3. Dependence of site-specific incidence of nonsense mutation on site-specific base damage incidence: (6-4) lesions (●, ◆) and cyclobutane pyrimidine dimers (○, ◇) at C → T substitution sites (—) and at C → A substitution sites (---). Site A23 is an apparent exception. Abscissa: % scission. Ordinate: mutants isolated.

has biased the results. Transition mutations at TT and CT sequences cannot be scored in this selection. This is particularly serious as all UV-induced mutations that arise at high frequency and that yield chain-terminating codons are transitions. Moreover, at any given TC and CC sequences, the rates of formation of dimers generally correlate with the rate of formation of the alkali-labile lesions; where the rate of one is high, so is the rate of the other. At such sites, mutation frequency correlates as well with dimer damage as it does with the alkali-sensitive lesion.

Four sites in the region of the *lacI* gene examined represent pronounced exceptions to this observation. At these sites, the rate of formation of the alkali-labile lesion is much lower than that of the dimers. In both cases, the mutation frequency is low, favoring the hypothesis that the alkali-labile lesions are premutagenic.

The possibility that the alkali-labile lesions might be premutagenic led us to investigate the chemical structure of the damage. Relevant to such analysis is the sequence specificity of these lesions. The DNA scissions always occur at the 3' position of adjacent pyrimidines. High frequency events occur at TC and CC sequences exclusively. Low frequency events arise at TT and also at some TC and CC sequences. These observations indicate that there is a chemical polarity to the lesions at bipyrimidine sites and that 3'-cytosines are potentially more reactive than are 3'-thymines.

The observation that the highest frequency of events occurred at TC sequences raised the possibility that the lesion that gave rise to sites of alkali-lability was the same as that which produced the precursor of the product Thy(6-4)Pyo identified by Wang and his co-workers upon acid hydrolysis of UV-irradiated DNA. The structure of this compound and postulated intermediates is shown in Fig. 4. The compound is red shifted in absorbance relative to nucleosides and is fluorescent.

To study the chemistry of potential precursor lesions, we used a series of pyrimidine dinucleotides as model compounds. A major product of irradiation of all four pyrimidine dinucleotides was found to have a maximum UV absorbance between 310 and 320 nm and was fluorescent. These fluorescent photoproducts were readily separated from their parent dinucleotides by the use of reverse-phase high-pressure liquid chromatography. Irradiation of the dinucleotide dTpdC with a high dose of UV light resulted in the formation of a fluorescent product in substantial yield (Fig. 5). The relative ratio of formation of the two photoproducts for all four nucleotides is shown in Fig. 6. Only for the case of $T<>T$ is the rate of formation of the dimer higher than

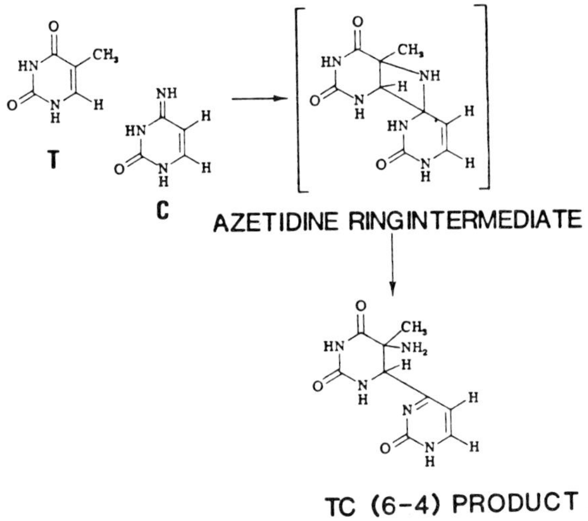

Fig. 4. Proposed mechanism for the formation of the thymine–cytosine (6-4) UV-induced photoproduct—the TC (6-4) product. The product is formed via an azetidine ring intermediate as shown.

it is for the fluorescent compounds. These unusual compounds form the major UV photoproducts, even at low UV doses, for the other three dinucleotides.

The fluorescent compounds obtained upon UV irradiation of the dinucleotides were subjected to acid hydrolysis with trifluoroacetic acid. The acid-hydrolyzed TC product was found to be identical to the product Thy(6-4)Pyo, as it had identical UV absorption and fluorescence spectra. The product also co-eluted with an authentic sample of Thy(6-4)Pyo on reverse phase and anion-exchange high-pressure liquid chromatographic separations.

If these compounds are the precursors of the alkali-labile lesions, they should be unstable under conditions of hot alkali. Treatment of all four compounds with hot alkali resulted in a rapid loss of the original compound and formation of a compound that migrated as if it were much more polar on the reverse phase column. The original fluorescent products were unaffected by treatment with either bacterial alkaline phosphatase or the 3'-phosphatase activity of polynucleotide kinase. Such treatment of the alkali- treated compounds resulted in a shift in the elution profile. From these observations, we concluded that the precursors to the alkali-labile lesions are the (6-4) class of photoproducts. Therefore, we call these precursor products UV-induced

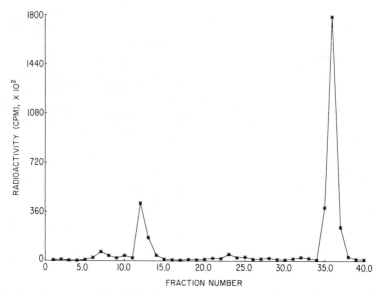

Fig. 5. Formation of the TC (6-4) product as the major photoproduct upon UV irradiation of the dinucleotide dTpdC. The dinucleotide dTpdC was 5' labelled with ^{32}P and was exposed to a total UV dose of 50,000 J/m^2. The exposed sample was then injected onto a reverse phase column and separated by high pressure liquid chromatography. Fractions were collected and were counted by Cerenkov counting. The figure shows a radiochromatogram of the irradiated dinucleotide [^{32}P]dTpdC. Four major peaks are seen in this separation. The first peak, at fraction 7, represents the solvent front; the second large peak at fraction 12 is the TC (6-4) product; the small peak at fraction 23 is the T<>C cyclobutane dimer; the large peak at fraction 37 is the radiolabeled dinucleotide.

pyrimidine–pyrimidone (6-4) photoproducts [abbreviated (6-4) products]. We suspect that the structure of these compounds is that illustrated in Fig. 4.

Comparison of the relative rates of formation of the (6-4) products at different pyrimidine–pyrimidine sites in DNA and in dinucleotides presented a puzzling problem. The ratio of (6-4) products formed in dinucleotides were comparable to one another. However, in DNA no such products were observed at CT sequences, and the rate of formation of these products at TT sequences was much lower than that at most TC and CC sites. One difference between dinucleotides and DNA is the extent of rotational freedom permitted the bases. To address the effect of structural constraints, we analyzed the spatial relationship of dipyrimidine sequences in B DNA. The reaction distances for formation of the postulated intermediate for the (6-4) products are

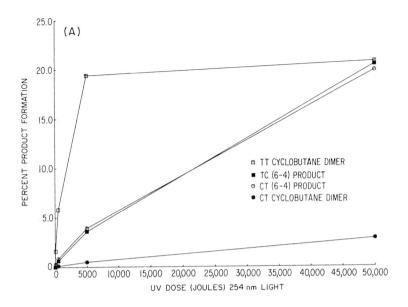
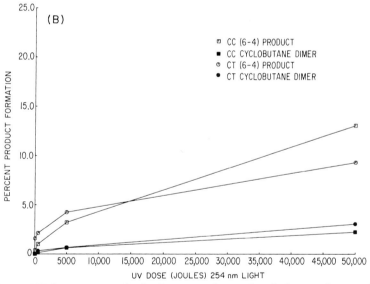

Fig. 6. UV dose responses for the formation of UV-induced photoproducts. Radiolabeled dinucleotides were exposed to UV doses ranging from 50 to 50,000 J/m² and were separated by reverse phase HPLC. The percent photoproduct formation at each dose was determined by dividing the amount of ³²P radioactivity contained in the photoproduct fraction by the total radioactivity of the photoproduct and parent radiolabeled dinucleotide, multiplied by 100. (A) Dose response for the formation of dimers and (6-4) products in *pdTpdC and *pdTpdT. (B) Dose response for the formation of dimers and (6-4) products in *pdCpdC and *pdCpdT.

only favorable if the four position of the 3'-pyrimidine participates in the exocyclic reactions. This consideration should determine the polarity of the reaction.

Consideration of the geometry of base stacking does not explain the discrepancy between the results for TC and TT in DNA with respect to those obtained with dinucleotides. For example, the difference cannot be due to a rate difference in the formation of oxetane versus azetidine ring intermediates as (6-4) products form readily in TT dinucleotides. We suggest that the exclusion of T in the 3' position of (6-4) products is due to interference by the 5-methyl group of the 3'-pyrimidine and by the additional degrees of rotational freedom in dinucleotides as compared to polynucleotides. This could explain why such products form at a high rate in dinucleotides, but not in DNA. Consistent with the hypothesis of steric hindrance by 5-methyl groups is the observation that no alkali-labile lesions arise at the sequence C-5 methyl C in the *lacI* gene, although C<>C dimers were formed at a high rate at the same site (11).

We conclude that the unusual sequence specificity observed in the (6-4) lesion is best explained by two parameters: (1) the structure of B DNA and (2) the interference of formation of the reactive intermediate by 5-methyl groups of the 3' base.

If the (6-4) products are premutagenic, it is expected that UV-induced mutations will arise at a much higher frequency at TC than at CT sequences. Hutchinson and his co-workers at Yale found a dramatic preference for mutations at TC as compared to CT sequences in UV-induced missense mutations in the *cI* gene of phage lambda. Such a preference is not expected for cyclobutane dimers as both TT and CT sequences are equally reactive for formation of such compounds. In the same collection of missense mutations, it was also noted that there was a strong preference for transition mutations at the 3' position of pyrimidines at potential dimer sites. A 5' to 3' polarity is anticipated from the structure of the (6-4) lesion, provided that one assumes that the structure of the DNA lesion is informative for the mutational specificity.

The experiments presented above, raised a clear possibility that the (6-4) lesion is premutagenic. However, before a firm conclusion can be reached, the physiology of such lesions must be examined. As mentioned above, it is known that the major repair functions of both bacteria and mammalian cells exert profound influences on both the lethal and mutagenic effects of UV light. Therefore, the ability of these repair pathways to excise (6-4) lesions must be studied. Furthermore, photoreactivation with visible light is known to reverse most of the lethal and mutagenic effects of sunlight damage. The effect of photoreactivation on (6-4) lesions must therefore be investigated as well.

SUMMARY

Methods for analysis of the sequence specific DNA damage by carcinogens and antitumor agents are presented. The methods developed are applicable to a study of DNA damage as it occurs in DNA sequences within intact human cells.

The example of UV light is given. UV light creates both cyclobutane dimers and (6-4) products at adjacent pyrimidine sites. The mutational specificity of UV light in both the *lacI* gene of *E. coli* and the *cI* gene of phage lambda is better explained by the assumption that the (6-4) lesion, rather than the cyclobutane dimer, is the major premutagenic damage, at least for mutations of the base substitution type.

REFERENCES

1. Wang, S. Y., ed. (1976) "Photochemistry and Photobiology of Nucleic Acids," Vol. 1. Academic Press, New York.
2. Hanawalt, P. C., Cooper, P. K., and Smith, C. A. (1979) *Annu. Rev. Biochem.* **48**, 783–836.
3. Freidberg, E. C., Ehmann, U. K., and Williams, J. I. (1979) *Adv. Radiat. Biol.* **8**, 86–173.
4. Witkin, E. M. (1966) *Radiat. Res., Suppl.* **6**, 30–47.
5. Witkin, E. M. (1976) *Bacteriol. Rev.* **40**, 869–907.
6. Witkin, E. M. (1966) *Science* **152**, 1345–1353.
7. Haseltine, W. A., Gordon, L. K., Lindan, C. P., Grafström, R. H., Shaper, N. L., and Grossman, L. (1980) *Nature (London)* **285**, 643–641.
8. Gordon, L. K., and Haseltine, W. A. (1980) *J. Biol. Chem.* **255**, 12047–12050.
9. Lippke, J. A., Gordon, L. K., Brash, D. E., and Haseltine, W. A. (1981) *Proc. Natl. Acad. Sci. U.S.A.* **78**, 3388–3392.
10. Maxam, A. M., and Gilbert, W. (1980) "Methods in Enzymology" (L. Grossman and K. Moldave, eds.), Vol. 65, Part 1, pp. 499–560.
11. Brash, D. E., and Haseltine, W. A. (1982) *Nature (London)* **298**, 189–192.
12. Coulondre, C., and Miller, J. H. (1977) *J. Mol. Biol.* **117**, 525–567.
13. Coulondre, C., and Miller, J. H. (1977) *J. Mol. Biol.* **117**, 577–606.

Membrane-Mediated Chromosomal Damage

P. A. CERUTTI,* I. EMERIT,† AND P. AMSTAD*

*Department of Carcinogenesis
Swiss Institute for Experimental Cancer Research
Lausanne, Switzerland
and
† Laboratory of Experimental Cytogenetics
Institut Biomédical des Cordeliers
Université Pierre et Marie Curie
Paris, France

INTRODUCTION

It is generally accepted that malignant transformation is a multistep process. The simplest model proposes an initiation step followed by promotion. Initiation can be accomplished in a single irreversible reaction with the carcinogen, whereas promotion usually requires multiple treatments and is reversible in its early phases (1–5). There are agents that possess predominantly initiator or promotor activity, but it is debatable whether agents exist that act exclusively in only one capacity. Complete carcinogens strike an equilibrium between both activities, which makes them particularly potent for malignant transformation. Initiation is usually believed to entail the structural modification of DNA, starting with DNA damage and resulting in mutation, gene amplification, gene transposition, or gene (in)activation. Until recently, it appeared that promotors, in particular the mouse skin promotor phorbol-myristate-acetate (PMA), exerted their pleiotropic effects via epigenetic mechanisms without causing chromosomal damage. It is now evident that PMA is a potent clastogen for human leukocytes (6,7) and mouse epidermal cells (8), which exerts its DNA damaging effect via indirect action (9), i.e., via the intermediacy of active oxygen species.

Active oxygen species inflicting ubiquitous damage to the cellular macromolecules including DNA may play a crucial role in promotion.

They may represent "necessary but insufficient" intermediates (10). *Therefore, all damaging agents operating by indirect action (9) are candidates for promotional activity.* The following groups of biologically important indirect agents that produce chromosomal damage can be distinguished: (A) *Physical agents,* e.g., ionizing radiation causing water radiolysis (11,12), near-ultraviolet light via photosensitization and photodynamic action (13,14). (B) *Radiomimetic drugs:* Fenton-type reagents and peroxides (15), e.g., ascorbic acid-Cu/Fe (16), bleomycin (17), benzoylperoxide (18); intermediates of microsomal drug metabolism: e.g., quinoid systems from mitomycin C (19), adriamycin (20), streptonigrin (21), 4-nitroquinolin-N-oxide (22), certain polycyclic aromatic hydrocarbons (23); CCl_4 (24). (C) *Membrane active agents* stimulating the arachidonic acid (AA) cascade and an oxidative burst, e.g., the mouse skin promotors PMA (25–29) and teleocidin (30); certain complete carcinogens, e.g., benzo(*a*)pyrene, aflatoxin B_1 (26,31–33); viruses (34); components of the immune system (35,36); particulates (37,38), e.g., asbestos, silica.

In this article, we are concentrating on membrane-active indirect DNA damaging agents of the last group. We are proposing that they produce lipid–hydroperoxides, active oxygen, and aldehydic components because they stimulate the AA-cascade, elicit an oxidative burst, and disturb the structural integrity of the membranes. Lipid–hydroperoxides and aldehydic components are likely candidates as mediators between the initial events at the membrane and the genome. Lipid–hydroperoxides may be considered "active oxygen carriers and stabilizers," which release active oxygen and degrade to aldehydic compounds, e.g., malondialdehyde (39,40) and 4-hydroxy-2,3-*trans*-nonenal. Lipid–hydroperoxides and their degradation products are expected not only to induce chromosomal damage in the stimulated cell itself but also in the neighboring tissue, i.e., they may represent clastogenic factors, CF (see below). A simple model of "membrane-mediated" chromosomal damage is illustrated in Fig. 1. We can further speculate that agents operating only by indirect action possess mostly promotional activity, agents operating only by direct action (9), in general producing carcinogen DNA adducts, are mostly initiators, and agents operating by both mechanisms are complete carcinogens.

It is worthwhile to recall the following observations concerning the mechanism of action of the classical tumor-promotor PMA (1–5), which represents the "prototype" of a membrane active agent of the third group of indirect DNA-damaging agents. In an early step, it interacts with membrane receptors in a hormone-like fashion. Receptor binding, probably via changes in Ca^{2+} influx, has a profound effect on

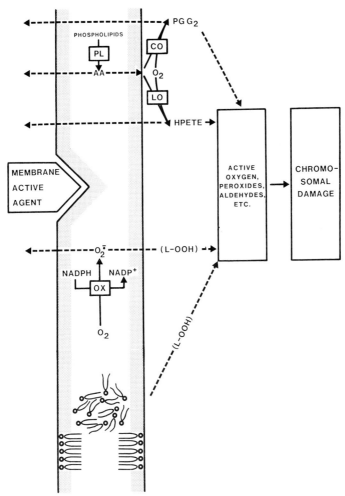

Fig. 1. Model of membrane-mediated chromosomal damage. Membrane-active agents can stimulate the arachidonic acid (AA)-cascade and via the intermediacy of AA-hydroperoxides produce active oxygen (upper part). Alternatively, they can produce active oxygen and lipid–hydroperoxides by an oxidative burst (middle part), or they may facilitate the oxidation of phospholipids because they disrupt the structural integrity of the membrane (lower part). Active oxygen, lipid–hydroperoxides, and aldehydic components produced by either pathway inflict chromosomal damage in the stimulated cell itself and in the surrounding tissue. PL, phospholipase; CO, cyclooxygenase; LO, lipoxygenase; OX, oxidase; PGG_2, prostaglandin G_2; HPETE, hydroperoxy-arachidonic acid; L-OOH, lipid–hydroperoxides.

phospholipid metabolism. The release of free AA and its oxidative conversion to prostaglandins (PG), prostacyclins, thromboxanes (TX), and hydroxy-AA (HETE) are stimulated (25,26–29,41). The latter products are formed via the intermediacy of highly reactive organic hydroperoxides, such as PGG_2 and 5-, 12-, and 15-hydroperoxy-AA, which decay under formation of active oxygen (42–44). As already mentioned, related compounds are formed in specialized cells as a consequence of an oxidative burst or in radical chain reactions initiated by radiation (39,45) and radiomimetic drugs (20).

The mechanism of action of a variety of structurally different antipromotors and anticarcinogens may involve the inhibition of a step in the formation of membrane-mediated chromosomal damage. For example, the anti-inflammatory steroids dexamethasone and fluocinolone-acetonide exert their antipromotional effect (46–48) probably via inhibition of phospholipase A_2 (49,50). They counteract many of the effects of PMA, e.g., PMA-induced epidermal DNA synthesis and cell proliferation (51), plasminogen activator production (52), and PG synthesis (49,50). Similarly, nonsteroidal anti-inflammatory agents that inhibit AA-cyclooxygenase and/or lipoxygenase and phospholipase A_2 antagonize PMA, and they appear to possess antipromotional activity under certain conditions. The best studied representative is indomethacin (see, e.g., ref. 5). It inhibits PMA-induced DNA synthesis stimulation (51,53) and cell proliferation (51) in mouse skin, induction of plasminogen activator formation (51), induction of ornithine decarboxylase (54), stimulation of PG-, hydroperoxy-, and hydroxy-AA-synthesis (31,55,56). Antipromotor and anticarcinogenic activity of superoxide dismutase (SOD), catalase, glutathioneperoxidase, organic antioxidants, and radical scavengers, (57,58) are expected on the basis of our model, because they, therefore, inactivate active oxygen species and/or organic radical intermediates and protect the cellular macromolecules from damage induced by indirect action. Exogenous SOD and catalase may exert their protective effect without being taken up by the cell via the destruction of superoxide radicals O_2^- and H_2O_2 in the extracellular medium and in the proximity of the plasma membrane, which represents the major site of the promotor induced AA-cascade.

Are there indications for analogous mechanisms in human pathobiology? Active oxygen causing ubiquitous macromolecular damage may play a role in the collagen diseases systemic lupus erythematosus (SLE), periartheritis nodosa, progressive systemic sclerosis, Crohn's ileocolitis (59), and the cancer-prone chromosomal breakage disorders Bloom syndrome (BS) and Ataxia telangiectasia (AT) (60). Involvement of active oxygen is suggested in SLE, BS, and AT, because patients with these diseases or their cells in culture are hypersensitive to radiation and radiomimetic drugs operating via indirect action. SLE and BS are sensitive to solar radiation (61), and AT to X-rays (62,63) and bleomycin (64). Near-ultraviolet light was also shown to produce excessive DNA strand breakage in BS-fibroblasts (65). Patients with the diseases listed above release clastogenic factors (CF) into their serum (66,67) and BS- (68) and AT-fibroblasts (67) produce CF in culture. CF are low molecular weight components of yet unidentified structural identity, which break chromosomes in lymphocyte cultures of healthy donors (see Emerit, 1982, for review (69)). The observation that Cu-Zn SOD inhibited the action of CF from BS indicates that O_2^--radicals are formed as intermediates in their clastogenic process. It is interesting to note that elevated oxygen tension led to excessive chromosomal breakage in cultures of lymphocytes from Fanconi anemia patients relative to normal donors. We are considering the working hypothesis that an abnormality in the formation or detoxication of active oxygen contributes to the pathology of these and related diseases (68,60). Increased stationary levels of active oxygen (i.e., "pro-oxidant" conditions) may produce a state of chronic inflammation in the collagen diseases and a state of chronic promotion in the cancer-prone diseases BS, AT, Fanconi's anemia, and Crohn's disease.

Below we are describing experiments with PMA as a representative of a "pure" promotor which together with observations from the literature support the model of "membrane-mediated chromosomal damage." We demonstrate that PMA induces the formation of a CF in human lymphocytes, which may play a role as intra- and extracellular mediator for PMA-induced chromosomal damage. Lipid–peroxides are implicated since inhibitors of the oxidative metabolism of AA, radical scavengers, and antioxidants substantially diminish both the formation of CF in response to PMA as well as its clastogenic potency. The intermediacy of O_2^--radicals is indicated in the formation and action of PMA-induced CF since CuZn SOD is inhibitory. PMA is a weak mitogen for human lymphocytes, but a potent stimulator of phospholipase A_2 leading to the release of AA from the plasma membrane. Availability of free AA is a determining factor for the activity of the cyclooxygenase and lipoxygenase pathways.

THE CLASTOGENIC ACTION OF PMA IN HUMAN LYMPHOCYTES IS MEMBRANE-MEDIATED

Binding to specific membrane receptors is the first step in the sequence of reactions which lead to the pleiotropic effects of PMA. Because it induces chromosomal damage and profound changes in gene expression apparently without covalent binding to DNA, the central question arises as to the signal that mediates the initial membrane interaction to the genome. We demonstrated that PMA induces a low molecular weight CF in blood and in regular lymphocyte cultures (such cultures contain residual platelets and monocytes). Lipid hydroperoxides plus free AA and aldehydic compounds are likely candidates for the identity of the CF. The CF may represent an intracellular link between membrane and genome and may communicate clastogenic activity among neighboring cells.

PMA INDUCES A CLASTOGENIC FACTOR IN HUMAN LYMPHOCYTES

Blood cultures from normal donors were stimulated with PHA and treated with 10 or 100 ng/ml PMA for 72 hr. The data in the left-half of Table I show that PMA is strongly clastogenic for lymphocytes in such cultures in agreement with our earlier observations (6). The medium of these cultures was collected and purified by ultrafiltration as described previously for Bloom Syndrome fibroblast cultures (68). Under these conditions, media components in the molecular weight range from 1–10,000 are isolated and concentrated 20 times. Small aliquots of these preparations, which are essentially free of PMA (see below), were tested for their potency to induce aberrations in blood cultures. The data in the right-half of Table I show that ultrafiltrates of the media of these cultures contain CF. The concentration range of the ultrafiltrates that yielded significant clastogenicity and low toxicity was relatively narrow. Clastogenic activity was detectable in media preparations following treatment with as little as 1 ng/ml PMA (data not shown). No simple dose relationship was observed between the concentration of PMA and the potency of the clastogenic components which were induced. In contrast to PMA, the nonpromotor phorbol was not clastogenic *per se* and did not induce significant clastogenic activity.

It was crucial to demonstrate that the clastogenic activity of the media ultrafiltrates was not due to residual PMA contaminating the ultrafiltrates but rather to low molecular weight components released by the cells in response to the PMA treatment. For this purpose, several experiments were carried out with radioactive PMA, which allowed the determination of the concentration of PMA in the various fractions during the preparation of the clastogenic ultrafiltrates and in the final test culture. PHA-stimulated lymphocyte cultures were treated with [^{3}H]PMA for 72 hr, and the culture media was purified. Filtration through UM10 ultrafilters eliminated more than 99% of the radioactiv-

TABLE I
PMA-Induced Clastogenic Factor in Human Lymphocytes: Effect of Anti-Inflammatory Agents

% Mitosis with aberrations[a,b]			
PMA 29.6 ± 12.1 (5)		CF 18.2 ± 3.6 (8)	
Phorbol 4.0 (2)		CF 5.0 (2)	
PMA 36.7 ± 10.3 (3)	PMA + SOD 5.3 ± 1.1 (3)	CF 24.4 ± 4.6 (5)	CF + SOD 4.0 ± 3.7 (5)
PMA 31.0 ± 9.7 (6)	PMA + INDOMETHACIN 9.0 ± 4.9 (6)	CF 18.0 ± 3.6 (6)	CF + INDOMETHACIN 4.0 ± 3.1 (6)
PMA 25.2 ± 5.1 (5)	PMA + IMIDAZOL 14.5 ± 3.9 (5)	CF 17.0 ± 4.8 (4)	CF + IMIDAZOL 2.5 ± 3.0 (4)
PMA 26.4 ± 6.0 (12)	PMA + ETYA 17.9 ± 6.2 (12)	CF 15.3 ± 3.0 (3)	CF + ETYA 4.7 ± 4.2 (3)
CONTROLS 3.0 ± 2.6 (4)		CONTROLS 2.0 ± 2.1 (4)	

[a] 100 ng/ml PMA (or phorbol) in the absence or presence of 10 μg/ml SOD, 10 μg/ml indomethacin, 50 μg/ml imidazol, and 10 μg/ml ETYA in standard blood cultures (68). The values in parenthesis give the number of independent experiments; means with standard deviations are listed; controls, 0.1% acetone.

[b] 0.1 ml of 20 times UM2 concentrated UM10 ultrafiltrates from 100 ng/ml PMA-treated blood cultures added to standard blood cultures (68) in the absence or presence of drugs under the conditions mentioned under (1).

ity. Concentration of the ultrafiltrates with UM2 filters and sterilization by passage through Millipore filters further decreased the radioactivity content. The residual PMA concentrations in the media of the test cultures were calculated from their radioactivity content. The ultrafiltrates were strongly clastogenic at residual PMA concentrations as low as 0.005–0.008 ng/ml, and some clastogenicity was detectable at 3–5 times lower concentrations. PMA in this concentration range is no longer clastogenic for PHA stimulated human lymphocytes. It was concluded that the clastogenic activity of the ultrafiltrates is produced by the cells in response to PMA treatment and is not due to residual PMA, which contaminates the preparations (70).

ANTICLASTOGENIC ACTIVITY OF SUPEROXIDE DISMUTASE AND NONSTEROIDAL ANTI-INFLAMMATORY AGENTS

Based on our earlier observation that SOD inhibited the clastogenic activity of PMA (6), we tested the effect of this enzyme on the activity of CF and on the formation of CF. The data in Table I confirm our earlier conclusion, and it shows in addition that SOD also inhibits the activity of performed CF. Because bovine erythrocyte CuZn SOD used in these experiments has a molecular weight of approximately 30,000 (71), it is removed by our standard ultrafiltration procedure for the preparation of CF. Therefore, we also compared the clastogenicity of CF preparations induced by PMA in the absence and presence of SOD. The potency of the CF was decreased from $17.3 \pm 2.1\%$ to $5.0 \pm 3.6\%$ mitosis with aberrations when 10 µg/ml SOD was present together with 100 ng/ml PMA during its formation (means $\pm$ SD of three experiments). It was concluded that SOD inhibits the clastogenic activity of PMA, the activity of preformed CF, and the formation of CF (70). Precluding any unknown side effects of SOD, these results implicate O_2^--radicals in all three reactions. The clastogenic action of PMA itself may be mediated by the CF. If this is the case, the effect of SOD on PMA clastogenicity is due to the inhibition of CF formation. The observation that SOD also inhibited the activity of preformed CF is reminiscent of our results with a CF isolated from Bloom Syndrome fibroblast cultures (68). No consistent results could be obtained with catalase.

The nonsteroidal anti-inflammatory agents indomethacin, imidazol, and 5,8,11,14-eicosatetraynoic acid (ETYA) (1–5,72) were tested for their effect on the clastogenicity of PMA and the potency of PMA-induced CF. From the data in Table I, it is evident that all three drugs were anticlastogenic at concentrations where they had no significant effects on cell growth. They diminished the clastogenicity of PMA and inhibited the activity of CF (70). Because there is no simple procedure to remove these drugs quantitatively from the media, it could not be tested whether they also inhibited the formation of CF. These results suggest that metabolites of AA participate in PMA clastogenicity and in the action of PMA induced CF. Because the selectivity of these drugs for the cyclooxygenase or lipoxygenase pathways of AA is limited, no safe conclusions can be drawn about the identity of the clastogenic metabolites. The possibility cannot be excluded that they also act on other enzymes and that they may influence the radical scavenging potential of the cell (72–76). PGG_2 and hydroperoxy-AA are probably clastogenic, because they release active oxygen upon transformation to more stable derivatives (42–44). Alternatively, malondialdehyde (MDA), which is formed upon decomposition of PGG_2, might be involved. MDA is weakly mutagenic and has been proposed as an intermediate in radiation carcinogenesis (39,40). Aldehydes of higher molecular weight, such as 4-hydroxy-2,3-*trans*-nonenal, which are formed by phospholipid peroxidation, are expected to be clastogenic as well (77).

PMA STIMULATES THE RELEASE OF FREE AA FROM HUMAN LYMPHOCYTES

Regular human lymphocytes (i.e., Ficoll–Hypaque preparations from heparinized blood which were contaminated with a small number of platelets, monocytes, and other leukocytes) were prelabeled with [^{3}H]AA and incubated either with phytohemagglutinin M and P, PMA (10–50 ng/ml), or both for 48 hr. The long incubation time was chosen in analogy to our cytogenetic studies described above. During the last 20 hr, [^{14}C]thymidine was added in order to compare DNA synthetic activity under these conditions. The culture media were then acidified to pH 3.5 and extracted with ethylacetate according to published procedures (78). Table II compares the amount and composition of [^{3}H]AA label contained in the media extracts. Approximately 30% more [^{3}H]AA label was released by the PMA-treated lymphocytes relative to untreated controls and approximately twice the amount relative to PHA-stimulated cells. Analysis of the ethylacetate extracts by thin-layer chromatography showed that most of the additional radioac-

TABLE II
Release of [^{3}H]AA and Metabolites from [^{3}H]AA-Labeled Human Lymphocytes by PHA and PMA[a]

	Extractable radioactivity	START	PG	TX	X	H(P)ETE	AA
	%	%	%	%	%	%	%
Control	20.0	1.3	0.7	2.1	2.6	2.2	10.1
PHA	14.1	0.6	1.3	2.7	1.1	1.6	5.6
PHA + TPA (10 ng/ml)	11.9	2.6	0.3	1.0		1.7	5.1
TPA (10 ng/ml)	33.4	4.2	0.7	2.0	2.3	2.0	20.5
TPA (50 ng/ml)	30.2	5.1	0.3	1.6	1.0	2.4	19.1

[a] The original cultures contained 5 × 10^6 lymphocytes labeled with approximately 10^6 cpm [^{3}H]AA. Following 48-hr incubation, the cells were removed; the media was acidified to pH 3.5 and exhaustibly extracted with ethylacetate. Aliquots of the extracts were chromatographed on tlc (silicagel 60 F$_{254}$, Merck, developed with benzene: dioxane:acetic acid v/v 40:20:2) The structural assignments of the radioactivity peaks were made with the help of authentic markers and are tentative. (PG, mostly PGF$_1$ or PGF$_2$; TX, mostly TXB$_2$.) The data are expressed as % radioactivity in total ethylacetate extracts and individual tlc peaks relative to total radioactivity in original sample.

tivity in the PMA-treated sample was AA, whereas the content in PG, TX, and H(P)ETE was comparable. More radioactivity accumulated at the start of the chromatogram following PMA treatment relative to controls and PHA-treated cultures. This highly polar material remains unidentified. These results were corroborated by high-pressure liquid chromatography (HPLC) (P. Amstad and P. Cerutti, unpublished). Residual platelets and, to a lesser degree, monocytes in our lymphocyte preparations were expected to contribute to the oxidative metabolism of free AA (78). DNA synthesis stimulation was only approximately $\frac{1}{5}$ for PMA alone relative to PHA and PHA plus PMA (P. Amstad and P. Cerutti, unpublished; cf ref. 79). It was concluded that PMA is a potent stimulator of membrane phospholipid degradation in human lymphocytes but a weak mitogen. PHA reduced the amount of free AA by PMA to the level of PHA alone. It is likely that AA was reincorporated into phospholipids as a consequence of the mitogenic stimulation by PHA. The HPLC fractions are presently being tested for clastogenicity in human blood cultures.

CONCLUDING REMARKS

According to our model of membrane-mediated chromosomal damage, chemically unrelated membrane-active agents that stimulate the AA-cascade elicit an oxidative burst and disturb membrane integrity so that phospholipids become more susceptible to oxidation can induce chromosomal damage and may possess promotional activity. All these pathways produce lipid–peroxides. The AA-cascade is initiated by the stimulation of membrane phospholipase, which results in the release of AA. Increased amounts of free AA become available to microsomal oxidative metabolism in the stimulated cell itself as well as in neighboring cells. For the case of PMA, the participation of neighboring cells is suggested by the observation that the potency of CF was higher when it was derived from platelet-rich rather than platelet-free lymphocyte cultures. Therefore, we speculate that PMA induced CF from regular lymphocyte preparations consists of free AA plus lipid hydroperoxides and aldehydic compounds and that platelets assist in the metabolism of AA. It is not yet known whether PMA and other membrane-active agents induce the formation of a CF in polymorphonuclear leukocytes and macrophages. This appears likely because they are known to stimulate an oxidative burst and possibly the AA-cascade in these cells.

In order to test the generality of our model of *membrane-mediated chromosomal damage,* we have recently extended our studies to the complete carcinogen AFB_1. Our results show that AFB_1 stimulates the release of free AA from human lymphocytes and induces chromosomal aberrations without forming detectable amounts of covalent DNA adducts (P. Amstad, I. Emerit and P. Cerutti, unpublished). We conclude tentatively that AFB_1 induces membrane-mediated chromosomal damage and speculate that this property of AFB_1 is related to its promotional activity as complete carcinogen.

Our studies have concentrated on human lymphocytes, and it is important that they be expanded to other types of cells. Indeed, it has been shown recently that PMA induced aberrations in cultured mouse epidermal cells (8), but these results have to be evaluated cautiously in view of the inherent genetic instability of these cells. PMA also induced mitotic aneuploidy in yeast (80) and gene-amplification in mouse cells (81) in further support of its action as chromosomal damaging agent.

ACKNOWLEDGEMENTS

This work was supported by grant 3'305.78 from the Swiss NSF, a grant from the Swiss Association of Cigarette Manufacturers, CNRS grant ER 149 and grant U 155 from Institut National de la Santé et de la Recherche Médicale.

REFERENCES

1. Diamond, L., O'Brien, G., and Baird, W. (1980) *Adv. Cancer Res.* 32, 1–74.
2. Slaga, T., Sivak, A., and Boutwell, R., eds. (1978). "Carcinogenesis," Vol. 2. Raven Press, New York.
3. Hecker, E., Fusenig, N., Kunz, W., Marks, F., and Thielmann, H., eds. (1982). "Carcinogenesis," Vol. 7. Raven Press, New York.
4. Weinstein, I., Mufson, R., Lee, L.-S., Fischer, P., Laskin, J., Horowitz, A., and Ivanovic, V. (1980) *In* "Carcinogenesis: Fundamental Mechanisms and Environmental Effects" (B. Pullman, P. Ts'o, and H. Gelboin, eds.), pp. 543–563. Reidel Publ., Dodrecht, Netherlands.
5. Levine, L. (1981) *Adv. Cancer Res.* 35, 49–79.
6. Emerit, I., and Cerutti, P. (1981) *Nature (London)* 293, 144–146.
7. Birnboim, H. (1982) *Science* 215, 1247–1249.
8. Fusenig, N., and Dzarlieva, R. (1982) *Carcinog.—Compr. Surv.* 7, 201–216.
9. Cerutti, P. (1978) *In* "DNA Repair Mechanisms" (P. Hanawalt, E. Friedberg, and C. Fox, eds.), pp. 717–722. Academic Press, New York.
10. Troll, W., Witz, G., Goldstein, B., Stone, D., and Sugimura, T. (1982) *Carcinog.—Compr. Surv.* 7, 593–597.

11. Draganic, I., and Draganic, Z. (1971) "The Radiation Chemistry of Water." Academic Press, New York.
12. Cerutti, P. (1974) *Naturwissenschaften* **61**, 51–59.
13. Foote, C. (1976) *In* "Free Radicals in Biology" (W. A. Pryor, ed.), Vol. 2, pp. 85–124. Academic Press, New York.
14. Cerutti, P., and Netrawali, M. (1979) *Radiat. Res. Proc. Int. Congr., 6th, 1979* pp. 423–432.
15. Fridovich, I. (1978) *Science* **201**, 875–880.
16. See, e.g., Wong, K., Morgan, A., and Paranchych, W. (1974) *Can. J. Biochem.* **52**, 950–958.
17. Lown, J., and Sim, S.-K. (1977) *Biochem. Biophys. Res. Commun.* **77**, 1150–1157.
18. Slaga, T., Triplett, L., Jotti, L., and Trosko, J. (1981) *Science* **213**, 1023–1025.
19. Tomasz, M. (1976) *Chem.-Biol. Interact.* **13**, 89–97.
20. Goodman, J., and Hochstein, P. (1977) *Biochem. Biophys. Res. Commun.* **77**, 797–803.
21. Bachur, N. R., Gordon S. L., and Gee, M. W. (1978) *Cancer Res.* **38**, 1745–1750.
22. Biaglow, J. E., Jabobson, B., and Nygaard, O. (1977) *Cancer Res.* **37**, 3306–3313.
23. Lesko, S., Ts'o, P., Jang, S.-U., and Cheng, R. (1982) *In* "Free Radicals, Lipid Peroxidation and Cancer" (D. McBrien and T. F. Slater, eds.), pp. 401–414. Academic Press, New York.
24. Slater, T. F. (1978) *In* "Biochemical Mechanisms of Liver Injury" (T. F. Slater, ed.), pp. 745–801. Academic Press, New York.
25. Brune, K., Glatt, M., Kälin, H., and Feskar, B. (1978) *Nature (London)* **274**, 261–263.
26. Levine, L., and Ohuchi, L. (1978) *Cancer Res.* **38**, 4142–4146.
27. Bresnick, E., Meunier, P., and Lamden, M. (1979) *Cancer Lett.* **7**, 121–125.
28. Wertz, P., and Muller, G. (1978) *Cancer Res.* **38**, 2900–2904.
29. Marks, F., Fürstenberger, G., and Kownatzki, E. (1981) *Cancer Res.* **41**, 696–702.
30. Goldstein, B., Witz, G., Amoruso, M., Stone, D., and Troll, W. (1981) *Cancer Lett.* **11**, 257–262.
31. Ohuchi, K., and Levine, L. (1978) *Prostaglandins Med.* **1**, 421–431.
32. Cerutti, P., and Remsen, J. (1977) *In* "DNA Repair Processes" (W. Nichols and D. Murphy, eds.), pp. 147–166. Symposia Specialists, Miami, Florida.
33. Ivanovic, V., and Weinstein, I. B. (1981) *Nature (London)* **293**, 404–406.
34. Peterhans, E. (1980) *Virology* **105**, 445–455.
35. Peterhans, E., Albrecht, H., and Wyler, R. (1981) *J. Immunol. Methods* **4**, 295–302.
36. Hafelman, D., and Lucas, Z. (1979) *J. Immunol.* **123**, 55–62.
37. Morley, J., Bray, M. A., Jones, R. W., Nugteren, D. H., and van Dorp, D. A. (1979) *Prostaglandins* **17**, 729–736.
38. Humes, J., Burger, S., Galavage, M., Kuehl, F. A., Jr., Wightman, P. D., Dahlgren, M. E., Davies, P., and Bonney, R. J. (1980) *J. Immunol.* **124**, 2110–2116.
39. Petkau, A. (1980) *Acta Physiol. Scand., Suppl.* **492**, 81–90.
40. Marnett, L., and Tuttle, M. (1980) *Cancer Res.* **40**, 276–282.
41. Fürstenberger, G., Richter, H., Fusening, N., and Marks, F. (1981) *Cancer Lett.* **11**, 191–198.
42. Egan, R., Paxton, J., and Kuehl, F. A. (1976) *J. Biol. Chem.* **251**, 7329–7335.
43. Rahimtula, A., and O'Brien, P. J. (1976) *Biochem. Biophys. Res. Commun.* **70**, 893–899.
44. Sujioka, K., and Nakano, M. (1976) *Biochim. Biophys. Acta* **423**, 203–216.

45. Slater, T. F. (1979) In "Drug Toxicity" (J. W. Garrod, ed.), pp. 123. Taylor & Francis, London.
46. Belman, S., and Troll, W. (1972) Cancer Res. 32, 450–454.
47. Scribner, J., and Slaga, T. (1973) Cancer Res. 33, 542–546.
48. Rivedal, E. (1982) Cancer Lett. 15, 105–103.
49. Flower, R. (1978) Adv. Prostaglandin Thromboxane Res. 3, 105.
50. Blackwell, G., Carnuccio, R., Di Rosa, M., Flower, R., Parente, L., and Persico, P. (1980) Nature (London) 287, 147–149.
51. Viaje, A., Slaga, T., Wigler, M., and Weinstein, I. (1977) Cancer Res. 37, 1530–1536.
52. Wigler, M., and Weinstein, I. (1976) Nature (London) 259, 232–233.
53. Fürstenberger, G., and Marks, F. (1978) Biochem. Biophys. Res. Commun. 84, 1103–1111.
54. Verma, A., Ashendel, C., and Boutwell, R. (1980) Cancer Res. 40, 308–315.
55. Fürstenberger, G., and Marks, F. (1980) Biochem. Biophys. Res. Commun. 92, 749–756.
56. Anderson, T., and Voorhees, J. (1979) J. Invest. Dermatol. 73, 180.
57. Oberley, L., and Buettner, G. (1979) Cancer Res. 39, 1141–1149.
58. Wattenberg, L. (1980) Carcinog.—Compr. Surv. 5, 85–98.
59. Emerit, I. (1980) Z. Rheumatol. 39, 84–90.
60. Cerutti, P. (1982) Prog. Mutat. Res. 4, 203–214.
61. Zbinden, I., and Cerutti, P. (1981) Biochem. Biophys. Res. Commun. 98, 579–587.
62. Taylor, A., Harnden, D., Arlett, C., Harcourt, S., Lehmann, A., Stevens, S., and Bridges, B. (1975) Nature (London) 258, 427–429.
63. Paterson, M. C., and Smith, P. J. (1979) Annu. Rev. Genet. 13, 291–318.
64. Taylor, A., Rosney, C., and Campbell, J. (1979) Cancer Res. 39, 1046–1050.
65. Hirschi, M., Netrawali, M., Remsen, J., and Cerutti, P. (1981) Cancer Res. 41, 2003–2007.
66. Emerit, I., Jalbert, P., and Cerutti, P. (1982) Hum. Genet. 61, 65–67.
67. Shaham, M., Becker, Y., and Cohen, M. (1980) Cytogenet. Cell Genet. 27, 1–7.
68. Emerit, I., and Cerutti, P. (1981) Proc. Natl. Acad. Sci. U.S.A. 78, 1868–1872.
69. Emerit, I. (1982) Prog. Mutat. Res. 4, 61–74.
70. Emerit, I., and Cerutti, P. (1982) Proc. Natl. Acad. Sci. U.S.A. 79, 7509–7513.
71. McCord, J., and Fridovich, I. (1969) J. Biol. Chem. 244, 6049–6055.
72. Flower, R. (1974) Pharmacol. Rev. 26, 33.
73. Needleman, P., Bryan, B., Wyche, A., Bronson, S., Eakens, J., Ferrendelli, A., and Minkes, M. (1977) Prostaglandins 13, 897.
74. Collins, F., and Raiskums, B. (1973) Proc. Aust. Biochem. Soc. 6, 28.
75. Weber, P., Scherer, B., and Larson, C. (1977) Eur. J. Pharmacol. 41, 329–332.
76. Kaplan, L., Weiss, J., and Elsbach, P. (1978) Proc. Natl. Acad. Sci. U.S.A. 75, 2955–2958.
77. Parker, C., Stenson, W., Huber, M., and Kelley, J. (1979) J. Immunol. 122, 1572–1577.
78. Cf. Abb, J., Bayliss, G., and Deinhardt, F. (1979) J. Immunol. 122, 1639–1642.
79. Parry, J. M., Parry, E. M., and Barrett, J. (1981) Nature (London) 294, 263–265.
80. Varshavsky, A. (1981) Cell 25, 561–572.

Molecular Mechanism of Mutagenesis by the Ultimate Carcinogen *N*-Acetoxy-*N*-2-Acetylaminofluorene

ROBERT P. P. FUCHS, NICOLE KOFFEL-SCHWARTZ,
AND MICHEL P. DAUNE

*Institut de Biologie Moléculaire
et Cellulaire du CNRS
Strasbourg, France*

INTRODUCTION

The hypothesis that carcinogenesis by chemicals is a multistep process involving mutagenesis as an early event has been proposed for many years. Supporting this view is the fact that (a) in general a tumor has a clonal origin and that (b) most of the chemical carcinogens are also mutagens in bacterial systems (1). The premutational event is the covalent binding of the ultimate carcinogen to the DNA bases or phosphate groups. The chemical structure of the adducts formed and, to a lesser extent, the structural changes induced in the DNA double helix in the neighborhood of the adducts have been extensively studied for the last 10 years. An unanswered but nevertheless crucial question is to know how the chemically altered base (the premutational lesion) is converted into a stable mutation. It is clear that most of the adducts are recognized by an excision repair-type of mechanism that processes the lesion in an error-free way so that the genetic information is preserved. In some cases however a stable mutation is fixed at the site or in the vicinity of the premutational lesion as a consequence of what is known as error-prone repair or SOS repair in bacteria (2–4). *N*-Acetoxy-*N*-2-acetylaminofluorene (*N*-AcO-AAF) is a model ultimate metabolite of the strong rat liver carcinogen 2-acetylaminofluorene

(AAF). *In vivo* studies have shown the mutagenicity of AAF and its derivatives in bacterial (5,6) and eukaryotic systems (7). N-AcO-AAF reacts *in vitro* with DNA leading mainly to the formation of a guanine adduct (8), N-2-(Deoxyguanosin-8-yl)acetylaminofluorene (80%) and also to at least three minor adducts (N. Koffel-Schwartz, R. P. P. Fuchs, and M. P. Daune, unpublished results), one of which is characterized as 3-(deoxyguanosin-N^2-yl)acetylaminofluorene (9).

Studies from our group led to the general conclusion that binding of N-AcO-AAF to DNA resulted in a local distortion of the DNA helix around the C-8 adduct (10–12). We have called this structural alteration the insertion–denaturation model (13). A similar model has been proposed by Grunberger and Weinstein (14).

In this paper we describe the analysis of forward mutations induced in the tetracycline-resistance gene of pBR322 by directing the chemical reaction of the carcinogen to a small restriction fragment (*Bam*HI,*Sal*I) inside the antibiotic-resistance gene. A preliminary report of this work has recently been published (15).

MATERIAL AND METHODS

The *Esherichia coli* strains used were AB 1157 or AB 1886 (16). N-AcO-AAF (^{3}H-ring) was synthetized as described previously (17) (specific activity: 196 mCi/mmole). N-AcO-AAF (^{3}H-ring) reaction with supercoiled plasmid DNA was performed in Tris 10 mM, EDTA 1 mM pH 8 buffer (TE buffer) containing 5% of ethanol (DNA concentration: 50 µg/ml). Removal of unbound fluorene derivatives was achieved by four successive ethanol precipitations. The number of AAF residues bound per plasmid molecule was determined as previously described (18).

Samples of pBR322 reacted with N-AcO-AAF to various extents (ranging from 0 to 2.5% of modified bases) were digested with *Bam*HI, and *Sal*I restriction enzymes (Boehringer, Mannheim). The large fragment (16 S fragment) and the small fragment (6 S fragment) were separated and purified either by velocity sedimentation on sucrose gradient (5 to 20%) or by electrophoresis on 0.8% agarose or on 8% polyacrylamide gels followed by electroelution. T4 DNA ligase (Biolabs) was used to ligate the unmodified 16 S fragment with either the unmodified 6 S fragment or the 6 S fragments obtained from the various AAF-modified pBR322 samples (6S-AAF). The ligation was performed under the conditions specified by the T4 DNA ligase manufacturer.

UV IRRADIATION OF THE CELLS PRIOR TO TRANSFORMATION

In some cases, the $E.\ coli$ cells were UV irradiated prior to the transformation procedure. This treatment was used to induce the cellular SOS response. The cells were UV irradiated as a suspension in 0.01 M $MgSO_4$ with a germicidal lamp (15 W, Phillips) at a dose giving about 50% survival (i.e., 60 J/m^2 for the wild-type strain, AB 1157; 6 J/m^2 for the $uvrA$ strain, AB 1886). The cells were then incubated in LB medium for 30 min at 37°C to allow expression of the SOS functions.

$E.\ coli$ TRANSFORMATION AND SELECTION OF THE AMPICILLIN-RESISTANT (Ap^R) AND TETRACYCLINE-SENSITIVE (Tc^S) CLONES

The $E.\ coli$ cells were prepared for transformation using the classic $CaCl_2$ treatment procedure (19). The different ligation mixtures were diluted by a factor of 100 in 10 mM Tris, 10 mM $CaCl_2$, and 10 mM $MgCl_2$ (pH 7) and used to transform the competent cell suspension by mixing 1 volume of the DNA solution with 2 volumes of the concentrated $E.\ coli$ suspension. Following the transformation procedure, the cells were spread on LB plates containing ampicillin (50 μg/ml) and incubated at 37°C overnight. The clones were then replated on LB plates containing tetracycline (20 μg/ml). Clones that grew on Ap but not on Tc were scored as Ap^R, Tc^S mutants. Such individual mutant clones were then grown further in LB medium plus ampicillin for preparation of the plasmid DNA contained in these clones. Plasmid DNA was purified either on a small scale (10 ml of culture) by an adaptation of the method of Clewell and Helinski (20) or on a larger scale (1 liter culture) by a NaCl/SDS lysis procedure followed by a CsCl/ethidium bromide centrifugation step (21).

DNA SEQUENCE ANALYSIS OF THE MUTANTS

Plasmids were digested with BamHI and SalI restriction enzymes and ^{32}P end-labeled at their 5' extremities using calf intestine phosphatase and T4 DNA kinase (Boehringer, Mannheim). Strand separation and sequencing were performed according to the method of Maxam and Gilbert (22).

RESULTS

pBR322 plasmid DNA carries the resistance genes for two antibiotics: tetracycline (Tc) and ampicillin (Ap). The tetracycline resistance gene contains two unique restriction sites (*Bam*HI and *Sal*I) at nucleotide position 375 and 650, respectively [numbering starts clockwise at the unique *Eco*RI site (23)]. This small *Bam*HI–*Sal*I restriction fragment (275 bp) was taken as the target for the *in vitro* mutagenesis experiments.

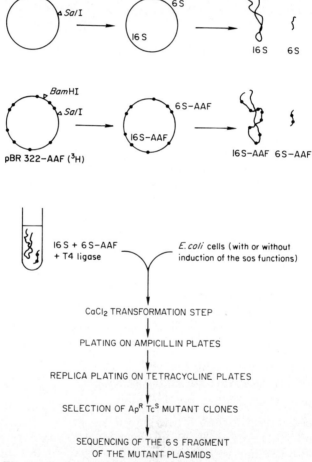

Fig. 1. Strategy for the site-directed mutagenesis experiment.

The strategy that was used to obtain mutants in the tetracycline resistance gene, within the small restriction fragment (BamHI,SalI), is outlined in Fig. 1. This restriction fragment (6 S fragment) modified to various extents with N-AcO-AAF (^{3}H-ring) (0 to 14 AAF residues/6 S fragment) was reinserted by *in vitro* ligation into the nonreacted large (BamHI,SalI) restriction fragment (16 S fragment). This large fragment contains both the gene coding for the β-lactamase (Ap resistance gene) and the origin of replication. The ligation mixture was used to transform $CaCl_2$ treated *E. coli* recipient cells. Mutants are selected for ampicillin (Ap) resistance and tetracycline (Tc) sensitivity.

MUTATION FREQUENCY AND RESTRICTION ENZYME ANALYSIS OF THE Ap^RTc^S MUTANTS

To obtain relatively high levels of mutation, it was necessary to induce the SOS function in the host cell prior to transformation. This was achieved by UV irradiation of the bacteria according to the protocol described in Material and Methods.

Mutant clones having the Ap^RTc^S phenotype were subcultured in 2 ml-sized cultures; their plasmid DNA was prepared and analyzed by agarose gel electrophoresis for its length and for the presence of the BamHI and SalI restriction sites. Two classes of mutants could easily be identified at this point.

1. Class I mutants: Such a mutant is defined by the fact that either its size is markedly different from the size of wild-type pBR322 DNA (as seen by its migration on 0.8% agarose gels), and (or) it has lost the BamHI and (or) the SalI restriction site. Such mutants are in general shorter than wild-type pBR322 by 0.2 to 0.8 kb. Moreover, they have lost both BamHI and SalI restriction sites. We propose that these mutants arise by head to tail dimerization of the large 16 S fragment followed by *in vivo* monomerization through recombination. An other subclass of mutants belonging to what we call class I mutants have the same size as wild-type pBR322 but lack either the BamHI or the SalI restriction site. These mutants must probably arise through mutagenesis at one of the restriction sites by an exonucleolytic contamination of the restriction enzyme preparation.

2. Class II mutants: These mutants have the pBR322 wild-type size and retain both BamHI and SalI restriction sites.

The class I mutation frequency does not depend on the induction of

TABLE I
Mutation Frequency in Strain AB 1157 with and without Induction of the SOS Functions[a]

	Class I mutants		Class II mutants	
	Number of mutants/total number of clones	Mutation frequency	Number of mutants/total number of clones	Mutation frequency
Without SOS induction	20/6488	3×10^{-3}	2/6488	3×10^{-4}
With SOS induction	11/4271	2.5×10^{-3}	18/4271	4.2×10^{-3}

[a] The numbers given in this table are the sum of all the data obtained for the different levels of AAF modification.

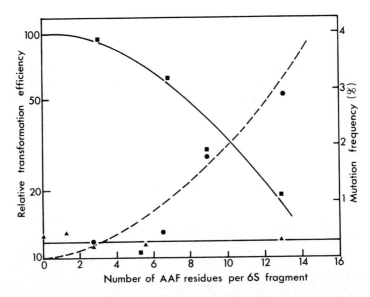

Fig. 2. Relative transformation efficiency and mutation frequencies as a function of the number of AAF residues bound to the 6 S fragment. The transformation of *E. coli* AB 1157 strain has been performed after UV irradiation of the cells (60 J/m²) as described in Materials and Methods. Selection of the transformants was done on LB plates containing ampicillin (50 µg/ml). Mutants were selected by replica plating on LB plates containing tetracycline (20 µg/ml). (■) Relative transformation efficiency (log scale); (▲) class I mutants; (●) class II mutants. From R. P. P. Fuchs *et al.*, see ref. (15), with permission from the publisher.

TABLE II
Properties of Class I and Class II Mutants

	Class I mutants	Class II mutants
Size	Different or equal to pBR322	Equal to pBR322
BamI and SalI restriction sites	Loss of one or both restriction sites	Presence of both restriction sites
Mutation frequency	AAF independent	Increases with the AAF modification level
	No effect of the SOS functions	Increases when the SOS functions are induced

the SOS functions (Table I), and it is not a function of the level of AAF modification (Fig. 2). Class I mutation frequency is constant and equal to about 0.2–0.3% throughout all our experiments. On the other hand, class II mutation frequency is inducible (Table I) and increases in a dose-dependent way with the level of AAF modification (Fig. 2). The mutation frequency increases by a factor of 10–20 when the SOS functions are induced, and reaches about 3% for the highest AAF modification level tested (i.e., 14 AAF residues/6 S fragment). From these data, it is clear that only class II mutants are due to the AAF modification and are therefore further analyzed for changes in their nucleotide sequence. The properties of class I and class II mutants are given in Table II. With increasing levels of AAF modification it is shown that the relative transformation efficiency (i.e., relative to nonmodified DNA) is decreasing (Fig. 2). The extent of this AAF-dependent inhibition of the transformation efficiency is strongly related to the general repair genotype of the recipient cell (24).

SEQUENCE ANALYSIS OF CLASS II MUTANTS

Twenty-one class II mutant plasmids were isolated from either the wild-type *E. coli* strain, AB 1157, or the corresponding *uvrA* mutant strain, AB 1886. The double *Bam*HI/*Sal*I digested DNA was ^{32}P-end labeled at the 5' extremities and sequenced according to the Maxam and Gilbert technique (22). The sequence of the wild-type 6 S fragment was found identical to the sequence published by Sutcliffe (23) except for an additional C within the run of the three C's at position 526–528. A new ATG initiation codon for the tetracycline resistance protein is therefore proposed (Fig. 4) with an 1188-nucleotide long open-reading frame. In all of the twenty-one class II mutants, we found a mutation located within the 6 S fragment. In most cases, (20/21) only one sequence alteration was found within the 6 S fragment. In one other case, two sequence alterations were observed within the 6 S fragment. Except for one case, all of the 21 sequenced mutations affected GC base pairs (deletion of one, two, or three adjacent GC base pairs or the addition of one GC base pair). Moreover, the mutations are clustered at a small number of sites that appear to be mutational hot spots. Nine mutants are described in Table III. The description of the other mutants, and their distribution along the 6 S fragment, will be described elsewhere (25). Two hot-spot sequences for mutagenesis were found within the collection of the nine mutants described in Table III.

HOT SPOT SEQUENCE 1

As shown in Table III, four out of the nine mutants (mutants number 4, 30, 41, and 45) show a deletion of a single G residue at position 520 or 521. (Numbering starts clockwise from the unique *Eco*RI site.) It should be noted that the four mutants have arisen under quite different conditions (i.e., in two different strains, with or without UV irradiation of the SOS functions, with very different AAF modification levels).

HOT SPOT SEQUENCE 2

Mutants 34 (as shown in Fig. 3) and 36 exhibit a -2 deletion within the alternating GCGC sequence at positions 435–438 (deletion of a GC sequence at position 435–436 or 437–438, or of a GC sequence at position 436–437). Mutant 33 also exhibits a -2 deletion within a GCGC sequence at position 548–551. The mutation in mutant 35, although being different (-1 deletion of G 416 and double transition at 414–415) also takes place within the same six-nucleotide long sequence, GGCGCC, sequence that is in common to all four mutants 34, 36, 33, 35. This given sequence, which is found 3 times within the 6 S fragment, can therefore be considered as a mutational hot spot.

DISCUSSION

N-Acetoxy-*N*-2-acetylaminofluorene is known to react covalently with guanine residues (8,9). All the mutations that are found affect GC base pairs and can therefore be considered as targeted mutations. It is

TABLE III
Description of the Different Mutants[a,b]

Description of the mutation	Mutant number	Strain	Induction of SOS functions	Average number of bound AAF residues/6 S fragment	Sequence in the neighborhood of the mutation
−1 deletion of G 520 or 521	4	AB 1886	+	2.0	511　　　　　520　　　　　529 G G G T A T G G T G G C A G G C C C G 　　　　　　　　　　−1
	30	AB 1886	−	2.8	
	41	AB 1157	+	13.8	
	45	AB 1157	+	2.8	
−1 deletion of G 389 or 390	32	AB 1157	+	8.8	381　　　　　390　　　　　399 T C T A C G C C G G A C G C A T C G T 　　　　　　　　　−1
−2 deletion of GC 435-436 or 437-438 or CG 436-437	34	AB 1157	+	6.6	428　　　　　　　　　　446 G T T G C T G G C G C C T A T A T C G 　　　　　　　−2
	36	AB 1157	+	8.8	
−2 deletion of GC 548-549 or 550-551 or CG 549-550	33	AB 1157	+	13.8	540　　　　　　　　　　558 A C T G T T G G C G C C A T C T C C 　　　　　　　　−2
−1 deletion of G416 and double transition at 414-415	35	AB 1157	+	8.8	406　　　　　　　　　　424 C A T C A C C C G C G C C A C A G G T 　　　　　　double　↑　−1 　　　　　transition A T

[a] The sequences appearing in this table are the wild-type sequences with the numbering defined by Sutcliffe (23). The bases involved in the deletion mutations are boxed with dotted lines. The different possibilities to obtain a given mutated sequence are shown. Mutant 35 has got a −1 deletion of a G at position 416 and a double transition, GC → AT, at position 414–415.

[b] From R. P. P. Fuchs et al., see ref. (15), with permission from the publisher.

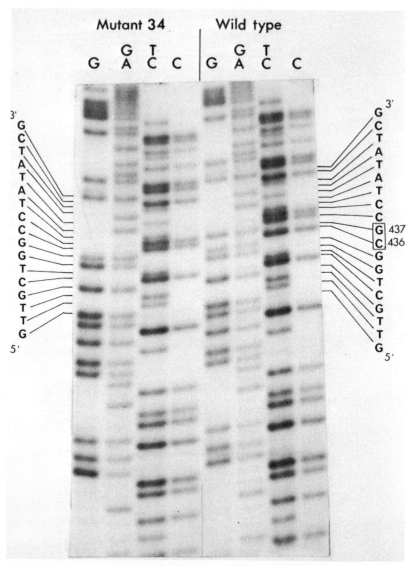

Fig. 3. Part of the sequence of mutant 34 and the wild-type DNA showing the −2 deletion of a CG doublet within the hot spot sequence GGCGCC.

not known however which of the guanine adducts [C(8) or N(2)] is the premutagenic lesion of the observed mutations. Let us consider that any of the 190 guanine residues of the 6 S fragment is a potential target for the premutagenic lesion. The probability that a mutation takes place at the same G residue 4 times within a small collection of mu-

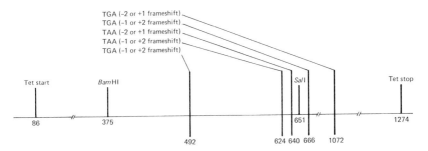

Fig. 4. Occurrence of stop codons in the tetracycline resistance gene by shifting the reading frame. The numbering has been altered to take into account the additional C within sequence 526–528.

tants is as low as 1.4×10^{-7}. It is therefore justified to refer to both sequences that are described above as hot spot sequences. The question remains to understand why these particular sites are hot spots for mutagenesis. First, we had to examine the possibility that the observed hot spots could arise as a bias in the selection procedure. The identification *in vivo* of the tetracycline resistance gene product (26) allows the determination of the reading frame. Among the 21 mutants that we have sequenced to date, 18 are frameshift mutants (25). From the wild-type sequence of the tetracycline resistance gene, it is easy to show that any frameshift (either ±1, or ±2) that takes place within the 6 S fragment results in an early nonsense codon TGA or TAA (Fig. 4). This observation rules out the possibility of a bias in the selection procedure of these frameshift mutants.

The reason why such particular sequences are hot spots for mutagenesis is not clear. Whether or not such sequences are hot spots for

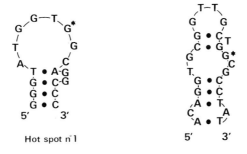

Fig. 5. Hypothetical hairpin structure at hot spot sequences 1 and 2. According to the insertion–denaturation model proposed by Fuchs and co-workers (10,13) there is a local denaturation of the helix around the guanine–AAF adduct that might favor the hairpin structure shown in this figure.

the AAF binding reaction itself is presently under investigation. Alternatively, it is more likely that the processing of the premutational lesions is strongly sequence dependent. It is noteworthy that at both hot spots the sequence is quasipalindromic (Fig. 5). The potential importance of quasipalindromic sequences in mutational hot spots has recently been stressed (27,28). Such hairpins are probably too short to be stable by themselves but might be high stabilized by the conformational change that -AAF introduces when bound to C-8 of guanine. As stated by the insertion–denaturation model proposed by Fuchs and co-workers (10,13), there is a local denaturation of the helix around the guanine–AAF adduct that might favor the hairpin structure shown in Fig. 5. Due to the multicopy state of plasmid pBR322 and to the recessivity of the mutations that are scored in our system, the conversion of the premutagenic lesion into a stable mutation most likely occurs simultaneously in both strands prior to replication.

The molecular mechanism by which the mutation is being fixed remains to be elucidated.

SUMMARY

The covalent binding of an ultimate carcinogen to the DNA bases or phosphate groups creates a premutational lesion that *in vivo* is processed by the repair, replication, and recombination enzymes, and eventually may be converted into a mutation. Being interested in the way that an initial premutational event is converted into a stable heritable mutation, we sequenced stable mutations in a gene that formed covalent adducts *in vitro* with N-acetoxy-N-2-acetylaminofluorene (N-AcO-AAF, a model for the ultimate metabolite of the rat liver carcinogen 2-acetylaminofluorene, AAF). *In vivo* studies have shown the mutagenicity of AAF and its derivatives in both bacterial and eukaryotic systems. N-AcO-AAF reacts *in vitro* with DNA leading mainly to the formation of a guanine adduct, N-2-(deoxyguanosin-8-yl)acetylaminofluorene (80%) and to at least three minor adducts (N. Schwartz, R. P. P. Fuchs, and M. P. Daune, unpublished results). Studies of our group showed that binding of N-AcO-AAF to DNA resulted in a local distortion of the DNA helix around the C-8 adduct (the insertion–denaturation model). We described the analysis of forward mutations induced in the tetracycline-resistance gene of pBR322 by directing the chemical reaction of the carcinogen to a small restriction fragment (*Bam*HI-*Sal*I) inside the antibiotic-resistance gene. Mutants were selected for ampicillin (Ap) resistance and tetracycline (Tc) sensitivity.

The plasmid DNA of such mutants was analyzed for sequence changes in the fragment where the AAF binding had been directed. We have shown that the mutations are mainly frameshifts involving GC base pairs and that certain base pairs (hotspots) are affected at high frequencies.

ACKNOWLEDGMENTS

This work has been supported by a Grant 79.7.0664 from the D.G.R.S.T. (Délégation Générale à la Recherche Scientifique et Technique).

REFERENCES

1. McCann, J., Choi, E., Yamasaki, E., and Ames, B. N. (1975). *Proc. Natl. Acad. Sci. U.S.A.* **72**, 5135–5139.
2. Radman, M. (1975) In "Molecular Mechanisms for Repair of DNA" (P. Hanawalt and R. B. Setlow, eds.), pp. 355–367. Plenum, New York.
3. Witkin, E. M. (1976) *Bacteriol. Rev.* **40**, 869–907.
4. Little, J. W., and Mount, D. W. (1982) *Cell* **29**, 11–22.
5. Ames, B. N., Gurney, E. G., Miller, J. A., and Bartsch, H. (1972) *Proc. Natl. Acad. Sci. U.S.A.* **69**, 3128–3132.
6. Santella, R. M., Fuchs, R. P. P., and Grunberger, D. (1979) *Mutat. Res.* **67**, 85–87.
7. Landolph, J. R., and Heidelberger, C. (1979) *Proc. Natl. Acad. Sci. U.S.A.* **76**, 930–934.
8. Kriek, E., Miller, J. A., Juhl, V., and Miller, E. C. (1967) *Biochemistry* **6**, 177–182.
9. Westra, J. G., Kriek, E., and Hittenhausen, H. (1976) *Chem.-Biol. Interact.* **15**, 149–164.
10. Fuchs, R. P. P., and Daune, M. P. (1972) *Biochemistry* **11**, 2659–2666.
11. Fuchs, R. P. P., and Daune, M. P. (1974) *Biochemistry* **13**, 4435–4440.
12. Fuchs, R. P. P. (1975) *Nature (London)* **257**, 151–152.
13. Fuchs, R. P. P., Lefèvre, J. F., Pouyet, J., and Daune, M. P. (1976) *Biochemistry* **15**, 3347–3351.
14. Grunberger, D., and Weinstein, I. B. (1979) In "Chemical Carcinogens and DNA" (P. Grover, ed.), Vol. 2, pp. 59–94. CRC Press, Boca Raton, Florida.
15. Fuchs, R. P. P., Schwartz, N., and Daune, M. P. (1981) *Nature (London)* **294**, 657–659.
16. Howard-Flanders, P., Boyce, R. P., and Theriot, L. (1966) *Genetics* **53**, 1119–1136.
17. Lefèvre, J. F., Fuchs, R. P. P., and Daune, M. P. (1978) *Biochemistry* **17**, 2561–2567.
18. de Murcia, G., Lang, M. C., Freund, A. M., Fuchs, R. P. P., Daune, M. P., Sage, E., and Leng, M. (1979) *Proc. Natl. Acad. Sci. U.S.A.* **76**, 6076–6080.
19. Cohen, S. N., Chang, A. C. Y., and Hsu, L. (1972) *Proc. Natl. Acad. Sci. U.S.A.* **69**, 2110–2114.
20. Clewell, D. B., and Helinski, D. R. (1969) *Proc. Natl. Acad. Sci. U.S.A.* **62**, 1159–1166.

21. Katz, L., Kingsbury, D. K., and Helinski, D. R. (1973) *J. Bacteriol.* **114**, 557–591.
22. Maxam, A. M., and Gilbert, W. (1977) *Proc. Natl. Acad. Sci. U.S.A.* **74**, 560–564.
23. Sutcliffe, J. G. (1979) *Cold Spring Harbor Symp. Quant. Biol.* **43**, 77–90.
24. Fuchs, R. P. P., and Seeberg, E. S. (1983). In preparation.
25. Schwartz, N., Daune, M. P., and Fuchs, R. P. P. (1983). In preparation.
26. Sancar, A., Heck, A. M., and Rupp, W. D. (1979) *J. Bacteriol.* **137**, 692–693.
27. Ripley, L. S. (1982) *Proc. Natl. Acad. Sci. U.S.A.* **79**, 4128–4132.
28. Todd, P. A., and Glickmann, B. W. (1982) *Proc. Natl. Acad. Sci. U.S.A.* **79**, 4123–4127.

Nitrated Polycyclic Aromatic Hydrocarbons: A Newly Recognized Group of Ubiquitous Environmental Agents

HERBERT S. ROSENKRANZ,* ELENA C. McCOY,* AND
ROBERT MERMELSTEIN†

*Center for the Environmental Health Sciences
and Department of Epidemiology and Community Health
School of Medicine
Case Western Reserve University
Cleveland, Ohio
and
†Joseph C. Wilson Center for Technology
Xerox Corporation
Webster, New York

Because nitrated chemicals are widely used as therapeutic agents in human and veterinary medicine (e.g., nitrofurans, nitroimidazoles, choramphenicol) and as intermediates in the chemical industry (nitrobenzenes, nitrotoluenes, nitroxylenes), their toxicology and their metabolic fate have been studied in some detail (see ref. 1). On the other hand, the nitrated polycyclic aromatic hydrocarbons (nitroarenes) have largely been laboratory curiosities, although the metabolism, carcinogenicity, and metabolism of some of the simpler ones (e.g., nitronaphthalenes) have been explored (see ref. 2). However, two recent developments have focused attention on the biological properties of the nitrated tri, tetra, and pentacyclic aromatic hydrocarbons. Thus it was found that nitropyrenes were among the most potent direct-acting mutagens for *Salmonella typhimurium* (3–5). In addition, the recognition that nitroarenes may have a ubiquitous environmental distribution has added some urgency to studies to determine their bio-

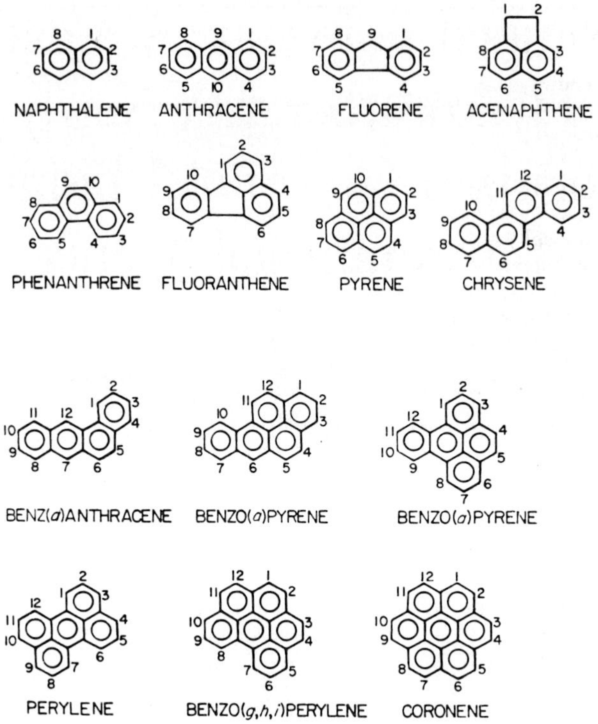

Fig. 1. Structures of polycyclic aromatic hydrocarbons.

logical properties. In addition, they are produced during incomplete combustion processes (2), notably in the diesel engine, and it has been predicted that, by 1990, diesel-powered cars may account for 25 to 50% of all passenger cars (6,7).

Although the nitropyrenes appear to be the most abundant of the nitroarenes in the environment (1-nitropyrene is present in approximately 100 ppm in diesel emissions), in excess of 60 other nitroarenes have been described in diesel exhausts. They occur in various condensed ring structures (Fig. 1) and can be mono- as well as polynitrated and may in addition contain a number of methyl or other alkyl substituents.

The nitroarenes present us with a fascinating biochemical problem as their mutagenicity is dependent upon structural features such as configuration of the rings and the number and position of the nitro substituents. Thus, isomeric nitroarenes may exhibit widely differing mutagenic activities (Table I).

Although the mutagenicity of nitroarenes has been studied most ex-

TABLE I
Mutagenicity of Nitroarenes for *Salmonella typhimurium* TA98

Chemical	Revertants per nanomole		Reference
	−S9	+S9	
1-Nitronaphthalene	0.05		8
2-Nitronaphthalene	0.2		8
2-Methyl-1-nitronaphthalene	0		9
1-Methyl-2-nitronaphthalene	0.2		9
3-Methyl-2-nitronaphthalene	1.0		9
1,3-Dinitronaphthalene	0.9		8
1,5-Dinitronaphthalene	3.3		8
1,8-Dinitronaphthalene	0		8
1,3,6,8-Tetranitronaphthalene	0.2		8
5-Nitroacenaphthene	2.5		10
2-Nitrofluorene	14		11
3-Nitro-9-fluorenone	383		12
2,7-Dinitrofluorene	471		11
2,7-Dinitro-9-fluorenone	1,459		11
2,4,7-Trinitro-9-fluorenone	2,125		11
2,4,5,7-Tetranitro-9-fluorenone	860		11
2-Nitroanthracene	892		13
9-Nitroanthracene	0.5		14
2-Nitrophenanthrene	<0.5		14
1-Nitrofluoranthene	74		14
3-Nitrofluoranthene	5,439		14
7-Nitrofluoranthene	544		14
8-Nitrofluoranthene	11,125		14
1-Nitropyrene	453		5
2-Nitropyrene	2,225		14
1,3-Dinitropyrene	144,760		5
1,6-Dinitropyrene	183,570		5
1,8-Dinitropyrene	254,000		5
1,3,6-Trinitropyrene	40,700		5
1,3,6,8-Tetranitropyrene	15,590		5
2-Nitrochrysene	<0.6	27	14
5-Nitrochrysene	<0.6	<2.7	15
6-Nitrochrysene	269		14
7-Nitrobenz[a]anthracene	0.3	1.4	14
6-Nitrobenzo[a]pyrene	0	466	16
1-(or 3-)Nitrobenzo[a]pyrene	1,567		16
3-(or 1-)Nitrobenzo[a]pyrene	1,070		16
1-Nitrobenzo[e]pyrene	39		14
3-Nitrobenzo[e]pyrene	890		14
3-Nitroperylene	<30	~1,784	14
4-Nitrobenzo[g,h,i]perylene	<0.6	~1,925	14
7-Nitrobenzo[g,h,i]perylene	<0.3	<0.6	14
1-Nitrocoronene	2.8	7.0	14
1-Nitrocarbazole	0	0	17
2-Nitrocarbazole	10.3		17
3-Nitrocarbazole	0.2	0.8	17
4-Nitrocarbazole	0.01	0.06	17

TABLE II
Gene Mutations in Cultured Mammalian Cells

Chemical	Response	Activation	Cell	Reference
2-Nitrofluorene	+	Not required	Mouse lymphoma L5178Y	18
2,4,7-Trinitro-9-fluorenone	+	—	Mouse lymphoma L5178Y	19
1-Nitropyrene	−	Absent	Chinese hamster ovary (CHO)	20
1-Nitropyrene	+	Required	Mouse lymphoma L5178Y	20, 21
1-Nitropyrene	−	—	Chinese hamster lung fibroblast	22
1,3-Dinitropyrene	+	—	Chinese hamster lung fibroblast	22
1,6-Dinitropyrene	+	—	Chinese hamster lung fibroblast	22
1,8-Dinitropyrene	+	—	Chinese hamster lung fibroblast	22
1,3,6-Trinitropyrene	+	—	Chinese hamster lung fibroblast	22
1,3,6,8-Tetranitropyrene	−	—	Chinese hamster lung fibroblast	22
1,8-Dinitropyrene	+	—	Mouse lymphoma L5178Y	23
1,8-Dinitropyrene	−	Absent	(Human) Xeroderma pigmentosum	24

tensively in *Salmonella,* recent findings indicate that they exert mutagenicity and genotoxicity for cultured mammalian cells as well (Tables II and III).

The biological activity of the nitroarenes depends upon their metabolism to ultimate mutagens. The arylhydroxylamines appear to be the ultimate or penultimate agents (2,31,32). They are capable of reacting at the C-8 position of the DNA guanine to form aminoaryl adducts. Thus, reduction of 1-nitropyrene followed by reaction with DNA results in the formation of the *N*-(deoxyguanosin-8-yl)-1-aminopyrene adduct (33). The specific expression of the mutagenicity, genotoxicity and DNA adduct formation, as well as presumably the carcinogenicity, depends upon the enzymic complement of the cell. A number of pathways have been demonstrated, and they, in turn, appear to be influenced by the substrate specificity of the enzyme involved. An overall scheme for the bioconversion of nitroarenes is presented (Fig. 2).

TABLE III
Genotoxic Effects of Nitroarenes

Chemical	Genetic endpoint[b]	System	Activation	Result	Reference
1-Nitropyrene	UDS	Primary rat hepatocyte	None	+	20
1-Nitropyrene	UDS	Human bronchus	None	+	25
1,3-Dinitropyrene	UDS	Human bronchus	None	+	25
1,6-Dinitropyrene	UDS	Human bronchus	None	+	25
Nitrated pyrenes[a]	UDS	Hela	None	+	26
Nitrated fluoranthenes[a]	UDS	Hela	None	+	26
Nitrated perylenes[a]	UDS	Hela	None	+	26
Nitrated chrysenes[a]	UDS	Hela	None	+	26
Nitrated anthracenes[a]	UDS	Hela	None	+	26
Nitrated benzo[a]pyrenes[a]	UDS	Hela	None	+	26
Nitrated benzo[e]pyrenes[a]	UDS	Hela	None	+	26
Nitrated benz[a]anthracenes[a]	UDS	Hela	None	+	26
Nitrated benzo[g,h,i]perylenes[a]	UDS	Hela	None	+	26
1,6-Dinitropyrene	Chromosome aberrations	Rat epithelial cell line	None	+	27
1,8-Dinitropyrene	Chromosome aberrations	Rat epithelial cell line		+	27
1,8-Dinitropyrene	Preferential toxocity	Human XP cells	None	−	24
2-Nitrofluorene	Inhibition of DNA synthesis	Hela	None	+	28
2-Nitrofluorene	SCE	CHO	Rat S9	+	29
1-Nitropyrene	SCE	CHO	Rat S9	+	29
1,8-Dinitropyrene	SCE	CHO	Rat S9	+	29
2,4,7-Trinitro-9-fluorenone	SCE	CHO	Rat S9	+	19
2-Nitrofluorene	Leukemia viral enhancement		None	±	30

[a] Mixtures
[b] Key to abbreviations: SCE, sister chromatic exchanges; CHO, Chinese hamster ovary cells; UDS, unscheduled DNA synthesis.

In bacteria (*Salmonella typhimurium* as well as the anaerobic flora of the colon), the aromatic structure of the nitroarenes is maintained intact. The initial event is reduction of the nitro moiety (Steps 1 and 3, Fig. 2), which is catalyzed by a family of nitroreductases differing in substrate specificity (31,32,34–36). *Salmonella* tester strains deficient in nitroreductase have been constructed (34,37,38). They are immune

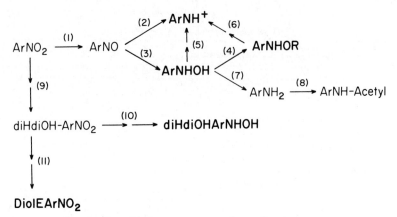

Fig. 2. Possible bioconversion scheme for nitroarenes.

to the mutagenic action of nitroarenes; however, this block can be bypassed by using the corresponding nitroso and hydroxylamino derivatives (Table IV, see also ref. 39). These results suggest that either the nitrosoarenes are the ultimate mutagens or that they are reduced further enzymically to the corresponding arylhydroxylamines.

However, because the nitrosoarenes tested (e.g., 2-nitrosofluorene) do not react extensively with DNA (Table V), it must be hypothesized either that they dismutate further nonenzymatically to form the reactive nitrenium ion (Step 2, Fig. 2) or that they are converted by a new enzyme(s) to the corresponding arylhydroxylamines (Table V) (Step 3, Fig. 2). These hydroxylamines have been shown capable of reacting

TABLE IV
Mutagenicity of Nitroarenes and Their Derivatives for Nitroreductase-Deficient *Salmonella typhimurium*

Chemical	Revertants per nanomole			
	TA100	TA100NR[a]	TA98	TA98NR
1-Nitronaphthalene	1.0	0		
N-Hydroxy-1-aminonaphthalene	10.6	11.8		
2-Nitronaphthalene	1.3	0.1		
N-Hydroxy-2-aminonaphthalene	15.0	16.1		
2-Nitrofluorene			84	2
2-Nitrosofluorene			542	504
N-Hydroxy-2-aminofluorene			1175	1266

[a] NR denoted strains deficient in the "classical" nitroreductase.

TABLE V
Effect of 2-Nitrofluorene and Derivatives on
the Thermal Helix-to-Coil Transition of DNA[a]

Treatment	T_m (°C)	ΔT_m
None	70.2	0
2-Nitrofluorene	69.7	−0.5
2-Nitrosofluorene	69.3	−0.9
N-Hydroxy-2-aminofluorene	63.2	−7.0
N-Acetoxy-2-acetylaminofluorene	65.4	−4.8

[a] Calf thymus DNA (1 mg/ml SSC/10) was reacted for 32 hr at 56°C in the absence or presence of the indicated fluorene derivatives (final concentration 0.3 mM), whereupon the DNA was precipitated and extensively washed with ethanol. The DNA fibers were redissolved in SSC/10 and processed for the determination of helix-to-coil transition profiles. The T_m is the temperature of the midpoint of the transition. SSC: 0.15 M NaCl in 0.015 M sodium citrate.

directly with DNA (Table V and refs. 40–42) and of forming nitrenium ions as well (Step 5, Fig. 2) (40,43–47). More frequently, however, the arylhydroxylamines are converted to the corresponding esters, which are reactive electrophiles capable of reacting with and forming DNA adducts (Step 4, Fig. 2) (43,48–52). This reaction has been thought to be rather nonspecific, involving a number of possible pathways.

Recently evidence was obtained that indicates that in *Salmonella*, at least, esterification appears to be under the control of a specific gene (53). This finding is of considerable interest and significance as arylhydroxylamines are also the postulated intermediates in the oxidation of arylamines (43). Evidence for the intermediacy of a specific gene product responsible for esterification was discovered during the characterization of a bacterial tester strain that was isolated, originally, as resistant to the mutagenicity of 1,8-dinitropyrene (35,54). This strain, TA98/1,8-DNP$_6$, is also resistant to the mutagenicity of 2-nitrofluorene but is fully sensitive to the mutagenicity of quercetin and of 7-bromomethyl-12-methylbenz(a)anthracene (Table VI). This would seem to indicate that the lesion in strain TA98/1,8-DNP$_6$ involves the classical nitroreductase (Step 1, Fig. 2). However, unlike the results obtained with TA98NR, the strain deficient in the classical nitroreductase (Table IV), the block in TA98/1,8-DNP$_6$ was not overcome by 2-nitrosofluorene (Table VI). This result would tend to suggest a

TABLE VI
Mutagenicity of Nitroarenes and Derivatives
for *Salmonella typhimurium* TA98/1,8-DNP$_6$

Chemical	Revertants per nanomole	
	TA98	TA98/1,8-DNP$_6$
BMBA[a]	14	17
Quercetin	0.9	1
1,8-Dinitropyrene	254,000	5,840
2-Nitrofluorene	84	2
2-Nitrosofluorene	542	79
N-Hydroxy-2-aminofluorene	1,175	31
N-Acetoxy-2-acetylaminofluorene	10	9

[a] BMBA, 7-Bromomethyl-12-methylbenz[a]anthracene.

*nitroso*reductase (Step 3, Fig. 2) as the deficiency. If this were the case, however, the block should be overcome by N-hydroxy-2-aminofluorene. The results (Table VI) clearly show that the blockage was not overcome in TA98/1,8-DNP$_6$. On the other hand, the defect in TA98/1,8-DNP$_6$ was overcome when N-acetoxy-2-acetylaminofluorene was used as the probe (Table VI), thus suggesting that the block involves Step 4 (Fig. 2) and further implying the existence of a specific esterification enzyme. [The low specific mutagenicity of N-acetoxy-2-acetylaminofluorene when compared to N-hydroxy-2-aminofluorene (Table VI) may reflect the fact that formation of the acetylaminofluorene-guanine adduct forces the DNA into the syn conformation, which does not result in a strongly mutagenic response(41,55–58)].

In mammalian cells, as well, the genotoxicity of nitroarenes appears to be primarily dependent upon the reduction of the nitro function (Steps 1 and 3, Fig. 2). Thus 1-nitropyrene can be activated to a genotoxicant by enzymes present in postmitochondrial preparations as well as by xanthine oxidase(33,59–61). This leads to the formation of the appropriate adduct (33) as well as of further reduction to amino- and N-acetylaminopyrene (Steps 7 and 8, Fig. 2) (20). The same end products have been detected in the culture medium of *Salmonella* exposed to this chemical (62,63). On the other hand, in mammalian cells (64), the nitro function may also "direct" the oxidation of the ring moiety (Step 9, Fig. 2), which is then followed by reduction of the nitro function to the corresponding hydroxylamine (Step 10, Fig. 2) and possibly esterification (Step 4). Thus, in addition to reduction to N-hydroxy-1-aminopyrene (33), 1-nitropyrene can also be converted, by cell-free extract, to 4,5-dihydro-4,5-dihydroxy-1-nitropyrene (4)(Step 9), which subsequently is reduced to the corresponding arylhydroxylamine (Step 10).

Similarly, 5-nitroacenaphthene is oxidized to the 1-hydroxy,2-hydroxy,1-oxo- and 2-oxo-5-nitroacenaphthene derivatives (Step 9), which subsequently are reduced to the corresponding hydroxylamines (64,65) (Step 10), which presumably are the penultimate mutagens.

Finally, although 6-nitrobenzo(a)pyrene is not mutagenic for *Salmonella* (Table I), in mammalian cells, or extracts thereof a series of intermediates (1-hydroxy- and 3-hydroxy-6-nitrobenzo(a)pyrene, 6-nitrobenzo(a)pyrene-1,9- and -3,9-hydroxyquinone and the desnitrobenzo(a)pyrene-3,6-quinone) are formed (Step 9) (66,67). These appear to be capable of further metabolism to mutagens (and/or carcinogens) without concomitant reduction of the nitro function (Step 11, Fig. 2).

In addition to the mutagenic and genotoxic effects on mammalian cells (Tables II and III), nitroarenes also possess transforming activity (Table VII) and demonstrated carcinogenicity in rodents (Table VIII). Six of the seven nitroarenes listed produced evidence of carcinogenicity in at least one species.

In considering the potential health effects of these nitroarenes, we may ask to what conclusions do these findings lead us? The most

TABLE VII
Cell Transforming Activity of Nitroarenes

Chemical	System	Activation required	Result	Reference
2-Nitrofluorene	Hamster BHK-21	+S9	+	68
	Syrian hamster embryo	Hamster hepatocytes	+	69
2-Nitronaphthalene	Syrian hamster embryo	None	+	69
1-Nitropyrene	Mouse BALB 3T3	None	−	70
1,3-Dinitropyrene	Mouse BALB 3T3	None	−	70
1,6-Dinitropyrene	Mouse BALB 3T3	None	−	70
1,8-Dinitropyrene	Mouse BALB 3T3	None	−	70
1,3,6-Trinitropyrene	Mouse BALB 3T3	None	−	70
1,3,6,8-Tetranitropyrene	Mouse BALB 3T3	None	−	70
1-Nitropyrene	Normal human fibroblasts	Anaerobiosis	+	33
6-Nitrobenzo[a]pyrene	Normal human fibroblasts	Anaerobiosis	+	33

TABLE VIII
Carcinogenicity of Nitroarenes

Chemical	Carcinogenicity	Reference
1-Nitronaphthalene	−	71
2-Nitronaphthalene	+	72
5-Nitroacenaphthene	+	71
2-Nitrofluorene	+	73
1-Nitropyrene	+	74
3-Nitrofluoranthene	+	74
6-Nitrochrysene	+ (Tumor Initiator)	79
3-Nitroperylene	+ (Tumor Initiator)	79

abundant source of nitroarenes are the diesel engines. It can be calculated that by 1990 the annual emission rate in the United States by passenger cars of 1-nitropyrene alone (one of the least mutagenic of the group) will be approximately 15,000 kg (7). This emission of 1-nitropyrene will be accompanied by the emission of approximately 60 other nitroarenes, some occurring in lesser amounts but with approximately 500 times the mutagenic potency of 1-nitropyrene. The nitroarenes present in diesel particulates seem to be bioavailable as evidenced by the fact that they are desorbed from diesel emission particulates by physiological solvents (75,76).

Although difficult and fraught with uncertainty, it would seem irresponsible to disregard this potent mutagenicity in the extrapolation of mutagenicity in *Salmonella* to health hazard in man. Moreover, some of the nitroarenes have indeed been shown to be carcinogenic for rodents (Table VIII). This then presents us with a rather unique situation. Those nitroarenes that have already been tested have been shown to possess unwanted properties (e.g., mutagenicity and carcinogenicity). There are at least another 60 or so nitroarenes that have not been evaluated yet. Hence, the exact magnitude of the possible health hazards cannot be quantitated. On the other hand, it has been demonstrated that the nitroarenes are not essential to the fuel-efficient diesel process. They are generated in the afterburn when the polycyclic aromatic hydrocarbons emitted come into contact with oxides of nitrogen and traces of acid (see ref. 7). Hence, the possibility exists of altering the technology in such a way as to minimize the formation of these biologically potent molecules. As a matter of fact, prototype afterburn units exist that can accomplish this (7). It would therefore seem wise to encourage industry to develop these devices further and to place them into production.

In evaluating the potential hazards of nitroarenes to the environment, it should be mentioned that they differ from other combustion products in that the nonspecific nitroreductases, which are universal enzymes (e.g., xanthine oxidase), are capable of activating them to biologically active chemicals. This, then, places them in a different class than the potent mutagens that are generated during the food preparation process (77,78) but which are dependent upon activation by liver enzymes for their biological potency. It would appear, then, that the nitroarenes could present a problem to the biosphere in general and especially to the germplasm of the highly inbred food-producing plants that could be subjected to genetic change as a result of the ubiquitous distribution of the nitroarenes.

ACKNOWLEDGMENTS

This investigation was supported by the National Institute of Environmental Health Sciences and the U.S. Environmental Protection Agency.

REFERENCES

1. Rickert, D. E., ed. (1983) "The Toxicity of Nitroaromatic Compounds." Hemisphere Publ. Co., New York (in press).
2. Rosenkranz, H. S., and Mermelstein, R. (1983) *Mutat. Res.* **114** (in press).
3. Löfroth, G., Hefner, E., Alfheim, I., and Møller, M. (1980) *Science* **209**, 1037–1039.
4. Rosenkranz, H. S., McCoy, E. C., Sanders, D. R., Butler, M., Kiriazides, D. K., and Mermelstein, R. (1980) *Science* **209**, 1039–1043.
5. Mermelstein, R., Kiriazides, D. K., Butler, M., McCoy, E. C., and Rosenkranz, H. S. (1981) *Mutat. Res.* **89**, 187–196.
6. Ingalls, M. N., and Bradow, R. L. (1981) *Air Pollut. Control Assoc. 74th Annu. Meet., Abstract*, 81–56.2.
7. Rosenkranz, H. S. (1982) *Mutat. Res.* **101**, 1–10.
8. McCoy, E. C., Rosenkranz, E. J., Petrullo, L. A., Rosenkranz, H. S., and Mermelstein, R. (1981) *Environ. Mutagen.* **3**, 499–511.
9. El-Bayoumi, K., Lavoie, E. J., Hecht, S. S., Fow, E. A., and Hoffmann, D. (1981) *Mutat. Res.* **81**, 143–153.
10. Rosenkranz, H. S., McCoy, E. C., and Mermelstein, R. (1983) *In* "Third Symposium on Application of Short-Term Bioassays in the Fractionation and Analysis of Complex Environmental Mixtures" (M. D. Waters *et al.*, eds.). Plenum, New York (in press).
11. McCoy, E. C., Rosenkranz, E. J., Rosenkranz, H. S., and Mermelstein, R. (1981) *Mutat. Res.* **90**, 11–20.
12. Pederson, T. C., and Siak, J.-S. (1981) *J. Appl. Toxicol.* **1**, 54–60.
13. Scribner, J. D., Fisk, S. R., and Scribner, N. K. (1979) *Chem.-Biol. Interact.* **26**, 11–25.
14. Nilsson, L., Löfroth, G., Toftgard, R., and Greibrokk, T. (1982) *Mutat. Res.* **97**, 208.
15. Tokiwa, H., Nakagawa, R., and Ohnishi, Y. (1981) *Mutat. Res.* **91**, 321–325.

16. Pitts, J. N., Jr., Lokensgard, D. M., Harger, W., Fisher, T. S., Majia, V., Schuler, J. J., Scorziell, G. M., and Katzenstein, Y. A. (1982) *Mutat. Res.* **103**, 241–249.
17. LaVoie, E. J., Govil, A., Briggs, G., and Hoffmann, D. (1981) *Mutat. Res.* **90**, 337–344.
18. Amacher, D. E., Paillet, S. C., and Turner, G. N. (1979) *Banbury Rep.* **2**, 277–289.
19. Burrell, A. D., Andersen, J. J., Jotz, M. M., Evans, E. L., and Mitchell, A. D. (1981) *Environ. Mutagen.* **3**, 360.
20. Ball, L. M., Kohan, M. J., Claxton, L., and Lewtas, J. (1982) *Abst. Chem. Ind. Inst. Toxicol. Conf. Toxicol.: Toxic. Nitroaromatic Compd.*, 5th 1982 p. 6.
21. Lewtas, J., Nishioka, M. G., and Petersen, B. A. (1982) *Abstr. Chem. Ind. Instit. Toxicol. Conf. Toxicol.: Toxic, Nitroaromatic Compd.*, 5th, 1982 p. 7.
22. Nakayasu, M., Sakamoto, H., Wakabayashi, K., Terada, M., Sugimura, T., and Rosenkranz, H. S. (1982) *Carcinogenesis (London)* **8**, 917–922.
23. Cole, J., Arlett, C. F., Lowe, J., and Bridges, B. A. (1982) *Mutat. Res.* **93**, 213–220.
24. Arlett, C. (1983) *In* "The Toxicity of Nitroaromatic Compounds" (E. C. Rickert, ed.). Hemisphere Publ. Co., New York (in press).
25. Kawachi, T. (1983) *In* "The Toxicity of Nitroaromatic Compounds" (D. E. Rickert, ed.). Hemisphere Publ. Co., New York (in press).
26. Campbell, J., Crumplin, G. C., Garner, J. V., Garner, R. C., Martin, C. N., and Rutter, A. (1981) *Carcinogenesis (London)* **2**, 559–565.
27. Danford, N., Wilcox, P., and Parry, J. M. (1982) *Mutat. Res.* **105**, 349–355.
28. Painter, R. B., and Howard, R. (1982) *Mutat. Res.* **92**, 427–437.
29. Nachtman, J. P., and Wolff, S. (1982) *Environ. Mutagen.* **4**, 1–5.
30. Yoshikura, H., Kuchino, T., and Matsushimu, T. (1979) *Cancer Lett.* **7**, 203–208.
31. Mermelstein, R., McCoy, E. C., and Rosenkranz, H. S. (1982) *In* "The Genotoxic Effects of Airborne Agents," pp. 369–396. Plenum, New York.
32. Rosenkranz, H. S., Karpinsky, G. E., Anders, M., Rosenkranz, E. J., Petrullo, L. A., McCoy, E. C., and Mermelstein, R. (1983) *In* "Induced Mutagenesis: Molecular Mechanisms and their Implications for Environmental Protection" (C. W. Lawrence, L. Prakash, and F. Sherman, eds.) Plenum, New York (in press).
33. Howard, P. C., and Beland, F. A. (1982) *Biochem. Biophys. Res. Commun.* **104**, 727–732.
34. Rosenkranz, H. S., and Mermelstein, R. (1980) *In* "The Predictive Value of Short-Term Screening Tests in the Evaluation of Carcinogenicity" (G. M. Williams, R. Kroes, H. W. Waaijers, and K. W. van de Poll, eds.), pp. 5–26. Elsevier/North-Holland, Amsterdam.
35. McCoy, E. C., Rosenkranz, H. S., and Mermelstein, R. (1981) *Environ. Mutagen.* **3**, 421–427.
36. Bryant, D. W., McCalla, D. R., Leeksma, M., and Laneuville, P. (1981) *Can. J. Microbiol.* **27**, 81–86.
37. Rosenkranz, H. S., and Speck, W. T. (1975) *Biochem. Biophys. Res. Commun.* **66**, 520–525.
38. Rosenkranz, H. S., and Poirier, L. A. (1979) *JNCL, J. Natl. Cancer Inst.* **62**, 873–892.
39. Wirth, P. J., Alewood, P., Calder, I., and Thorgeirsson, S. S. (1982) *Carcinogenesis (London)* **3**, 167–170.
40. Kriek, E. (1965) *Biochem. Biophys. Res. Commun.* **20**, 793–799.
41. Sage, E., and Leng, M. (1980) *Proc. Natl. Acad. Sci., U.S.A.* **77**, 4597–4601.
42. Spodheim-Maurizot, M., Saint-Ruf, G., and Leng, M. (1980) *Carcinogenesis (London)* **1**, 807–812.
43. Miller, J. A. (1970) *Cancer Res.* **30**, 559–576.

44. Frederick, C. B., Mays, J. B., Ziegler, D. M., Guengerich, F. P., and Kadlubar, F. F. (1982) *Cancer Res.* **42**, 2671–2677.
45. Kadlubar, F. F., Unruh, L. E., Flammang, T. J., Sparks, D., Mitchum, R. K., and Mulder, G. J. (1981) *Chem.-Biol. Interact.* **33**, 129–147.
46. Sakai, S., Reinhold, C. E., Wirth, P. J., and Thorogeirsson, S. S. (1978) *Cancer Res.* **38**, 2058–2067.
47. Schut, H. A. J., Wirth, P. J., and Thorgeirsson, S. S. (1978) *Mol. Pharmacol.* **14**, 682–692.
48. King, C. M., and Phillips, B. (1968) *Science* **159**, 1351–1353.
49. Weisburger, J. H., Yamamoto, R. S., Williams, G. M., Grantham, P. H., Matsushima, and Weisburger, E. K. (1972) *Cancer Res.* **32**, 491–500.
50. Lotlikar, P. D., and Luha, L. (1971) *Mol. Pharmacol.* **7**, 381–388.
51. Weeks, C. E., Allaben, W. T., Tresp, N. M., Louie, S. C., Lazear, E. J., and King, C. M. (1980) *Cancer Res.* **40**, 1204–1211.
52. Weeks, C. E., Allaben, W. T., Louie, S. C., Lazear, E. J., and King, C. M. (1978) *Cancer Res.* **38**, 613–618.
53. McCoy, E. C., McCoy, G. D., and Rosenkranz, H. S. (1982) *Biochem. Biophys. Res. Comm.* **108**, 1362–1367.
54. Rosenkranz, E. J., McCoy, E. C., Mermelstein, R., and Rosenkranz, H. S. (1982) *Carcinogenesis* (London) **3**, 121–123.
55. Grunberger, D., and Santella, R. M. (1981) *J. Supramol. Struct. Cell. Biochem.* **17**, 231–244.
56. Leng, M., Ptak, M., and Rio, P. (1980) *Biochem. Biophys. Res. Commun.* **96**, 1095–1102.
57. Santella, R. M., Kriek, E., and Grunberger, D. (1980) *Carcinogenesis (London)* **1**, 897–902.
58. Santella, R. M., Grunberger, D., Broyde, S., and Hingerty, B. E. (1981) *Nucleic Acids Res.* **9**, 5459–5467.
59. Claxton, L. D., Kohan, M., and Lewtas, J. (1982) *Environ. Mutagen.* **4**, 382.
60. Kohan, M., and Claxton, L. (1981) *EPA Diesel Emission Symp., 1981* Abstracts.
61. Pederson, T. C., and Siak, J.-S. (1981) *J. Appl. Toxicol.* **1**, 61–66.
62. Quilliam, M. A., Messier, F., Lu, C., Andrews, P. A., McCarry, B. E., and McCalla, D. R. (1982) "Polynuclear Aromatic Hydrocarbons: Physical and Biological Chemistry," Vol. 6 (M. Cooke, A. J. Dennis and G. L. Fisher, eds.) pp. 667–672. Battelle Press.
63. Messier, F., Lu, C., Andrews, P., McCarry, B. E., Quilliam, M. A., and McCalla, D. R. (1981) *Carcinogenesis (London)* **2**, 1007–1011.
64. El-Bayoumi, K. (1982) In "The Toxicity of Nitroaromatic Compounds" (D. E. Rickert, ed.). Hemisphere Publ. Co., New York (in press).
65. El-Bayoumi, K., and Hecht, S. S. (1982) *Cancer Res.* **42**, 1243–1248.
66. Fu, P. P., Chou, M. W., Yang, S. K., Beland, F. A., Kadlubar, F. F., Casciano, D. A., Heflich, R. H., and Evans F. E. (1982) *Biochem. Biophys. Res. Commun.* **105**, 1037–1043.
67. Tong, S., and Selkirk, J. K. (1982) *Proc. Am. Assoc. Cancer Res.* **23**, 80.
68. Styles, J. A. (1981) "Mammalian Cell Transformation by Chemical Carcinogens" (N. Mishra, V. Dunkel, and M. Mehlman, eds.), pp. 85–131. Senate Press, Inc., Princeton Junction, New Jersey.
69. Pienta, R. J. (1980) "The Predictive Value of Short-Term Screening Tests in Carcinogenicity Evaluation" (G. M. Williams, R. Kroes, H. W. Waaijers, and K. W. Van de Poll, eds.), pp. 149–169. Elsevier/North-Holland, Amsterdam.

70. Tu, A. S., Sivak, A., and Mermelstein, R. (1982) *Abst., Chem. Ind. Inst. Toxicol. Conf. Toxicol.: Toxic. Nitroaromatic Compd., 5th, 1982* p. 8.
71. Griesemer, R. A., and Cueto, C., Jr. (1980) *IARC Sci. Publ.* **27**, 259–281.
72. Poirier, L. A., and Weisburger, E. K. (1979) *J. Natl. Cancer Inst.* **62**, 833–840.
73. Weisburger, E. K., and Weisburger, J. H. (1958) *Adv. Cancer Res.* **5**, 331–431.
74. Ohgaki, H., Matsukura, N., Morino, K., Kawachi, T., Sugimura, T., Morita, K., Tokiwa, H., and Hirota, T. (1982) *Cancer Lett.* **15**, 1–7.
75. King, L. C., Tejada, S. B., and Lewtas, J. (1981) *EPA Diesel Emission Symp., Oct. 5–7, 1981* Abstract.
76. King, L. C., Kohan, M. J., Austin, A. C., Claxton, L. D., and Huisingh, J. L. (1981) *Environ. Mutagen.* **3**, 109–121.
77. Sugimura T., and Nagao, M. (1979) *CRC Crit. Rev. Toxicol.* **6**, 189–209.
78. Nagao, M., Wakabayashi, K., Kasai, H., Nishimura, S., and Sugimura, T. (1981) *Carcinogenesis (London)* **2**, 1147–1149.
79. El-Bayoumi, K., Hecht, S. S., and Hoffmann, D. (1982) *Cancer Lett.* **16**, 333–337.

PART II

MULTISTAGE EVENTS

Cellular Events in Multistage Carcinogenesis

I. BERNARD WEINSTEIN, ANN HOROWITZ,
ALAN JEFFREY, AND VESNA IVANOVIC

Division of Environmental Sciences
and
Cancer Center/Institute of Cancer Research
Columbia University
New York, New York

MULTISTAGE CARCINOGENESIS

There is increasing appreciation of the fact that, in experimental animals and in humans, carcinogenesis is a multistage process that can proceed over a considerable fraction of the life-span of the individual and that the evolution of a fully malignant tumor is subject to a variety of promoting, as well as inhibitory, factors. These basic phenomena are also becoming apparent during the transformation of cells in culture, whether the process is induced by chemical carcinogens or by certain oncogenic viruses. It seems likely, therefore, that a full understanding of the molecular mechanisms underlying the carcinogenic process will require an understanding of these complex events (1).

Multistage carcinogenesis and tumor progression are often thought of in terms of a series of successive mutations and selections, eventuating in the clonal outgrowth of a fully malignant tumor. This may, however, be an oversimplification. As discussed in detail elsewhere (1,2), we think that a more suitable model is the multistage process that occurs during normal embryologic development, in which new stem cell populations emerge and develop into specialized cells and tissues. There is considerable evidence that the successive stages in carcinogenesis may involve qualitatively different events, that they can be enhanced or inhibited by quite different types of environmental and host factors, and that at least the early stages are often reversible (1–3).

One of the most instructive experimental models of multistage carcinogenesis has been the system of two-stage mouse skin carcinogenesis, which has been extensively analyzed by Berenblum (4), Van Duuren (5), Boutwell (6), and Hecker (7). These studies clearly define two qualitatively different stages, initiation and promotion, and indicate that each stage can be brought about by quite different types of agents. Recent studies indicate that the process of tumor promotion can also be divided into a least two phases (6,8). Elsewhere, we have contrasted the various properties of initiators and promoters (1,2,9). Tumor promoters can be defined as compounds that have very weak or no carcinogenic activity when tested alone, but markedly enhance tumor yield when applied repeatedly following a low or suboptimal dose of a carcinogen (initiator). At the biochemical level, it appears that the major difference between initiators and promoters is that initiators (or their metabolites) bind covalently to cellular DNA, whereas the primary site of action of the phorbol ester tumor promoters appears to be cell membranes (1,2).

Although current evidence suggests that the covalent binding of certain initiating carcinogens or their metabolites to cellular DNA is a critical event in the initiation process, the subsequent series of biochemical and genetic events that lead to the conversion of a normal cell to a cancer cell are not known (1). The initiating event is often thought of as a simple random point mutation resulting from errors in DNA replication at the sites of carcinogen damage. Several features of the carcinogenic process, however, including the high efficiency of carcinogen-induced cell transformation when compared to specific locus mutations, the lengthy latency period needed for the expression of the transformed state, and the multistep nature of carcinogenesis, are not consistent with this mechanism (1). An alternative mechanism would be that carcinogen-induced DNA damage might induce complex genomic changes, including, for example, gene rearrangements or gene amplification (1,2,10). Consistent with this hypothesis is the increasing evidence that gene amplification and gene rearrangements occur in several eukaryotic systems undergoing development and differentiation. It is possible, therefore, that carcinogen–DNA interactions might disrupt the cellular and molecular mechanisms that control such genetic events, and thus produce aberrations in the control of gene expression. We have begun a series of studies to explore this possibility (10–12), and these are discussed in the article by Kirschmerer et al. in this volume.

MECHANISM OF ACTION OF TUMOR PROMOTERS

Within the past few years cell culture systems have provided a wealth of information on the action of 12-O-tetradecanoyl phorbol-13-acetate (TPA) and related phorbol ester tumor promoters (2,13–15). These compounds can exert highly pleiotropic effects on the growth, function, and differentiation of a variety of cell types. We have found it convenient to classify these effects into three categories: A) mimicry and enhancement of transformation, B) modulation of differentiation, C) and membrane effects (2).

A number of the cell culture effects of the phorbol ester tumor promoters led us to suggest that they act by binding to and usurping the function of membrane-associated receptors that are normally utilized by an endogenous growth factor (2,9,16). In this sense, tumor promotion induced by the phorbol esters may be closely related to the more general phenomenon of hormonal carcinogenesis. We have recently found, for example, that when rodent fibroblast cultures are infected with an oncogenic adenovirus or are exposed to X-ray, and then grown in the presence of epidermal growth factor (EGF), this growth factor enhances cell transformation to an extent comparable to that obtained with TPA (17). Like TPA, EGF also enhances the *in vitro* progression of adenovirus transformed rat embryo fibroblasts (18). Subsequent studies have indicated that EGF also markedly enhances the *in vitro* transformation of murine granulosa cells by the Kirsten strain of murine sarcoma virus (19). One could readily imagine, therefore, that in the intact animal the endogenous level of growth factors and/or the responsiveness of a given tissue to growth factors would be important host factors in determining whether exposure of that tissue to a chemical, physical, or viral agent will eventuate in neoplasia. Exogenous or endogenous growth factors might also influence tumor progression and the development of variant cell populations in fully established tumors.

Utilizing [^{3}H]phorbol dibutyrate (PDBu), several laboratories have recently obtained direct evidence for specific high affinity saturable "phorboid receptors" in membrane preparations and intact cells from avian, rodent, and human tissues (2,20–23). Membrane-associated phorboid receptors have also been found in diverse tissues and cell types (2,20–23). We have determined the binding constants and receptor numbers in several different types of tissue culture cells (24). In general, one or two classes of [^{3}H]PDBu receptors are found in each cell type. In most cases, a high affinity site, with a K_D in the range of 3 to 17 nM is present at $0.2-4 \times 10^5$ sites per cell. Where a second, lower affinity site is observed the K_D is 300 nM or greater, and the number of sites is greater than 1×10^6 per cell. Some variation in receptor number occurs within a cell line depending on the growth state of the cells at the time of assay. When only one binding site is detected the affinity for PDBu appears to be lower (K_D of 12–41 nM), but the number of sites is, in general, higher. It is of interest that in two rat brain tumor cell lines displaying glial-like properties as many as 2×10^6 high affinity sites per cell are present. This observation is in accord with the results of Nagle et al. (25), showing that brain tissue contains a very large number of phorboid receptors. We have also demonstrated PDBu receptors in normal human keratinocytes and in normal and transformed human melanocytes (26).

In general, the abilities of a series of phorbol esters to compete with [³H]PDBu for binding to cell surface receptors correlates with their known potencies in cell culture and with their activities as tumor promoters on mouse skin. These results provide evidence that the phorboid receptors mediate the biologic action of the phorbol esters. We have also found that the antileukemic plant diterpenes gnidilatin and gnilatimacrin are potent inhibitors of [³H]PDBu binding (Horowitz and Weinstein, unpublished studies). As discussed below, it appears that the phorboid receptors also mediate the TPA-like effects of two new classes of tumor promoters, teleocidin (and structurally related indole alkaloids) and aplysiatoxin.

Although the putative endogenous ligand for the phorboid receptors has not yet been identified, we have partially purified a factor present in normal human and rodent sera that inhibits [³H]PDBu-receptor binding (22,27). The serum factor inhibits [³H]PDBu binding in both intact monolayer cultures of the rat embryo cell line CREF N and in a subcellular system containing membranes from these cells. Inhibition occurred at both 37° and 4°C and was rapid and reversible. An analysis of [³H]PDBu binding in the presence of the serum factor indicated that inhibition of [³H]PDBu binding by the serum factor was noncompetitive. Using gel filtration to separate the serum factor from free [³H]PDBu, we obtained evidence that the serum factor does not act by binding or trapping the [³H]PDBu. Unlike the phorbol ester tumor promoters, the serum factor alone did not stimulate the release of choline or arachidonic acid from cellular phospholipids, nor did it inhibit the binding of ^{125}I-labeled epidermal growth factor to cellular receptors. The factor did, however, antagonize the inhibition of epidermal growth factor binding induced by PDBu. Sera from pregnant women were, in general, more inhibitory of [³H]PDBu binding than were those from nonpregnant women, which were more inhibitory than those from men. Taken together, our results indicate that the serum factor inhibits [^{3}H]PDBu binding by a direct physical effect at the level of the phorboid receptors or their associated membranes. It would appear that if this factor acts *in vivo*, then it might antagonize certain effects of this class of tumor promoters. The normal physiologic role of this factor and its possible effects on carcinogenesis remain to be determined (27).

We have postulated that the phorboid receptor system could play a role during embryogenesis by enhancing the outgrowth of new stem cell populations (1,2). In the adult, this same system might enhance stem cell replication during hyperplasia, wound healing, and regeneration. During tumor promotion, aberrant stem cells (generated during initiation) might undergo preferential clonal expansion as a result of stimulation of the phorboid receptor system. If our hypothesis is correct, then it might be possible to develop analogs of the phorbol esters that could be used as pharmacologic agents to enhance normal tissue repair or to enhance the repletion of tissues with stem cells following trauma, radiation, or drug toxicity. Alternatively, it might be possible to design agents that would block the phorboid receptors and thus protect the host from certain endogenous or exogenous agents that affect tumor promotion or progression.

Until recently, the phorbol esters were the only class of skin tumor promoters that showed marked structure–function specificity and that were active at nanomolar concentrations. T. Sugimura, H. Fujiki, R. E. Moore, and their colleagues (28–30) have found, however, that the indole alkaloids teleocidin (isolated from *Streptomyces*) and dihydroteleocidin and a polyacetate compound aplysiatoxin (isolated from a marine algae) are as potent as TPA in inducing ornithine decarboxylase in mouse skin (28–30). Dihydroteleocidin and aplysiatoxin have also been demonstrated to be potent tumor promoters on mouse skin (27–29; see also Fujiki *et al.*, this volume). Teleocidin and aplysiatoxin also induce many of the same effects as TPA in cell culture, including effects on growth, differentiation, and membrane structure and function (27–29,31). The chemical structures of teleocidin and aplysiatoxin are quite different from those of the phorbol esters (Fig. 1). However, the fact that these compounds share similar biologic effects suggested to us that they might act by binding to the phorboid receptors. Indeed, in collaborative studies with Drs. H. Fujiki, T. Sugimura, K. Umezawa, and R. E. Moore, we have found that both teleocidin and aplysiatoxin are potent inhibitors of the binding of [^{3}H]PDBu to membrane receptors (32,33). These two compounds are also equipotent with TPA in inhibiting [^{125}I]EGF receptor binding and in stimulating

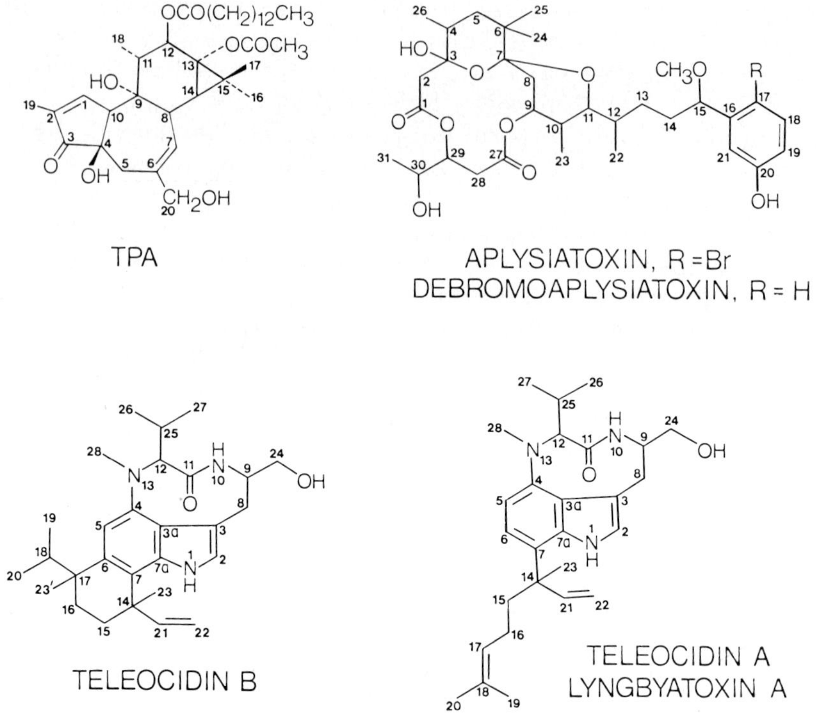

Fig. 1. Chemical structures of TPA, teleocidin B, and aplysiatoxin.

the release of [³H]arachidonic acid and [³H]choline from prelabeled cellular phospholipids (33).

These results prompted us to study the stereochemistry of these compounds to see if they might display structural similarities (33). All three types of compounds are amphipathic since they have both hydrophobic and hydrophilic domains (Fig. 1). In the case of the phorbol esters, there is evidence that all of the biologically active compounds have a highly hydrophobic residue on the 12 position, although the precise chemical structure of this residue is not critical. Presumably this region of the molecule is required for a relatively nonspecific hydrophobic interaction with a region on the phorboid receptor or the adjacent lipid microenvironment. The saturated six-membered ring of teleocidin and the side chain attached to the polyacetate ring of aplysiatoxin might play analogous roles. Extensive structure–activity studies of the phorbol esters on mouse skin and in cell culture (3–7) indicate that the region of the molecule containing the 3-keto, 4-OH,

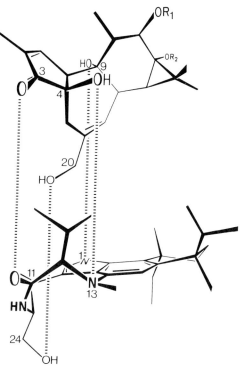

Fig. 2. Perspective drawings of TPA (top) and dihydroteleocidin B (bottom). The dotted lines connect heteroatoms whose spatial positions correspond with one another and represent residues that could form hydrogen bonds with a putative receptor.

and 6-CH_2OH residues displays marked structural and steric specificity. Our model building studies indicate that the nine-membered lactam region of teleocidin and the macrocyclic polyacetate ring system of aplysiatoxin (Fig. 1) can assume conformations which are remarkably similar to the corresponding region present in the biologically active phorbol esters (33). We postulate, therefore, that the respective regions of the phorbol esters, teleocidin, and aplysiatoxin form highly specific chemical bonding and/or steric interactions with the phorboid receptors. These relationships for a phorbol ester and teleocidin are displayed in Fig. 2. Our model is consistent with published data on the stereochemistry of these compounds. Precise model building studies with aplysiatoxin can not be done at the present time because its stereochemistry has not been elucidated. Additional compounds are being examined to obtain further data relevant to this hypothesis. Information of this type could make it possible to rationally design compounds that would either act as agonists or blockers of the phorboid receptor system.

PROMOTER-LIKE ACTIVITIES OF POLYCYCLIC AROMATIC HYDROCARBONS

Because with repeated applications benzo[a]pyrene (BP) and certain other polycyclic aromatic hydrocarbon (PAH) carcinogens can act as complete carcinogens on mouse skin, it seemed possible that PAHs might induce certain effects on membranes that are similar to those induced by the phorbol ester tumor promoters. One of the earliest and most sensitive membrane effects of the phorbol ester tumor promoters is inhibition of the binding of epidermal growth factor (EGF) to its cell surface receptors (16). This same effect has also been seen with teleocidin and aplysiatoxin (32,33). It was of interest, therefore, to determine whether BP and related PAHs might induce similar effects on EGF receptor binding, thus mimicking the action of certain tumor promoters. We have indeed found that the exposure of C3H 10T1/2 cells to BP and certain other PAHs leads to a loss of EGF-receptor binding (36–38). A Scatchard analysis revealed that BP treatment acts mainly by reducing the effective numbers, rather than affinities, of both high and low affinity EGF receptors. To further understand the mechanism of action, we analyzed the possible role of BP metabolites in BP-mediated inhibition of EGF binding. Confluent 10T1/2 cells were exposed to 1 μM [^{3}H]BP for 24 hr, and the ethyl acetate extract of the medium was analyzed by high-pressure liquid chromatography (HPLC). The HPLC profile indicated that about 90% of the recovered radioactivity was unmetabolized BP. The remaining radioactivity represented the following metabolites: BP-9,10-diol, BP-7,8-diol, BP-3,6-quinone, BP-9-phenol, and BP-3-phenol. Equimolar concentrations (1–4 μM) of these metabolites, and of several other BP phenols and BP quinones, exhibited less inhibition of EGF–receptor binding than that obtained with BP (38). In view of the latter results, the relatively small amounts of BP metabolites compared to the parent compound in 10T1/2 cultures, and the inability of the ultimate metabolite BP-7,8-diol-9,10-oxide to inhibit EGF binding, it seems likely that BP itself rather than a metabolite(s) is responsible for the inhibition of EGF binding. Progesterone, 17β-estradiol, benzo[e]pyrene, cholesterol, phenobarbital, 1,1-bis(p-chlorophenyl)-2,2,2-trichloroethane, hexachlorobenzene, or pregnenlone-16α-carbonitrile, did not inhibit EGF binding. On the other hand, several known inducers of P_1-450 were very effective inhibitors of EGF binding. These included: dimethylbenz[a]anthracene, 3-methylcholanthrene, benzo[a]pyrene, benz[a]anthracene, β-naphthoflavone, and α-naphthoflavone. These and other results have led us to propose that binding of certain compounds to the cytosolic "Ah" receptor induces a pleiotropic program that includes not only increases in certain drug metabolizing enzymes but also changes in membrane structure and function (36). Thus certain PAHs might be complete carcinogens because they not only are converted to metabolites that bind covalently to cellular DNA, but also because they induce membrane changes that alter growth and differentiation. Consistent with this hypothesis are recent studies indicating that 2,3,7,8-tetrachlorodibenzo-p-dioxin (TCDD), which binds to the Ah receptor, acts as a potent mouse skin tumor promoter in an appropriate strain of mice (this volume and Ref. 39). It is of interest that the glucorticoid hormones, which are extremely effective inhibitors of tumor promotion on mouse skin (40,41) are potent antagonists of some of the membrane-related effects of both TPA and BP (37). Thus reciprocal effects of certain agents on membranes may explain their tumor promoter or tumor inhibitory properties. Additional inhibitors of tumor promotion might be designed based on this rationale.

CHROMOSOMAL EFFECTS AND ACTIVATED OXYGEN

Although we have stressed the membrane and epigenetic effects of tumor promoters, other investigators have suggested that tumor promoters act by inducing chromosomal aberrations and/or segregation (42). A few studies have shown that TPA can induce sister chromatid exchange (SCE) as well as various chromosomal aberrations (34,42,43). These effects have not, however, been highly reproducible between laboratories, often require high concentrations of the TPA, are sometimes seen only with certain batches of TPA, and do not occur with all cell types. Moreover, we know that when a population of cells is exposed to TPA the entire population can undergo phenotypic changes within minutes or hours, and the changes are usually reversible when the agent is removed (3). In addition, the papillomas on mouse skin often regress when application of the tumor promoter is stopped (4). Thus, chromosome aberrations or segregation could not account for the early effects of tumor promoters. It is true that the malignant tumors that appear much later on mouse skin are autonomous and often show chromosomal abnormalities. Thus, if the *in vitro* chromosomal effects of TPA have any significance we suggest that they relate more to late stages in the process, such as tumor progression, rather than to tumor promotion. Consistent with this possibility is the fact that, whereas the early stages of neoplasia and benign tumors are often associated with a diploid karyotype, tumor progression and highly malignant tumors are often associated with chromosomal abnormalities (44).

Troll et al. (45) have demonstrated that TPA and teleocidin induce oxygen radicals and peroxides in polymorphonuclear leukocytes. The stimulation of arachidonic acid metabolism and lipid turnover (2) by tumor promoters could generate highly reactive forms of activated oxygen (35) that could lead to lipid peroxidation and other toxic effects, including chromosomal damage. Very recent studies indicate that TPA (10^{-8} M) can induce DNA strand breakage in human peripheral blood polymorphonuclear leukocytes. The authors also suggest that this is due to formation of activated oxygen species as part of the respiratory burst induced in these phagocytic cells by TPA (46). It remains to be determined whether this TPA-induced DNA damage is peculiar to these specialized cells or can also occur in other cell types.

To summarize this aspect, we would stress the need to determine whether or not the chromosomal effects reported with TPA are confined to only certain cell types, whether they occur only under conditions of toxicity, and whether they can actually be associated with tumor promotion or progression on mouse skin. This subject is discussed in further detail by Cerutti et al. (34) in this volume.

SUMMARY

Carcinogenesis is a multistep process resulting from a complex interaction between multiple factors, both environmental and endogenous. A number of environmental chemicals initiate the carcinogenic process by generating metabolites that bind covalently to cellular DNA. On the other hand, recent studies on the mechanism of action of certain tumor-promoting agents indicate that they act by binding to specific cell surface receptors, thus altering membrane structure and functions. Presumably, these alterations in membrane properties lead indirectly to changes in the expression of genes controlling growth and differentiation. Certain stereochemical requirements for the interaction of diverse compounds with this receptor system have been elucidated. In addition, evidence has been obtained that certain polycyclic aromatic hydrocarbon carcinogens can induce, via more indirect mechanisms, alterations in cell membranes that are somewhat similar to those produced by the phorbol ester tumor promoters. The latter effects may explain the abilities of these compounds to serve as complete carcinogens. Further studies are required to assess the effects of tumor promoters on chromosome structure and the possible significance of these effects on tumor promotion and progression. The diverse cellular targets and biologic effects of tumor initiators and promoters suggest that both genetic and epigenetic events are involved in multistep carcinogenesis.

ACKNOWLEDGMENTS

This research was supported by DHS, NCI Grants CA 021111, and CA 26056. The authors wish to acknowledge the valuable collaboration with Drs. T. Sugimura and H. Fujiki of the National Cancer Center Research Institute, Tokyo, K. Umezawa of the Cancer Institute, Tokyo, and R. E. Moore of the University of Hawaii in the studies of teleocidin and aplysiatoxin. We thank Patricia Kelly for assistance in preparing this manuscript. A. H. was supported by a gift from the Dupont Company.

REFERENCES

1. Weinstein, I. B. (1981) *J. Supramol. Struct. Cell. Biochem.* **17**, 99–120.
2. Weinstein, I. B., Mufson, R. A., Lee, L. S., Fisher, P. B., Laskin, J., Horowitz, A., and Ivanovic, V. (1980) *In* "Carcinogenesis: Fundamental Mechanisms and Environmental Effects" (B. Pullman, P. O. P. Ts'o, and H. Gelboin, eds.), pp. 543–563. Reidel Publ., Dordrecht, Netherlands.
3. Slaga, T. J., Sivak, A., and Boutwell, R. K., eds. (1978) "Carcinogenesis," Vol. 2. Raven Press, New York.
4. Berenblum, I. (1975) *In* "Cancer: A Comprehensive Treatise" (F. F. Becker, ed.), pp. 323–344. Plenum, New York.
5. Van Duuren, B. L. (1969) *Prog. Exp. Tumor Res.* **11**, 31–68.
6. Boutwell, R. K. (1974) *CRC Crit. Rev. Toxicol.* **2**, 419–443.
7. Hecker, E. (1975) *In* "Handbuch der allgemeinen Pathologie" (E. Grundmann, ed.), Vol. IV, Part 6, pp. 651–676. Springer-Verlag, Berlin and New York.
8. Slaga, T. J., Fisher, S. M., Nelson, K., and Gleason, G. L. (1980) *Proc. Natl. Acad. Sci. U.S.A.* **77**, 3659–3663.
9. Weinstein, I. B., Wigler, M., and Pietropaolo, C. (1977) *Cold Spring Harbor Conf. Cell Proliferation* **4**, 751–772.
10. Kirschmeier, P., Gattoni-Celli, S., Perkins, H., Hsiao, W., and Weinstein, I. B., this volume.
11. Gattoni, S., Kirschmeier, P., Weinstein, I. B., Escobedo, J., and Dina, D. (1982) *Mol. Cell. Biol.* **2**, 42–51.
12. Kirschmeier, P., Gattoni-Celli, S., Dina, D., and Weinstein, I. B. (1982) *Proc. Natl. Acad. Sci. U.S.A.* **79**, 273–277.
13. Blumberg, P. M. (1980) *CRC Crit. Rev. Toxicol.* **8**, 153–197.
14. Blumberg, P. M. (1981) *CRC Crit. Rev. Toxicol.* **8**, 199–238.

15. Diamond, L., O'Brien, T. G., and Baird, W. M. (1980) *Adv. Cancer Res.* **32**, 1–74.
16. Lee, L. S., and Weinstein, I. B. (1978) *Science* **202**, 313–315.
17. Fisher, P. B., Mufson, R. A., Weinstein, I. B., and Little, I. B. (1981) *Carcinogenesis* **2**, 183–188.
18. Fisher, P. B., Bozzone, J. H., and Weinstein, I. B. (1979) *Cell* **18**, 695–705.
19. Harrison, J., and Auersperg, N. (1981) *Science* **213**, 218–219.
20. Delclos, K. B., Nagle, D. S., and Blumberg, P. M. (1980) *Cell* **19**, 1025–1032.
21. Driedger, P. E., and Blumberg, P. M. (1980) *Proc. Natl. Acad. Sci. U.S.A.* **77**, 567–571.
22. Horowitz, A. D., Greenebaum, E., and Weinstein, I. B. (1981) *Proc. Natl. Acad. Sci. U.S.A.* **78**, 2315–2319.
23. Shoyab, M., and Todaro, G. J. (1980) *Nature (London)* **288**, 451–455.
24. Horowitz, A. D., Nicolaides, M., Greenebaum, E., Woodward, K., Giotta, G., and Weinstein, I. B. (1983) Submitted for publication.
25. Nagle, D. S., Jaken, S., Castagna, M., and Blumberg, P. M. (1981) *Cancer Res.* **41**, 89.
26. Greenebaum, E. M., Nicolaides, M., Eisinger, M., Vogel, R. H., and Weinstein, I. B. (1983) *JNCI, J. Natl. Cancer Inst.* (in press).
27. Horowitz, A. D., Greenebaum, E., and Weinstein, I. B. (1982) *Mol. Cell. Biol.* **2**, 545–553.
28. Fujiki, H., Mori, M., Nakayasu, M., Terada, M., and Sugimura, T. (1979) *Biochim. Biophys. Res. Commun.* **90**, 976–983.
29. Fujiki, H., Mori, M., and Sugimura, T. (1981) *Proc. Natl. Acad. Sci. U.S.A.* **78**, 3872–3876.
30. Fujiki, H., Suganama, M., Nakayasu, M., Hoshino, H., Moore, R. E., and Sugimura, T. (1982) *Gann* **73**, 497–499.
31. Fisher, P. B., Miranda, A. F., Mufson, R. A., Weinstein, L. S., Fujiki, H., Sugimura, T., and Weinstein, I. B. (1982) *Cancer Res.* **42**, 2829–2835.
32. Umezawa, K., Weinstein, I. B., Horowitz, A. D., Fujiki, H., Matsuschima, T., and Sugimura, T. (1981) *Nature (London)* **290**, 411–413.
33. Horowitz, A., Fujiki, H., Weinstein, I. B., Jeffrey, A., Okin, E., Moore, R. E., and Sugimura, T. (1983). *Cancer Res.*, in press.
34. Cerutti, P. A., Emerit, I., and Amstad, P., this volume.
35. Fridovich, I. (1978) *Science* **201**, 875.
36. Ivanovic, V., and Weinstein, I. B. (1981) *Carcinogenesis* **3**, 505–510.
37. Ivanovic, V., and Weinstein, I. B. (1981) *Nature (London)* **293**, 404–406.
38. Ivanovic, V., Okin, E., and Weinstein, I. B. (1982) *Proc. Am. Assoc. Cancer Res.* **23**, Abstr. 221.
39. Poland, A., Palen, D., and Glover, E. (1982) *Nature (London)* **300**, 271–273.
40. Belman, S., and Troll, W. (1972) *Cancer Res.* **32**, 450–454.
41. Viage, A., Slaga, T. J., Wigler, M., and Weinstein, I. B. (1977) *Cancer Res.* **37**, 1530–1536.
42. Kinsella, A. R., and Radman, M. (1978) *Proc. Natl. Acad. Sci. U.S.A.* **75**, 6149.
43. Nagasawa, H., and Little, J. B. (1979) *Proc. Natl. Acad. Sci. U.S.A.* **76**, 1943.
44. Klein, G. (1979) *Proc. Natl. Acad. Sci. U. S. A.* **76**, 2442.
45. Troll, W., Witz, G., Goldstein, B., Stone, D., and Sugimura, T. (1982) *In* "Cocarcinogens and Biological Effects" (E. Hecker, ed.), pp. 593–597. Raven Press, New York.
46. Birnboim, H. C. (1982) *Science* **215**, 1247–1249.

Potent New Mutagens From Pyrolysates of Proteins and New Naturally Occurring Potent Tumor Promoters

HIROTA FUJIKI AND TAKASHI SUGIMURA

National Cancer Center Research Institute
Chuo-ku, Tokyo, Japan

INTRODUCTION

It has been claimed that most human cancers are caused by environmental factors (1,2). Since the two-step concept of chemical carcinogenesis in tumor formation in skin, liver, bladder, and other organs is now widely accepted (3–9), it seems reasonable to classify environmental factors as initiators and promoters. The detections of new initiators and new promoters are necessary for understanding the mechanism of development of human cancer and may eventually lead to methods for cancer prevention.

The first part of this paper is concerned with a series of potent mutagens discovered in pyrolysates of amino acids and proteins and in broiled meat and fish (10–16). These potent mutagens are heterocyclic amines, and their adducts with DNA have been isolated (17,18). Some of these mutagens have already been found to be carcinogenic in animal experiments (19).

The second part of this paper describes the discovery to two new classes of potent tumor promoters, indole alkaloids (20–22) and polyacetates (23,24). These two classes of compounds were found to exhibit similar tumor promoting activities to phorbol esters, such as 12-O-tetradecanoylphorbol-13-acetate (TPA), in mouse skin. It is suggested that the actions of these compounds may be mediated via the same receptor system as that for TPA (25–27).

NEW MUTAGENS FROM PYROLYSATES OF AMINO ACIDS, PROTEINS, AND PROTEINACEOUS FOODS

Like the typical carcinogen benzo[a]pyrene, many biologically active compounds are produced in the process of burning. Since Japanese food is often cooked over a naked flame, we made a special attempt to look for initiators in pyrolysis products formed in this way. Using the most convenient method for detecting mutagens, Ames's test, we found that many mutagens are formed when food is cooked in this way (28). Much mutagenic activity was produced by broiling sundried sardines, a favorite food in Japan (10). Pyrolysis at above 200°C produced various types of mutagens through successive radical reactions. The formation of mutagens in hamburger increased with the duration of cooking (10). This potent mutagenic activity was found to be mainly produced by pyrolyzing proteins. Since proteins consist mainly of various amino acids, we subjected various amino acids and related compounds to pyrolysis. As shown in Fig. 1, we found that tryptophan, glutamic acid, phenylalanine, lysine, and ornithine yielded several potent new mutagens (11). From a tryptophan pyrolysate, two amino γ-carboline derivatives, 3-amino-1,4-dimethyl-5H-pyrido[4,3-b]indole (Trp-P-1) and 3-amino-1-methyl-5H-pyrido[4,3-b]indole (Trp-P-2), were isolated (29,30). We also obtained two new dipyrido imidazole derivatives from a glutamic acid pyrolysate; i.e., 2-amino-6-methyldipyrido[1,2-a:3′,2′-d]imidazole (Glu-P-1) and 2-amino-dipyrido[1,2-

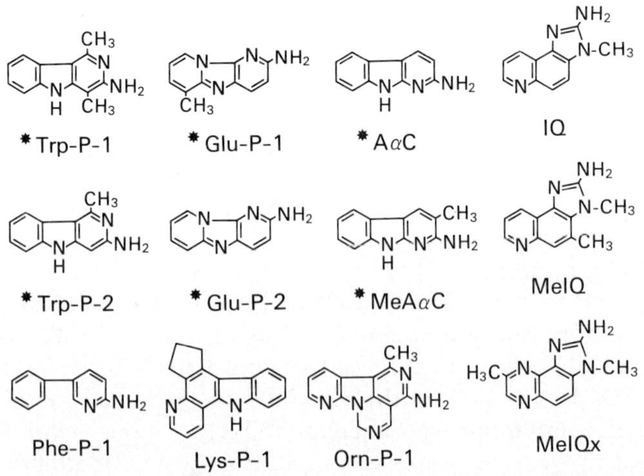

Fig. 1. Structures of new mutagens isolated from pyrolysates. The six compounds indicated by asterisks (*) were shown to be carcinogenic in in vivo animal experiments.

a:3′,2′-d]imidazole (Glu-P-2) (31). The structures of these compounds were determined by X-ray crystallography. These compounds have now been synthesized chemically.

In addition, Yoshida *et al.* (32) obtained two amino-α-carboline derivatives, 2-amino-9H-pyrido[2,3-b]indole (AαC) and 2-amino-3-methyl-9H-pyrido[2,3-b]indole (MeAαC) from pyrolysates of soy bean globulin. We also isolated three very strong mutagens, 2-amino-3-methylimidazo[4,5-f]quinoline (IQ) and 2-amino-3,4-dimethylimidazo[4,5-f]quinoline (MeIQ) from broiled sardines (33–35) and 2-amino-3,8-dimethylimidazo[4,5-f]quinoxaline (MeIQx) from broiled beef (36). The six compounds indicated by stars in Fig. 1 have been shown to be carcinogenic *in vivo* in animal experiments (19).

The specific mutagenic activities toward *Salmonella typhimurium* TA98 of all these new mutagens are shown in the left column of Fig. 2. The specific activities of MeIQ and Phe-P-1 differ greatly. The derivatives of amino imidazoquinolines, amino imidazoquinoxaline, amino γ-carbolines like Trp-P-1 and Trp-P-2, and Orn-P-1 and Glu-P-1, showed higher specific mutagenic activities toward TA98 than typical carcinogens such as AF-2, aflatoxin B_1 and 4NQO (37).

These heterocyclic amines are activated by microsomal enzymes,

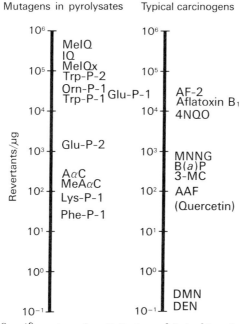

Fig. 2. Specific mutagenic activity toward *S. typhimurium* TA98.

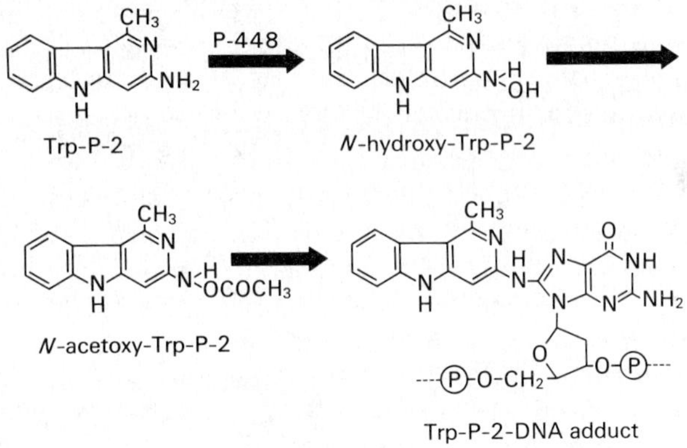

Fig. 3. Metabolic activation of Trp-P-2.

and especially cytochrome P-448, which can be induced in rat liver by PCB or 3-methylcholanthrene (38). The pathway for the metabolic activation of Trp-P-2 is given in Fig. 3. From Trp-P-2, an N-hydroxyamino compound is obtained through activation by P-448 (39). This compound may be further acylated to an N-acylhydroxyamino compound (17). The ultimate form is assumed to make an adduct with C-8 of guanine base. Other environmental heterocyclic amines might also react with DNA through the same mechanism (18).

The carcinogenicities of Trp-P-1, Trp-P-2, Glu-P-1, Glu-P-2, AαC, and MeAαC have already been demonstrated *in vivo* in animal feeding experiments (16,19). Interestingly, these six potent mutagens were found to be hepatocarcinogens in mice. In all cases, female mice were more susceptible than males for development of hepatomas. Moreover, four compounds, Glu-P-1, Glu-P-2, AαC, and MeAαC, also produced hemangioendotheliosarcomas in brown adipose tissue, which is located in the interscapular region (16). The three strongest mutagens produced during cooking of sardines and beef, MeIQ, IQ, and MeIQx, are now being examined in long-term *in vivo* animal experiments.

NEW NATURALLY OCCURRING POTENT TUMOR PROMOTERS

Even though several potent mutagens and carcinogens were found in food and in pyrolysis products, we realized that carcinogenesis may not be understood only through the study of initiators, if cancer is caused by a two-step mechanism. With this in mind, we began screening for new tumor promoters, based on the assumption that potent tumor promoters other than TPA may be present in our environment (40). In screening for tumor promoters, we used a series of screening tests (23).

The first test was an irritant test on mouse ear (41). Compounds were applied to the ear skin as solutions of various concentrations in 10 µl of acetone, and irritancy was estimated by redness of the ear 24 h later. Of 270 compounds tested in this manner, 43 induced significant redness.

Second, we tested the induction of ornithine decarboxylase (ODC) activity in mouse skin (42). ODC induction was determined 4 hr after application of the compound. Of the 43 compounds referred to above, 15 induced ODC activity with the same potency as TPA.

We realized recently that, by testing the effects of compounds in causing adhesion of human promyelocytic leukemia cells (HL-60), compounds can be classified as strong or weak tumor promoters (24). The value of the test of adhesion of HL-60 cells will be discussed later in the section on polyacetates. Up to now, 5 of the 15 compounds that induced ODC activity have been tested for adhesion of human pro-

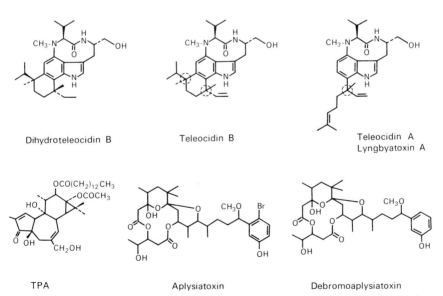

Fig. 4. Structures of various tumor promoters. The carbon atoms shown in circles have two variable positions, i.e., positions R and S.

myelocytic leukemia cells (HL-60) and also for carcinogenicity in animal experiments (21–24,43). Three of these (dihydroteleocidin B, teleocidin, which is a mixture of teleocidin A and B (23), and lyngbyatoxin A) are indole alkaloids (21), and the other two compounds, aplysiatoxin and debromoaplysiatoxin, are polyacetates (23). The chemical structures of these compounds are shown in Fig. 4.

INDOLE ALKALOIDS: DIHYDROTELEOCIDIN B, TELEOCIDIN, AND LYNGBYATOXIN A

Dihydroteleocidin B was obtained by catalytic hydrogenation of one of the isomers of teleocidin B (44). Teleocidin, a strong skin irritant, was originally isolated from mycelia of *Streptomyces mediocidicus* by Dr. Takashima in Japan (45). We also purified teleocidin from a crude extract of *Streptomyces*. A methanolic extract of the mycelia was subjected to LH-20 column chromatography and then rechromatographed on a silica gel column. The teleocidin thus obtained, giving a single spot on thin-layer chromatography, was used for experiments. We called this preparation as "teleocidin," but, when it was further subjected to high-performance liquid chromatography (HPLC), three peaks were obtained. The molecular weight of material in the first peak was 437, and the molecular weight of material in the second and third peaks was 451. To our surprise, each of these three peaks was further resolved into two peaks under different conditions of HPLC. In other words, our "teleocidin" preparation was a mixture of teleocidin A and teleocidin B. Teleocidin A has two isomeric forms, 14S- and 14R-, each with a molecular weight of 437, whereas teleocidin B has four isomeric forms, 14S- or 14R-, and 17S- or 17R-, each with a molecular weight of 451. The structures of these 6 isomers are now being elucidated by NMR and mass spectrophotometry (H. Fujiki, M. Suganuma, Z. Yamaizumi, H. Saito, and T. Sugimura, unpublished results).

Lyngbyatoxin A was first isolated from the blue-green alga *Lyngbya majuscula* by Dr. R. E. Moore of the University of Hawaii (46). Lyngbyatoxin A is known to be the causative agent of swimmer's itch in Hawaii. Recently we were struck as was Dr. Moore, by the fact that lyngbyatoxin A is identical with one of the two isomers of teleocidin A (H. Fujiki, M. Suganuma, and T. Sugimura, unpublished results).

Various biological effects of dihydroteleocidin B, teleocidin, and lyngbyatoxin A were compared with those of TPA (Table I). The ID_{50}^{24} in the irritant test represents the amount of the compound that reddened the ears of 50% of the mice in 24 hr. These compounds strongly

TABLE I
Effects of Various Tumor Promoters

Effect	TPA	Dihydroteleocidin B	Teleocidin	Lyngbyatoxin A	Aplysiatoxin	Debromoaplysiatoxin
Irritant test (ID_{50}^{24} nmoles/ear)	0.016	0.017	0.008	0.011	0.005	0.005
Induction of ODC (nmoles CO_2/5.0 μg compound)	1.45	1.55	1.89	2.05	2.15	2.05
Adhesion of HL-60 (ED_{50} ng/ml)	1.5	0.3	4.0	7.0	2.0	180
Phagocytosis of HL-60 (ED_{30} ng/ml)	2.5	1.4	3.6	2.5	1.7	100
Aggregation of NL-3 (ED_{50} ng/ml)	11.2	6.5	3.1	2.4	2.1	180
Inhibition of differentiation of Friend erythroleukemia cells (ED_{50} ng/ml)	1.0	0.2	2.0	0.4	—	150
Inhibition of specific binding of [^{3}H]PDBu (ED_{50} ng/ml)	3.0	—	5.0	8.0	4.0	52.0
Tumor incidence in week 30 (%)	100	90	100	80	87[a]	20[a]

[a] Tumor incidences in groups given DMBA plus aplysiatoxin and DMBA plus debromoaplysiatoxin were determined in week 20.

induced ODC activity (20,21). They also induced both adhesion and differentiation of HL-60 cells, enhanced aggregation of human lymphoblastoid cells (NL-3), and inhibited differentiation of Friend erythroleukemia cells, showing similar potencies to that of TPA (20,21,47,48).

IN VIVO CARCINOGENICITY TEST WITH DIHYDROTELEOCIDIN B, TELEOCIDIN, AND LYNGBYATOXIN A

Carcinogenesis was initiated in the skin of 8-week-old female CD-1 mice with 100 μg of 7,12-dimethylbenz[a]anthracene (DMBA). From 1 wk later, either 2.5 μg of dihydroteleocidin B or 2.5 μg of TPA was given twice a week until week 30, as a positive control. The group treated with DMBA plus dihydroteleocidin B showed 90% tumor incidence, and the group treated with DMBA plus TPA showed 100% (21,22).

Figure 5 shows the tumor incidence in the group treated with DMBA plus teleocidin or with DMBA plus TPA (49). These two groups showed 100% tumor incidence. No tumors were observed in the group treated with teleocidin alone or with a single application of DMBA. The group treated with DMBA plus teleocidin had 4.0 tumors per mouse, whereas the group given DMBA plus TPA had 9.8 per mouse. The tumors in these two groups were classified histologically into two types: squamous cell carcinomas and papillomas. Histological analysis revealed that teleocidin produced more cancers than TPA

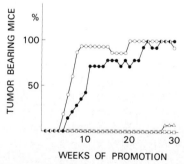

Fig. 5. Tumor incidences in mice treated with DMBA plus teleocidin (●) and DMBA plus TPA (○). One animal in the group treated with TPA alone had a tumor (△). No tumors were observed in groups treated with teleocidin alone or with DMBA alone (□).

did. The incidence of squamous cell carcinomas among all those examined histologically was 13.2% in the group treated with DMBA plus teleocidin and 2.2% in the group treated with DMBA plus TPA. Similar results were obtained in in vivo carcinogenicity tests with lyngbyatoxin A (23,43). Details of the experiments with lyngbyatoxin A will be described elsewhere.

POLYACETATES: APLYSIATOXIN AND DEBROMOAPLYSIATOXIN

In August 1980, over 400 people who swam at Kailua beach, Oahu island, suffered from acute vesicular dermatitis. Since Dr. Moore had already succeeded in isolating lyngbyatoxin A as a causative agent of swimmer's itch, he tried at this time to isolate the causative agent of this dermatitis (46). His work resulted in the identification of aplysiatoxin and debromoaplysiatoxin, a debrominated form of aplysiatoxin, as irritant substances present in an extract of a blue-green alga that grew on Kailua beach (Dr. R. E. Moore, unpublished results). Aplysiatoxin and debromoaplysiatoxin were first found in the digestive tract of the sea hare, *Stylocheilus longicauda*, by Drs. Kato and Scheuer at the University of Hawaii (50,51). Because the sea hare grazes on the blue-green alga, *Lyngbya majuscula*, the toxins in its digestive tract probably come from the blue-green alga (52). In collaboration with Dr. Moore's group, we carried out the various biological tests shown in Table I on aplysiatoxin and debromoaplysiatoxin. The effective concentrations of aplysiatoxin and debromoaplysiatoxin in the irritant test, and ODC induction were very similar to that of TPA (24). Moreover the concentrations of aplysiatoxin inducing adhesion and differentiation of HL-60 cells, and aggregation of NL-3 cells were similar to those of TPA (24). However, 100 times more debromoaplysiatoxin was required to achieve the same effects on HL-60 cells and NL-3 cells (21,23,24). On the basis of these results, we conclude that aplysiatoxin is a potent tumor promoter like TPA and indole alkaloids and that debromoapolysiatoxin is a weak tumor promoter. In week 20 of an *in vivo* carcinogenicity experiment now in progress on aplysiatoxin and debromoaplysiatoxin, 87% tumor incidence was observed in the group treated with DMBA plus aplysiatoxin, and 20% tumor incidence in the group treated with DMBA plus debromoaplysiatoxin (24). No tumors were observed in the group treated with only aplysiatoxin or only debromoaplysiatoxin.

It is noteworthy that aplysiatoxin, which is structurally unrelated to phorbol esters and indole alkaloids, showed very potent tumor promoting activity and also that removal of the bromine molecule from aplysiatoxin causes remarkable differences in its biological potencies.

Our group in Tokyo, in collaboration with Dr. Weinstein's group at Columbia University, obtained strong evidence that TPA, indole alkaloids, and aplysiatoxin all inhibited the specific binding of [^{3}H]PDBu to rat embryo fibroblasts in the same way (Table I) (25). Each promoter possesses both hydrophobic and hydrophilic domains. The bromine molecule in aplysiatoxin plays a crucial role in expansion of the hydrophobic region. Studies by computer graphics are in progress on whether these three different classes of tumor promoters have structural similarities (Dr. I. B. Weinstein, personal communication).

Recently several new interesting results with TPA and indole alkaloids were obtained by our collaborators. Dr. Sakiyama's group reported that the synthesis of 32,000-dalton protein increased approximately twofold within 2 h after exposure of BALB/c3T3 cells to TPA (53) and indole alkaloids (54). Dr. Hiai presented the evidence that pseudoemperipolesis (localization of leukemia cells under or between adherent cells), which is the basic cell interaction in symbiotic complexes, was inhibited by TPA (55) and teleocidin (56). Furthermore, Dr. Kakunaga's group found that dihydroteleocidin B markedly enhanced malignant cell transformation induced by 3-methylcholanthrene (57). Dihydroteleocidin B was at least 100 times more effective in enhancing transformation than TPA in experiments using cultured mouse cells (A31-1-1). The discovery of two new classes of tumor promoters now gives us a tool to study the biological mechanisms of tumor promotion.

ACKNOWLEDGMENTS

This work was supported in part by Grants-in-Aid for Cancer Research from the Ministry of Education, Science and Culture, the Ministry of Health and Welfare of Japan, and the Princess Takamatsu Cancer Research Fund. Work on the isolations of lyngbyatoxin A, aplysiatoxin, and debromoaplysiatoxin used in this study was supported by Grant CA 12632-09 to Dr. R. E. Moore at the University of Hawaii from the National Cancer Institute, Health and Human Services. We thank Drs. R. E. Moore and I. B. Weinstein for their valuable collaborative efforts.

REFERENCES

1. Doll, R. (1977) *Nature (London)* **265**, 589–596.
2. Wynder, E. L., Hoffmann, D., McCoy, G. D., Cohen, L. A., and Reddy, B. S. (1978)

In "Carcinogenesis, Mechanisms of Tumor Promotion and Cocarcinogenesis" (T. J. Slaga, A. Sivak, and R. K. Boutwell, eds.), Vol. 2, pp. 59–77. Raven Press, New York.
3. Berenblum, I. (1941) *Cancer Res.* **1**, 807–814.
4. Hecker, E. (1967) *Naturwissenschaften* **54**, 282–284.
5. Van Duuren, B. L. (1969) *Prog. Exp. Tumor Res.* **11**, 31–68.
6. Boutwell, R. K. (1977) *Cold Spring Harbor Conf. Cell Proliferation* **4**, (Book B), 773–783.
7. Peraino, C., Fry, R. J. M., and Staffeldt, E. (1971) *Cancer Res.* **31**, 1506–1512.
8. Hicks, R. M., Wakefield, J. St. J., and Chowaniec, J. (1975) *Chem.-Biol. Interact.* **11**, 225–233.
9. Narisawa, T., Magadia, N. E., Weisburger, J. H., and Wynder, E. L. (1974) *JNCI, J. Natl. Cancer Inst.* **55**, 1093–1097.
10. Sugimura, T., Nagao, M., Kawachi, T., Honda, M., Yahagi, T., Seino, Y., Sato, S., Matsukura, N., Matsushima, T., Shirai, A., Sawamura, M., and Matsumoto H. (1977) *Cold Spring Harbor Conf. Cell Proliferation* **4**, Book C, 1561–1566.
11. Nagao, M., Yahagi, T., Kawachi, T., Seino, Y., Honda, M., Matsukura, N., Sugimura, T., Wakabayashi, K., Tsuji, K., and Kosuge; T. (1977) *In* "Progress in Genetic Toxicology" (D. Scott, B. A. Bridges, and F. H. Sobels, eds.), pp. 259–264. Elsevier/North-Holland, Amsterdam.
12. Sugimura, T. (1979) *In* "Naturally Occurring Carcinogens-Mutagens and Modulators of Carcinogenesis" (E. C. Miller, J. A. Miller, I. Hirono, T. Sugimura, and S. Takayama, eds.), pp. 241–261. Jpn Sci Soc. Press, Tokyo.
13. Sugimura, T., Kawachi, T., Nagao, M., and Yahagi, T. (1981) *In* "Nutrition and Cancer: Etiology and Treatment" (G. R. Newell and N. M. Ellison, eds.), pp. 59–71. Raven Press, New York.
14. Sugimura, T. (1982) *In* "Molecular Interrelations of Nutrition and Cancer" (M. S. Arnott, J. van Eys, and Y.-M. Wang, eds.), pp. 3–24. Raven Press, New York.
15. Sugimura, T. (1982) *In* "Environmental Mutagens and Carcinogens" (T. Sugimura, S. Kondo, and H. Takebe, eds.), pp. 3–20. Alan R. Liss, Inc., New York.
16. Sugimura, T. (1982) *Cancer* **49**, 1970–1984.
17. Hashimoto, Y., Shudo, K., and Okamoto, T. (1979) *Chem. Pharm. Bull.* **27**, 1058–1060.
18. Hashimoto, Y., Shudo, K., and Okamoto, T. (1980) *Biochem. Biophys. Res. Commun.* **92**, 971–976.
19. Matsukura, N., Kawachi, T., Morino, K., Ohgaki, H., Sugimura, T., and Takayama, S. (1981) *Science* **213**, 346–347.
20. Fujiki, H., Mori, M., Nakayasu, M., Terada, M., and Sugimura, T. (1979) *Biochem. Biophys. Res. Commun.* **90**, 976–983.
21. Fujiki, H., Mori, M., Nakayasu, M., Terada, M., Sugimura, T., and Moore, R. E. (1981) *Proc. Natl. Acad. Sci. U.S.A.* **79**, 3872–3876.
22. Sugimura, T., Fujiki, H., Mori, M., Nakayasu, M., Terada, M., Umezawa, K., and Moore, R. E. (1982) *Carcinog.-Comp. Surv.* **7**, 69–73.
23. Fujiki, H., Sugimura, T., and Moore, R. E. (1982) *Environ. Health Perspect.* **50** (in press).
24. Fujiki, H., Suganuma, M., Nakayasu, M., Hoshino, H., Moore, R. E., and Sugimura, T. (1982) *Jpn. J. Cancer Res.* **73**, 497–499.
25. Umezawa, K., Weinstein, I. B., Horowitz, A., Fujiki, H., Matsushima, T., and Sugimura, T. (1981) *Nature (London)* **290**, 411–413.
26. Yamamoto, H., Katsuki, T., Hinuma, Y., Hoshino, H., Miwa, M., Fujiki, H., and Sugimura, T. (1981) *Int. J. Cancer* **28**, 125–129.

27. Solanki, V., and Slaga, T. J. (1981) *Proc. Natl. Acad. Sci. U.S.A.* **78**, 2549–2553.
28. Ames, B. N., McCann, J., and Yamasaki, E. (1975) *Mutat. Res.* **31**, 347–364.
29. Sugimura, T., Kawachi, T., Nagao, M., Yahagi, T., Seino, Y., Okamoto, T., Shudo, K., Kosuge, T., Tsuji, K., Wakabayashi, K., Iitaka, Y., and Itai, A. (1977) *Proc. Jpn. Acad.* **53**, 58–61.
30. Kosuge, T., Tsuji, K., Wakabayashi, K., Okamoto, T., Shudo, K., Iitaka, Y., Sugimura, T., Kawachi, T., Nagao, M., Yahagi, T., and Seino, Y. (1978) *Chem. Pharm. Bull.* **26**, 611–619.
31. Yamamoto, T., Tsuji, K., Kosuge, T., Okamoto, T., Shudo, K., Takeda, K., Iitaka, Y., Yamaguchi, K., Seino, Y., Yahagi, T., Nagao, M., and Sugimura, T. (1978) *Proc. Jpn. Acad., Ser. B* **54**, 248–250.
32. Yoshida, D., Matsumoto, T., Yoshimura, R., and Matsuzaki, T. (1978) *Biochem. Biophys. Res. Commun.* **83**, 915–920.
33. Kasai, H., Nishimura, S., Wakabayashi, K., Nagao, M., and Sugimura, T. (1980) *Proc. Jpn. Acad., Ser. B* **56**, 382–384.
34. Kasai, H., Yamaizumi, Z., Wakabayashi, K., Nagao, M., Sugimura, T., Yokoyama, S., Miyazawa, T., and Nishimura, S. (1980) *Chem. Lett.* pp. 1391–1394.
35. Kasai, H., Yamaizumi, Z., Wakabayashi, K., Nagao, M., Sugimura, T., Yokoyama, S., Miyazawa, T., Spingarn, N. E., Weisburger, J. H., and Nishimura, S. (1980) *Proc. Jpn. Acad., Ser. B* **56**, 278–283.
36. Kasai, H., Yamaizumi, Z., Shiomi, T., Yokoyama, S., Miyagawa, T., Wakabayashi, K., Nagao, M., Sugimura, T., and Nishimura, S. (1981) *Chem. Lett.* pp. 485–488.
37. Yokota, M., Narita, K., Kosuge, T., Wakabayashi, K., Nagao, M., Sugimura, T., Yamaguchi, K., Shudo, K., Iitaka, Y., and Okamoto, T. (1981) *Chem. Pharm. Bull.* **29**, 1472–1475.
38. Ishii, K., Ando, M., Kamataki, T., Kato, R., and Nagao, M. (1980) *Cancer Lett.* **9**, 271–276.
39. Yamazoe, Y., Ishii, K., Kamataki, T., Kato, R., and Sugimura, T. (1980) *Chem.-Biol. Interact.* **30**, 125–138.
40. Sugimura, T. (1982) *Jpn. J. Cancer Res.* **73**, 499–507.
41. Hecker, E. (1963) *Z. Krebsforsch.* **65**, 325–333.
42. O'Brien, T. G., Simsiman, R. C., and Boutwell, R. K., (1975) *Cancer Res.* **35**, 1662–1670.
43. Fujiki, H., Mori., and Sugimura, T. (1981) *Proc. Jpn. Cancer Assoc.* (40th Annu. Meet.), p. 50.
44. Takashima, M., Sakai, H., and Arima, K. (1962) *Agric. Biol. Chem.* **26**, 660–668.
45. Takashima, M., and Sakai, H. (1960) *Bull. Agric. Chem. Soc. Jpn.* **24**, 647–651.
46. Cardellina, J. H., II, Marner, F. J., and Moore, R. E. (1979) *Science* **204**, 193–195.
47. Nakayasu, M., Fujiki, H., Mori, M., Sugimura, T., and Moore, R. E. (1981) *Cancer Lett.* **12**, 271–277.
48. Hoshino, H., Miwa, M., Fujiki, H., and Sugimura, T. (1980) *Biochem. Biophys. Res. Commun.* **95**, 842–848.
49. Fujiki, H., Suganuma, M., Matsukura, N., Sugimura, T., and Takayama, S. (1982) *Carcinogenesis* (Oxford) **3**, 895–898.
50. Kato, Y., and Scheuer, P. J. (1974) *J. Am. Chem. Soc.* **96**, 2245–2246.
51. Kato, Y., and Scheuer, P. J. (1976) *Pure Appl. Chem.* **48**, 29–33.
52. Mynderse, J. S., Moore, R. E., Kashiwagi, M., and Norton, T. R. (1977) *Science* **196**, 538–540.
53. Hiwasa, T., Fujimura, S., and Sakiyama, S. (1982) *Proc. Natl. Acad. Sci. U.S.A.* **79**, 1800–1804.

54. Hiwasa, T., Sakiyama, S., and Fujimura, S. (1981) *Proc. Jpn. Cancer Assoc., 40th Annu. Meet.* p. 48.
55. Hiai, H., and Nishizuka, Y. (1981) *JNCI, J. Natl. Cancer Inst.* **67**, 1333–1340.
56. Kaneshima, H., Hiai, H., Fujiki, H., Iijima, S., Sugimura, T., and Nishizuka, Y. (1983) *Int. J. Cancer* (in press).
57. Hirakawa, T., Kakunaga, T., Fujiki, H., and Sugimura, T. (1982) *Science* **216**, 527–529.

Multistage Skin Tumor Promotion in Mouse Skin: Critical Protein Changes During Tumor Promotion

T. J. SLAGA AND K. G. NELSON[1]

University of Texas System Cancer Center
Science Park, Research Division
Smithville, Texas

INTRODUCTION

Skin tumors can be induced by the sequential application of a subthreshold dose of a carcinogen (initiation stage), followed by repetitive treatment with a weak or noncarcinogenic promoter (promotion stage). The initiation stage in mouse skin requires only a single application of either a direct acting carcinogen or a procarcinogen and is essentially an irreversible step which, as data suggests, probably involves a somatic mutation in some aspect of epidermal differentiation (1–3). There is a good correlation between the skin tumor-initiating activities of several polycyclic aromatic hydrocarbons (PAH) and their ability to bind covalently to epidermal DNA (3). Current information suggests that skin tumor promoters are not mutagenic and do not bind covalently to DNA; they bring about a number of important epigenetic changes in the skin (4).

The primary aims of this report are to (1) provide evidence for critical morphological and biochemical changes during skin tumor promotion, (2) provide evidence for the multistage nature of skin tumor promotion, (3) correlate promotion associated morphological and biochemical responses with specific stages of promotion, and (4) provide evidence for important protein changes that occur during the progression of papillomas to carcinomas.

---

[1] Present address: Department of Pathology, University of North Carolina, Chapel Hill, North Carolina.

TUMOR PROMOTION

Although the phorbol esters are the most potent of the mouse skin tumor promoters, a wide variety of other compounds have been shown to have skin tumor promoting activity, as shown in Table I. After the phorbol esters and dihydroteleocidin B, anthralin is the most potent tumor promoter known of the compounds listed in Table I. Van Duuren and co-workers have reported a fairly extensive structure–activity study with anthralin and derivatives (5). Likewise, Boutwell and co-workers (6) have reported a structure–activity study of a number of phenolic compounds that are weak promoters in comparison to the phorbol esters and anthralin. Although several of the other compounds shown in Table I have moderate to weak activity as tumor promoters, there have not been any extensive structure–activity studies performed. We have recently found that benzo(e)pyrene (7) and benzoyl peroxide (8) are relatively good tumor promoters. Other free radical-generating compounds like benzoyl peroxide, such as lauroyl peroxide and decanoyl peroxide, are good skin tumor promoters. These agents were found not to have skin tumor-initiating or complete carcinogenic activity (8).

In addition to causing inflammation and epidermal hyperplasia, the

TABLE I
Skin Tumor Promoters[a]

Promoters	Potency
Croton oil	Strong
Certain phorbol esters found in croton oil	Strong
Some synthetic phorbol esters	Strong
Certain euphorbia latices	Strong
Anthralin	Moderate
Certain fatty acids and fatty acid methyl esters	Weak
Certain long chain alkanes	Weak
A number of phenolic compounds	Weak
Surface active agents (sodium lauryl sufate, Tween 60)	Weak
Citrus oils	Weak
Extracts of unburned tobacco	Moderate
Tobacco smoke condensate	Moderate
Iodoacetic acid	Weak
1-Fluoro-2,4-dinitrobenzene	Moderate
Benzo(e)pyrene	Moderate
Dihydroteleocidin B	Strong

[a] See ref. (4) for individual references.

TABLE II
Morphological and Biochemical Responses of Mouse Skin to
Phorbol Ester and Other Tumor Promoters[a]

Responses
Induction of inflammation and hyperplasia Increase in DNA, RNA, and protein synthesis An initial increase in keratinization followed by a decrease Increase in phospholipid synthesis Increase in prostaglandin synthesis Increase in histone synthesis and phosphorylation
Increase in ornithine decarboxylase activity followed by increase in polyamines Increase in histidine and Dopa decarboxylase activity[b] Decrease in the isoproterenol stimulation of cAMP Decrease in the number of dexamethasone receptors[b] Decrease in SOD and catalase[b]

Induction of embryonic state in adult skin
  1. Induction of dark cells (primitive stem cells)
  2. Induction of embryonic proteins in adult skin
  3. Induction of morphological changes in adult skin resembling papillomas, carcinomas, and embryonic skin
  4. Decrease in histidase activity
  5. Increase in protease activity
  6. Decrease in response of G_1 chalone in adult skin
  7. Increase in cAMP independent protein kinase in adult skin resembling tumors and embryonic skin

[a] See ref. (4) for individual references.
[b] Slaga, unpublished results.

phorbol esters and other tumor promoters produce several other morphological and biochemical changes in skin as listed in Table II. Of the observed phorbol ester-related effects on the skin, the induction of epidermal cell proliferation, ornithine decarboxylase (ODC), and dark basal keratinocytes have the best correlation with promoting activity (9–14). In addition to the induction of dark cells, which are normally present in large numbers in embryonic skin, there are many other embryonic conditions that appear in adult skin after treatment with tumor promoters (Table II).

It is difficult to determine which of the many effects associated with phorbol ester tumor promotion are in fact essential components of the promotion process. A good correlation appears to exist between promotion and epidermal hyperplasia when induced by phorbol esters (10). However, other agents that induce epidermal cell proliferation do not necessarily promote carcinogenesis (15). However, it should be emphasized that all known skin tumor promoters do induce epidermal hyperplasia (4). O'Brien et al. (9) have reported an excellent correlation between the tumor promoting ability of various compounds (phorbol esters as well as non-phorbol ester compounds) and their ability to induce ODC activity in mouse skin. However, mezerein, a diterpene similar to TPA but with weak promoting activity, was found to induce ODC to levels that were comparable to those induced by TPA (16). Raick found that phorbol ester tumor promoters induced the appearance of "dark basal cells" in the epidermis, whereas ethylphenylpropiolate (EPP), a nonpromoting epidermal hyperplastic agent, did not (11–13,17). Wounding induced a few dark cells, which seemed to correlate with its ability to be a weak promoter (11–13). In addition, a large number of these dark cells are found in papillomas and carcinomas (12,13). Slaga et al. (14,18) reported that TPA induced about 3 to 5 times the number of dark cells as mezerein, which was the first major difference found between these compounds.

INHIBITORS OF TUMOR PROMOTION

Various modifiers of the tumor promotion process have been very useful in our understanding of the mechanism(s) of tumor promotion. Table III lists the potent inhibitors of phorbol ester tumor promotion in mouse skin. The anti-inflammatory steroid fluocinolene acetenoide (FA) was found to be an extremely potent inhibitor of phorbol ester tumor promotion in mouse skin (19). Repeated applications of as little as 0.01 μg almost completely counteracted skin tumorigenesis. FA also effectively counteracts the induced cellular proliferation associated with application of phorbol ester tumor promoters. Certain retinoids are also potent inhibitors of mouse skin tumor promotion (20). Verma and co-workers (20) have shown that the retinoids that inhibit skin tumor promotion are potent inhibitors of phorbol ester-induced epidermal ODC activity. We have found that a combination of FA and retinoids produces an inhibitory effect on skin tumor promotion greater than that produced by each separately (21).

The work of Belman and Troll indicated that protease inhibitors' cyclic nucleotides, dimethylsulfoxide, and butyrate also inhibit mouse skin tumor promotion by phorbol esters (22). In addition to butyric acid, acetic acid also inhibits tumor promotion (15,22). The phosphodiesterase inhibitor isobutylmethylxanthine was also found to inhibit tumor promotion, which gives further support to the inhibitory effect

TABLE III
Inhibitors of Phorbol Ester Skin Tumor Promotion[a]

Inhibitors

1. Anti-inflammatory steroids: cortisol, dexamethasone, and fluocinolone acetonide (FA)
2. Vitamin A derivatives
3. Combination of retinoids and anti-inflammatory agents
4. Protease inhibitors: tosyl lysine chloromethyl ketone (TLCK); tosyl arginine methyl ester (TAME); tosyl phenylalanine chloromethyl ketone (TPCK), antipain, and leupeptin
5. Cyclic nucleotides
6. Phosphodiesterase inhibitors; isobutylmethylxanthine (IBMX)[b]
7. Dimethylsulfoxide (DMSO)[b]
8. Butyrate, acetic acid
9. *Bacillus* Calmette-Guerin (BCG)
10. Polyriboinosinic: polyribocytidylic acid [poly(I:C)]
11. Prostaglandin synthesis inhibitor 5,8,11,14-eicosatetraynoic acid (ETYA) and RO-22-3582[a]
12. Aracidonic acid
13. Polyamine synthesis inhibitor difluoromethylornithine (DFMO)[b]
14. Butylated hydroxyanisole (BHA) and hydroxytoluene (BHT)[b]

[a] See ref. (4) for individual references.
[b] Slaga, unpublished results.

of cyclic nucleotides (Slaga and Weeks, unpublished results). Schinitsky and co-workers (23) reported the inhibitory effect of *Bacillus* Calmette-Guerin (BCG) vaccination on skin tumor promotion. It has been shown that poly(I:C) has an inhibitory effect on carcinogenesis and tumor promotion (24). This appears to be mediated by its inhibition of promoter and carcinogen-induced cell proliferation (24). Certain prostaglandin synthesis inhibitors, thromboxane synthesis inhibitors, and phospholipase A_2 inhibitors' also inhibit skin tumor promotion, which suggest that prostaglandins and thromboxane may be important in tumor promotion (25). Although the mechanism is not presently understood, arachidonic acid at high doses is a potent inhibitor of tumor promotion (25). α-Difluoromethylornithine (DFMO), a specific inhibitor of polyamine synthesis, also inhibits tumor promotion, which suggests that polyamines are also important (4). The mechanism(s) by which histamine and diphenhydramine inhibit tumor promotion is currently not known (S. M. Fischer, unpublished results). Although BHA, BHT, Disulfiram, and parahydroxyanisole are potent inhibitors of skin tumor promotion by both TPA and benzoyl peroxide, their mechanism of action is currently not known (T. J.

Slaga, unpublished results). It is possible that free radicals are important in tumor promotion, and thus these agents may prevent promotion by their free radical scavenging ability.

CRITICAL PROTEIN CHANGES DURING TUMOR PROMOTION

In order to gain some insight into possible critical protein changes during tumor promotion, we used two-dimensional gel electrophoresis to compare the changes in mouse epidermal proteins induced by the potent promoter TPA, by the moderate promoter mechanical abrasion, and by the weakly or nonpromoting hyperplastic agents mezerein and EPP (26). Evidence is presented that indicates that TPA caused many changes in the epidermal protein profiles especially related to the keratins, which are the major differentiation product of the epidermis. The keratin modification progresses with time after TPA treatment, resulting in a keratin pattern that resembles that of newborn mouse epidermis (Fig. 1). The criteria used for the identification of the keratins were extractability, isoelectric points, molecular weights, filament formation *in vitro*, immunological cross-reactivity, amino acid composition, and peptide mapping. Several other protein changes were evident in the more soluble epidermal proteins, which were also prominent in the newborn epidermis (Fig. 2). These protein alterations are observed not only early during the TPA induction of hyperplasia and inflammation at 48 and 72 hr, but also in 1- and 2-week samples in which the morphology of the epidermis has returned to normal. Mezerein and abrasion produced protein changes similar to those induced by TPA (Fig. 3). EPP-induced protein modifications not only occurred at later times compared with either mezerein or TPA, but also were less in magnitude (Fig. 3). However, although many of the protein modifications induced by TPA appear associated with the hyperplastiogenic properties of TPA, the major difference between a potent promoter like TPA and a weak promoter like EPP appeared to be related to the magnitude of the response and to the time of appearance of the protein changes.

The keratins, which are the major differentiation product of the epidermis, are thought to be involved in the regulation of cell shape and function as mechanical integrators of various cytoplasmic components (27–30). Some of the morphological and biochemical alterations attributed to TPA treatment may be due to the modification of the keratins, which may alter their normal cytoplasmic function.

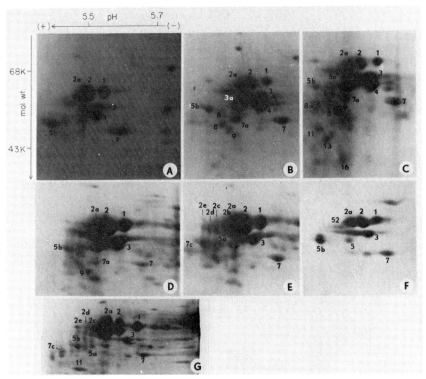

Fig. 1. Two-dimensional gel profiles of proteins extracted with SDS-DTT from adult SENCAR mouse epidermis at various times after treatment with a single dose of TPA or acetone. The pH gradient of the isoelectric focusing gels range between pH 4.0 and 6.0. Numbers and letters were assigned to individual proteins for ease of comparison. These proteins were predominantly the epidermal keratins. The time points investigated were (A) 12 hr, (B) 48 hr, (C) 72 hr, (D) 1 week, and (E) 2 weeks after a single application of TPA. The protein profiles of SENCAR mice treated with the delivery vehicle, acetone, did not change significantly at various times after treatment. Therefore, only the gel profiles of proteins extracted 48 hr after acetone treatment is shown (F). Protein patterns of SENCAR newborn mouse epidermis is also shown (G). Note similarities between newborn protein patterns and those obtained 2 weeks following TPA treatment of adult (E).

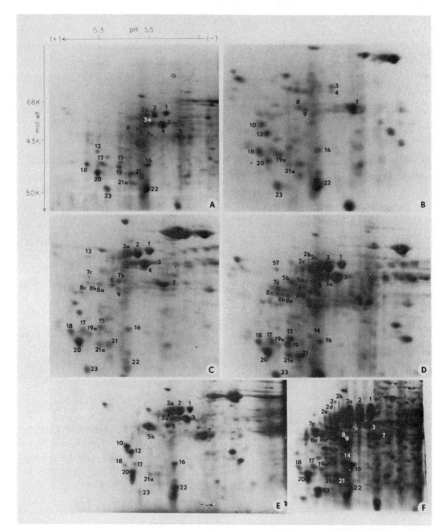

Fig. 2. Two-dimensional gel protein profiles of the 4 M urea solubilized proteins at various times following the application of TPA: (A) 48 hr, (B) 72 hr, (C) 1 week, and (D) 2 weeks after treatment. Gel profiles of proteins extracted 48 hr after acetone treatment (E).

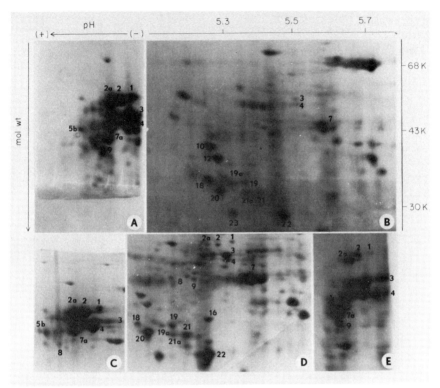

Fig. 3. Two-dimensional gel profiles of proteins isolated from mouse epidermis after treatment with the weakly promoting hyperplastiogenic agents. Gel profiles of the SDS-DTT protein extracts (A) mezerein, 48 hr after treatment; (C) EPP, 48 hr after treatment; and (E) abrasion, 4 days after treatment. Two 4 M urea-extracted protein profiles are presented: (B) mezerein, 48 hr after treatment; (D) EPP, 48 hr after treatment.

MULTISTAGE PROMOTION

As previously discussed, mezerein, a diterpene similar to TPA, was capable of causing most of the morphological and biochemical changes in skin and in cells in culture that TPA does, but TPA was at least 50 times more active as a tumor promoter (16). Mezerein is as potent or more potent than TPA, especially in the induction of epidermal ODC and epidermal hyperplasia. The effect of mezerein on ODC activity suggests that ODC induction is not a critical event in tumor promotion (17). It should be emphasized that this conclusion is also true for the other morphological and biochemical responses to mezerein.

Because of the many similarities in morphological and biochemical responses induced by TPA and mezerein, we felt that mezerein, although a weak promoter, would be a good candidate as a compound to be used in the second stage of a two-stage promotion protocol as originally reported by Boutwell (2). We recently reported that mezerein was a potent stage II promoter (31). A summary of the results on the use of mezerein as a second-stage promoter in two-stage promotion is shown in Table IV. As illustrated, TPA is about 50 times more active as a promoter than mezerein. When 2 µg of TPA are given twice weekly for only 2 weeks after DMBA initiation, no tumors are induced compared to twice weekly treatments for 18 weeks. However, when mezerein is given at a dose of either 1, 2, 4, or 6 µg twice weekly after the limited TPA treatment, it induced a significant tumor response in a dose-dependent manner.

TABLE IV
Two-Stage Promotion[a]

	Initiation	Promotion	Relative tumor response (%)
1. DMBA	1 week	TPA 32 ×	100
2. DMBA	1 week	Mezerein (4 µg) 32 ×	2

			Promotion	
	Initiation	Stage 1	Stage 2	Relative tumor response (%)
3. DMBA	1 week	TPA 4 ×	Acetone 28 ×	0
4. DMBA	1 week	TPA 4 ×	Mezerein (1 µg) 28 ×	35
5. DMBA	1 week	TPA 4 ×	Mezerein (2 µg) 28 ×	50
6. DMBA	1 week	TPA 4 ×	Mezerein (4 µg) 28 ×	85
7. DMBA	1 week	TPA 4 ×	Mezerein (6 µg) 28 ×	120
8. DMBA	1 week	4-O-Methyl TPA (80 µg) 4 ×	Mezerein (2 µg) 28 ×	40
9. DMBA	1 week	TPA 4 ×	4-O-Methyl TPA (80 µg) 28 ×	0
10. DMBA	1 week	A23187 (80 µg) 4 ×	Mezerein (2 µg) 28 ×	60
11. DMBA	1 week	TPA 4 ×	A23187 (80 µg) 28 ×	0
12. DMBA	1 week	EPP (14 mg) 32 ×		1
13. DMBA	1 week	TPA 4 ×	EPP (14 mg) 28 ×	2

[a] The mice were initiated with 10 nmol of DMBA and promoted with 2 µg of TPA or as shown above.

The ability of mezerein to act as a potent second-stage promoter was repeated in more than 15 separate experiment (4,31,32). Also shown in Table IV is the ineffectiveness of EPP as a complete promoter and as a second-stage promoter. In addition, we recently found that 4-O-methyl TPA, the calcium ionophore A23187, hydrogen peroxide, and wounding, which do not promote, are effective first-stage promoters (Tables IV and V). These compounds or wounding induce epidermal hyperplasia and increase the number of dark basal keratinocytes (4). Table V shows some of the characteristics of the first- and second-stages of promotion. Besides showing a good dose-response for TPA as a first-stage promoter, only a single application of TPA is necessary for stage I of promotion to be expressed after repeated applications of mezerein. In addition, stage I of promotion is partially irreversible for 4 weeks. As previously stated, stage II of promotion requires multiple

TABLE V
Characteristics of the First and Second Stages of Tumor Promotion[a]

Stage I
1. A good dose-response exists for TPA as a first-stage promoter.
2. Only one application of TPA is necessary.
3. Partially irreversible:
    A. Four weeks can separate first and second stages of promotion without a decrease in tumor response.
    B. There is a 80% decrease in tumor response if 10 weeks separate stage I and stage II of promotion.
4. The nonpromoting agents calcium ionophore (A23187), 4-O-methyl TPA, H_2O_2, and wounding can act as stage I promoters.
5. Increase in the number of dark basal keratinocytes (stem cells) are important. This occurs by stimulating existing dark cells to divide as well as converting basal cells to dark cells.
5. Prostaglandins are important because PGE_2 can enhance stage I by TPA.
7. Inhibited by anti-inflammatory steroid (FA) and protease inhibitor (TPCK).

Stage II
1. A good dose-response exists for mezerein as a second-stage promoter.
2. Multiple applications are required.
3. The nonpromoting agent DPtri-D can act as stage II promoter.
4. Reversible at first but becomes irreversible.
5. Polyamines are important because putrescine can enhance stage II by mezerein.
6. Inhibited by FA, RA, and DFMO.
7. Most of the morphological and biochemical events shown to be important in promotion occur in this stage.

[a] Part of the data was taken from refs. (4), (31), and (32) and part is unpublished data (Slaga).

applications and also shows a good dose-response with mezerein or 12-deoxyphorbol-13-2,4,6-decatrienoate (DPtri-D).

The effectiveness of some of the inhibitors of tumor promotion on two-stage promotion was recently reported by this laboratory (25). The effects of FA, retinoic acid (RA), DFMO, and tosyl phenylalanine chloromethylketone (TPCK) on two-stage promotion are shown in Table VI. FA was a potent inhibitor of stage I and II of promotion but to a greater degree for stage I than stage II. It should be emphasized that only four applications of FA with TPA were necessary to counteract the tumor response. RA was ineffective in stage I but was a potent inhibitor of stage II promotion, whereas TPCK specifically inhibited stage I but not stage II. These experiments were repeated several times and were very reproducible (4,32). Recently, Weeks and Slaga (unpublished results) found that DFMO was a potent specific inhibitor of stage II promotion.

Because the only major morphological or biochemical difference between the effects of TPA and mezerein on the skin is the ability of TPA to induce a large number of dark basal keratinocytes (4,14), we were interested in determining the effects of various inhibitors of promotion on the appearance of these dark cells. We reasoned that if these dark cells are critical in the first stage of promotion and if FA and TPCK are potent inhibitors of stage I and RA and DFMO of stage II,

TABLE VI
The Effects of Tumor Promotion Inhibitors on Two-Stage Promotion[a]

	Initiation	Promotion		Tumor response (% of control)
		Stage 1	Stage 2	
1. DMBA	1 week	TPA 4×	Mezerein 28×	100
2. DMBA	1 week	TPA + FA 4×	Mezerein 28×	0
3. DMBA	1 week	TPA 4×	Mezerein + FA 28×	20
4. DMBA	1 week	TPA + RA 4×	Mezerein 28×	95
5. DMBA	1 week	TPA 4×	Mezerein + RA 28×	20
6. DMBA	1 week	TPA + TPCK 4×	Mezerein 28×	25
7. DMBA	1 week	TPA 4×	Mezerein + TPCK 28×	94
8. DMBA	1 week	TPA + DFMO 4×	Mezerein 28×	98
9. DMBA	1 week	TPA 4×	Mezerein + DFMO 28×	45

[a] The mice were initiated with 10 nmol of DMBA and promoted with 2 μg of TPA and 2 μg of mezerein. FA (1 μg), RA (10 μg), TPCL (10 μg), and DFMO (2% in drinking H_2O) were applied simultaneously with TPA or mezerein.

TABLE VII
Effects of FA, RA, DFMO, and TPCK on Tumor Promotion and TPA-Induced Epidermal Hyperplasia, Dark Keratinocytes, and Polyamine Levels

Inhibitor	Relative ability to counteract (%)[a]			
	TPA promotion	TPA-induced hyperplasia	TPA-induced dark cells	TPA-induced ODC and polyamine levels
FA	100	100	100	20
RA	80	0	0	85
TPCK	70	0	70	10
DFMO[b]	55	0	0	95

[a] The ability of FA, RA, DFMO, and TPCK to counteract the various TPA responses are expressed from 100% (complete supression) to 0% (no effect). The effects of the inhibitors were determined from dose–response studies. See refs. (4) and (32) for details concerning these experiments.
[b] Weeks and Slaga, unpublished data.

then FA and TPCK should counteract the appearance of these cells, whereas RA and DFMO should not. The results of FA, RA, DFMO, and TPCK on the induction of dark basal keratinocytes by TPA are summarized in Table VII. As hypothesized, FA and TPCK were found to effectively counteract the appearance of the dark cells induced by TPA, whereas RA and DFMO had no effect (32).

Because TPCK inhibited stage I of promotion but not stage II and because TPCK counteracted the TPA-induced increase in the dark basal keratinocytes but did not have any effect on TPA-induced hyperplasia, we were interested in determining the effect of TPCK on TPA-induced ODC activity. As shown in Table VII, TPCK had very little effect on TPA- and mezerein-induced epidermal ODC activity.

The anti-inflammatory steroid FA not only counteracted the appearance of dark cells induced by TPA but also suppressed the hyperplasia induced by TPA. In fact, the skins from FA plus TPA-treated mice appeared as untreated skin. This is in agreement with our previously reported observations on the inhibitory effect of FA on TPA induced inflammation, hyperplasia, and DNA synthesis (19). However, FA had little effect on the TPA increased ODC activity (Table VII) as compared to its effect on inhibition of promotion.

It is also of interest to point out that although RA inhibited stage II of promotion, it had no inhibitory effect on the TPA or mezerein-induced hyperplasia (Table VII). However, certain retinoids have been found to be potent inhibitors of TPA- and mezerein-induced epidermal ODC activity (16). In this regard, DFMO is a specific irreversible

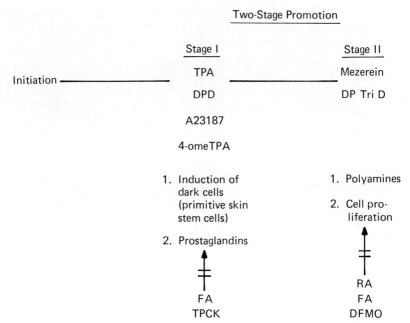

Fig. 4. A diagram of the various stages of skin carcinogenesis showing the important events in stage I and II of promotion and where FA, RA, DFMO, and TPCK inhibit promotion.

inhibitor of ODC activity. These data suggest that the induction of epidermal ODC activity followed by increased polyamines may be important in stage II of promotion. In this regard, FA and TPCK have either no effect or only a slight inhibitory effect on TPA or mezerein-induced ODC activity (32). FA does, however, significantly decrease the TPA induced spermidine levels in the epidermis (4). This effect plus FA's inhibitory effect on TPA-induced hyperplasia may be responsible for its inhibitory effect on stage II promotion. Figure 4 depicts the various stages of promotion, the important events in each stage, and where the various inhibitors are effective.

PROTEIN CHANGES IN PAPILLOMAS AND SQUAMOUS CELL CARCINOMAS

As pointed out earlier, TPA caused many protein modifications, especially the keratin proteins. These changes were even more pronounced in epidermal papillomas (33). In addition, the keratins of the

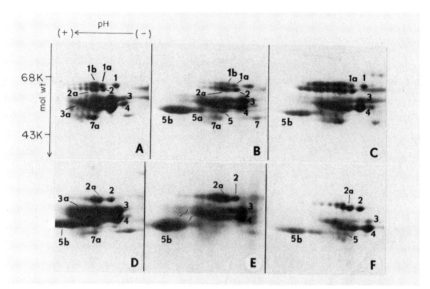

Fig. 5. Two-dimensional protein patterns of the SDS-DTT extracted proteins of papillomas arising during two-stage skin carcinogenesis. (A) Small papillomas obtained from initiated mice after 13 TPA treatments, (B) small papillomas obtained from 16 TPA treatments, (C) large papillomas obtained after 27 TPA treatments, (D) small papillomas obtained after 27 TPA treatments, (E) small papillomas obtained after 44 TPA treatments, (F) large papillomas obtained after 83 TPA treatments.

papillomas displayed greater charge heterogeneity, particularly among the high-molecular-weight keratins (60,000 to 62,000, see Fig. 5; proteins 1, 2, 3, and 4). As the experiment progressed, there appeared to be a selective loss of one group of high-molecular-weight keratins (62,000) in some of the papillomas (Fig. 5). Interestingly, the carcinomas that appeared at this time had significant reduction in both groups of high-molecular-weight keratins (Fig. 6; proteins 1 and 2). In fact, the keratin profiles of carcinomas were very similar to the patterns observed in basal cells after a single TPA treatment of adult epidermis (Fig. 6). This may indicate that the program of keratin expression of a carcinoma becomes permanently fixed at a basal cell pattern. Changes in keratin patterns may serve as a biochemical marker of malignant progression in mouse epidermis. Modification of the program of keratin expression in mouse epidermal basal cells may be one mechanism by which TPA promotes. TPA may induce changes that allow initiated basal cells to become fixed into a basal pattern of differentiating resulting in a permanent growth advantage.

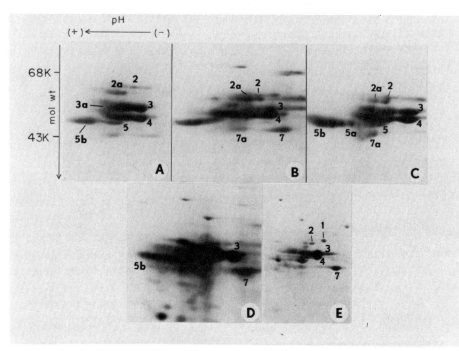

Fig. 6. Comparison by two-dimensional electrophoresis of the keratin patterns found in carcinomas during two-stage carcinogenesis (initiation and promotion) with those found in a carcinoma induced by complete carcinogenesis (a carcinogenic dose of DMBA) and with those found in basal cells after TPA treatment. (A) Carcinoma obtained from an initiated mouse after 44 TPA treatments, (B) carcinoma obtained after 57 TPA treatments, (C) carcinoma obtained after 80 TPA treatments, (D) carcinoma obtained after complete carcinogenesis with DMBA, (E) keratin patterns obtained from basal cells isolated from adult mouse epidermis 48 hr after a single TPA treatment.

CONCLUSION

Besides the phorbol ester tumor promoters, various peroxides such as benzoyl peroxide and lauroyl peroxide have been found to be effective skin tumor promoters. Skin tumor promoters have been shown to have many cellular and biochemical effects on the skin. Of all the observed promoter-related effects on the skin, the induction of epidermal cell proliferation, polyamines, prostaglandins, and dark basal keratinocytes, as well as other embryonic conditions, appears to correlate the best. We have found that skin tumor promoters also bring about many protein modifications, especially the keratin proteins that resemble the pattern found in fetal skin. Various inhibitors of the tumor promotion process have been very useful in our understanding of what cellular events are critical to tumor promotion. Besides the anti-inflammatory steroids, retinoids, protease inhibitors, and prostaglandin and polyamine biosynthesis inhibitors, antioxidants have been found to be effective inhibitors of skin tumor promotion. We have found that the tumor-promotion stage can be operationally and mechanistically further divided into at least two stages. Because hydrogen peroxide is an effective first-stage promoter, free radicals may be important in this stage of promotion. Dark basal keratinocytes and prostaglandins are important events in stage I of promotion, whereas polyamines and cell proliferation are important events in stage II of promotion. Protease inhibitors were found to be specific inhibitors of stage I of promotion and retinoic acid, and polyamine synthesis inhibiting are specific inhibitors of stage II of promotion. We have also found that the keratin protein changes found in epidermis after TPA treatment were even more pronounced in epidermal papillomas. Although the carcinoma keratin pattern was similar to that of papillomas, there was a significant reduction in high-molecular-weight keratins in carcinomas. In fact, the keratin profiles of carcinomas were very similar to the patterns observed in basal cells after TPA treatment of adult epidermis. Changes in keratin patterns may serve as a biochemical marker of malignant progression in skin.

REFERENCES

1. *Carcinog. Compr. Survey* (1978) **2**, 1–583.
2. Boutwell, R. K. (1964) *Prog. Exp. Tumor Res.* **4**, 207–250.
3. Slaga, T. J., Fischer, S. M., Weeks, C. E., and Klein-Szanto, A. J. P. (1980) In "Biochemistry of Normal and Abnormal Differentiation" (N. Seije and I. A. Bernstein, eds.), pp. 193–218. University of Tokyo Press, Tokyo.
4. Slaga, T. J., Fischer, S. M., Weeks, C. E., and Klein-Szanto, A. J. P. (1981) In "Reviews in Biochemical Toxicology), Vol. 3, (E. Hodgson, J. Bend, and R. M. Philpot, eds.), pp. 231–381. Elsevier-North Holland Amsterdam.
5. VanDuuren, B. L., and Goldschmidt, B. M. (1978) *Carcinog. Compr. Survey* **2**, 491–507.
6. Boutwell, R. K., and Bosch, D. K. (1959) *Cancer Res.* **19**, 413–419.
7. Slaga, T. J., Jecker, L., Bracken, W. M., and Weeks, C. E. (1979) *Cancer Letters* **7**, 51–59.
8. Slaga, T. J., Klein-Szanto, A. J. P., Triplett, L. L., Yotti, L. P., and Trosko, J. E. (1981) *Science* **213**, 1023–1025.
9. O'Brien, T. G., Simsiman R. C., and Boutwell, R. K. (1975). *Cancer Res.* **35**, 1662–1670.
10. Slaga, T. J., Scribner, J. D., Thompson S., and Viaje, A. (1974) *J. Natl. Cancer Inst.* **52**, 1611–1618.
11. Raick, A. N. (1973) *Cancer Res.* **33**, 269–286.

12. Raick, A. N. (1974) *Cancer Res.* **34**, 920–926.
13. Raick, A. N. (1974) *Cancer Res.* **34**, 2915–2925.
14. Klein-Szanto, A. J. P., Major S. M., and Slaga, T. J. (1980) *Carcinogenesis* **1**, 399–406.
15. Slaga, T. J., Bowden, G. T., and Boutwell, R. K. (1975) *J. Natl. Cancer Inst.* **55**, 983–987.
16. Mufson, R. A., Fischer, S. M., Verma, A. K., Gleason, G. L., Slaga, T. J., and Boutwell, R. K. (1979) *Cancer Res.* **39**, 4791–4795.
17. Raick, A. N., and Burdzy, K. (1973) *Cancer Res.* **33**, 2221–2230.
18. Slaga, T. J., Fischer, S. M., Weeks, C. E., and Klein-Szanto, A. J. P. (1980) *In* "Biochemistry of Normal and Abnormal Epidermal Differentiation" (M. Seije and I. A. Bernstein, eds.), pp. 193–218. University of Tokyo Press, Tokyo.
19. Schwartz, J. A., Viaje, A., Slaga, T. J., Yuspa, S. H., Hennings H., and Lichti, U. *Chem. Biol. Interact.* **17**, 331–347.
20. Verma, A. K., Rice, H. M., Shapas, B. G., and Boutwell, R. K. (1978) *Cancer Res.* **38**, 798–801.
21. Weeks, C. E., Slaga, T. J., Hennings, H., Gleason, G. L., and Bracken, W. M. (1979) *J. Natl. Cancer Inst.* **63**, 401–406.
22. Belman, S., and Troll, W. (1978) *Carcinog. Compr. Survey* **2**, 117–134.
23. Schinitsky, M. R., Hyman, L. R., Blazkovec, A. A., and Burkholder, P. M. (1973) *Cancer Res.* **33**, 659–663.
24. Gelboin, H. V., and Levy, H. B. (1970) *Science* **167**, 205–207.
25. Fischer, S. M., Gleason, G. L., Hardin, L. G., Bohrman, J. S., and Slaga, T. J. (1980) *Carcinogenesis* **1**, 245–248.
26. Nelson, K. G., Stephenson, K. B., and Slaga, T. J. (1982) *Cancer Res.* **42**, 4164–4175.
27. Fuchs, E., and Green, H. (1974) *Cell* **15**, 887–897.
28. Fuchs, E., and Green, H. (1980) *Cell* **19**, 1033–1042.
29. Sun, T-T., and Green, H. (1978) *J. Biol. Chem.* **253**, 2053–2060.
30. Sun, T-T., Shih, C., and Green, H. (1979) *Proc. Natl. Acad. Sci. U.S.A.* **76**, 2813–2817.
31. Slaga, T. J., Fischer, S. M., Nelson, K., and Gleason, G. L. (1980) *Proc. Natl. Acad. Sci. U.S.A.* **77**, 3659–3663.
32. Slaga, T. J., Klein-Szanto, A. J. P., Fischer, S. M., Weeks, C. E., Nelson, K., and Major, S. (1980) *Proc. Natl. Acad. Sci. U.S.A.* **77**, 2251–2254.
33. Nelson, K. G., and Slaga, T. J. (1982) *Cancer Res.* **42**, 4176–4179.

Tumor Promotion in the Skin of Hairless Mice by Halogenated Aromatic Hydrocarbons

ALAN POLAND,* JOYCE KNUTSON,* EDWARD GLOVER,* AND
ANDREW KENDE†

*McArdle Laboratory for Cancer Research
University of Wisconsin, Madison, Wisconsin
and
† Department of Chemistry
University of Rochester
Rochester, New York

INTRODUCTION

Over the past decade, the scientific community and the general public have developed an increasing awareness and concern over the widespread dispersion in our environment of halogenated chemicals. The halogenated aromatic hydrocarbons, i.e., the chlorinated dibenzo-p-dioxins, dibenzofurans, azo(xy)benzenes, biphenyls, and brominated biphenyls, have received considerable attention because of the extraordinary toxic potency of some members of this class of compounds and because their chemical stability permits their persistence in the environment (Fig. 1). All the biologically active halogenated aromatic hydrocarbons share a number of common properties. They are all approximate isostereomers, produce a similar and characteristic pattern of biochemical and toxic responses, and appear to act by a common mechanism (1–3). 2,3,7,8-Tetrachlorodibenzo-p-dioxin (TCDD) serves as the prototype of these compounds.

The administration of TCDD (and congeners) to laboratory animals induces the coordinate expression of a battery of enzymes primarily concerned with drug metabolism, including cytochrome P-450-mediated microsomal monooxygenase activity, in liver and a variety of other tissues (4). This pleiotropic response is mediated by the stereo-

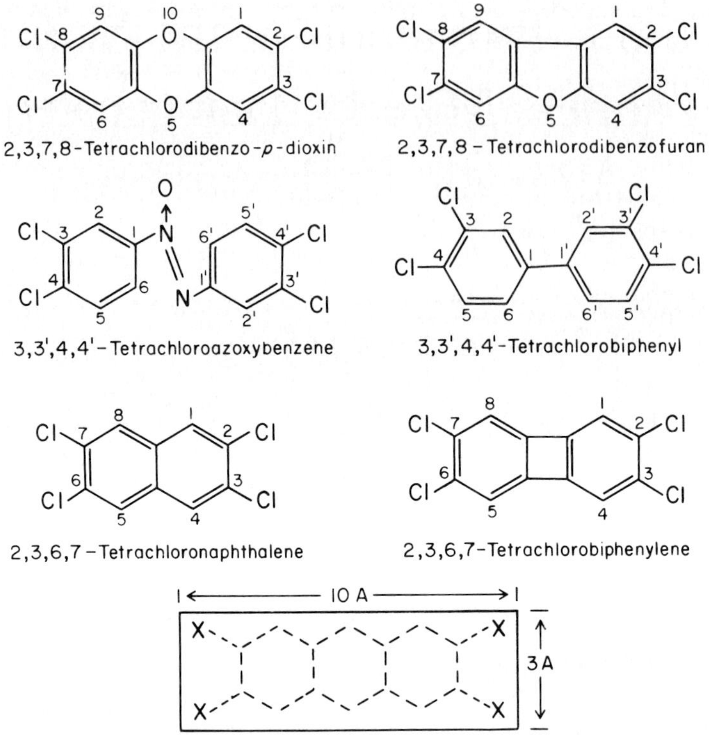

2,3,7,8-Tetrachlorodibenzo-*p*-dioxin

2,3,7,8-Tetrachlorodibenzofuran

3,3',4,4'-Tetrachloroazoxybenzene

3,3',4,4'-Tetrachlorobiphenyl

2,3,6,7-Tetrachloronaphthalene

2,3,6,7-Tetrachlorobiphenylene

Fig. 1. The isosteric tetrachloro- congeners and ring numbering systems of various halogenated aromatic hydrocarbons. The diagram at the bottom depicts the postulated recognition site on the cytosol receptor to which these isosteric compounds bind.

specific, reversible binding of TCDD to a cytosol receptor and translocation of the ligand–receptor complex to the nucleus (5–7). Two lines of evidence support this mechanism. (1) Among a large series of halogenated aromatic hydrocarbons, the structure–activity relationship for receptor binding corresponds to that for induction of aryl hydrocarbon hydroxylase (AHH) activity, a microsomal monooxygenase activity (1). (2) In mice there is a genetic polymorphism at the locus which determines the receptor, the *Ah* locus (8–10). Inbred strains of mice carrying the Ah^b allele (e.g., C57BL/6) have a high affinity receptor and are sensitive to the induction of AHH activity by TCDD, whereas other inbred strains possessing the Ah^d allele (e.g. DBA/2) have a lower affinity receptor and are less sensitive to enzyme induction by TCDD. Induction of AHH activity segregates with the *Ah* locus (8,9,11).

The halogenated aromatic hydrocarbons produce toxic responses, i.e., histopathologic changes, which are quite characteristic; however, many of the lesions are highly species specific (12–14). The administration of a lethal dose of TCDD or one of its congeners, produces a prolonged wasting syndrome prior to death, with loss of adipose tissue, involution of lymphoid organs, degeneration of the seminiferous tubules of the testicle, and embryotoxicity and/or teratogenicity in virtually all species tested. In contrast, only a limited number of species experience the more distinct lesions which involve proliferation and/or metaplasia of epithelial tissues, such as skin, stomach, intestines, and urinary tract. Similarly, the presence and severity of hepatic pathology vary considerably among species. The cause of death is unknown. Two lines of evidence indicate that the toxic responses produced by TCDD and congeners are mediated by their binding to the cytosol receptor: (1) the correlation of the structure–activity relationship for receptor affinity and for toxic potency (1,2,15); (2) the segregation with the *Ah* locus of several of the toxic responses produced by TCDD in mice (e.g., thymic involution, cleft palate formation in the fetus, an hepatic porphyria) (16,17).

CARCINOGENICITY STUDIES ON TCDD

The chronic administration of TCDD and other halogenated aromatic hydrocarbons to rats and mice is associated with an increased incidence of carcinoma of the liver and other tissues (for review, see references 18,19). Kociba *et al.* (20) reported that the lifetime dietary administration of TCDD, at a dose equivalent to 0.1 µg/kg/day produced an increased incidence of hepatocellular carcinomas, and squamous cell carcinomas of the lung, hard palate and nasal turbinates. However, TCDD has been found to be neither mutagenic *in vitro* (21, 22) nor to bind appreciably to DNA in rat liver *in vivo* (23). These results suggest that TCDD may not be a complete carcinogen, but rather that it may act as a tumor promoter, enhancing the tumorigenic expression of already initiated cells *in vivo*.

Pitot *et al.* (24) reported that TCDD was a promoter in a two-stage model of rat liver carcinogenesis. Following partial hepatectomy and a single suboptimal dose of diethylnitrosamine (DEN), repeated administration of TCDD (equivalent to 0.1 µg/kg/day) produced an increase in enzyme altered foci and hepatocellular carcinomas. Partial hepatectomy and DEN administration without subsequent TCDD treatment, or administration of TCDD alone, produced fewer enzyme-altered foci and no hepatic carcinomas.

In a two stage model of mouse skin tumorigenesis, TCDD did not act as a tumor promoter. Following a single application of 7,12-dimethylbenzanthracene (DMBA) to the skin, repeated topical administration of TCDD did not enhance skin papillomas in CD-1 (25) or Swiss-Webster mice (26). In the latter study, repeated administration of TCDD, with or without prior initiation with DMBA, produced an increase in integumental sarcomas, leading the authors to conclude that TCDD was carcinogenic.

In this chapter, we wish to review some recent studies of the effects of TCDD on mouse skin which define the genetic constitution required for the expression of epidermal toxicity and tumor promotion by halogenated aromatic hydrocarbons.

EPIDERMAL CHANGES PRODUCED BY HALOGENATED AROMATIC HYDROCARBONS

As noted above, the coordinate expression of the drug metabolizing enzymes and the histopathologic changes produced by TCDD appear to be receptor-mediated events. Many tissues *in vivo* and cells *in vitro* respond to TCDD with the induction of AHH activity, and hence possess the receptor, but display no evidence of a toxic response (11,27–29). Thus, although the receptor is essential for the expression of toxicity, the presence of the receptor is not sufficient. To illustrate this point, we will concentrate on the epidermal changes produced by halogenated aromatic hydrocarbons in mice.

TCDD and congeners produce a characteristic hyperplasia and hyperkeratosis of the epidermis and a squamous metaplasia of the sebaceous glands with the formation of keratinaceous comedones in humans (30), monkeys (31), and rabbits (32). Similar changes have not been observed in the skin of normal laboratory mice, rats, and guinea pigs (15,33). Inagami *et al.* (34), however, noted that *hairless* mice fed rice oil contaminated with polychlorinated biphenyls develop epidermal hyperplasia, hyperkeratosis, and sebaceous gland metaplasia.

Hairless is a recessive mutation in mice controlled by the *hr* locus on chromosome 14. Homozygous *hr/hr* mice develop a persistent alopecia after losing their first coat of hair around 4 weeks of age; heterozygous *hr/+* mice have a normal haired phenotype indistinguishable from the wild-type (+/+) (35,36). HRS/J is an inbred strain of mice segregating for the *hr* locus; HRS/J homozygous *hr/hr* and heterozygous *hr/+* mice are congenic, genetically identical except at the *hr* locus and other closely linked loci on chromosome 14 (37,38).

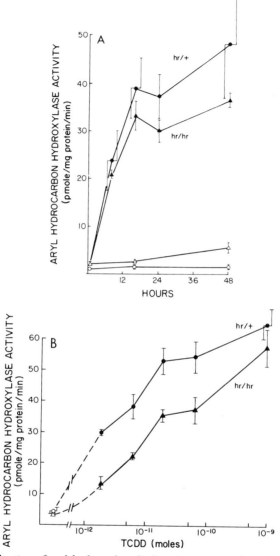

Fig. 2. Induction of aryl hydrocarbon hydroxylase activity by TCDD in the epidermis of HRS/J mice. (A) Time course of the induction of aryl hydrocarbon hydroxylase activity in epidermis of 5–7 week old HRS/J *hr/hr* and *hr/+* mice following a single application of TCDD (7.8 nmol). Control mice (open symbols) received only acetone. All animals were killed and the enzyme activity assayed on the same day. Each point is the mean ± standard error of the values of four mice. (B) Dose–response curves for the induction of aryl hydrocarbon hydroxylase activity in the epidermis of 4 to 8-week-old HRS/J *hr/hr* and *hr/+* mice 48 hours after TCDD application. The initial point is the value obtained from skin treated only with acetone. Each value represents the mean ± standard error from five mice.

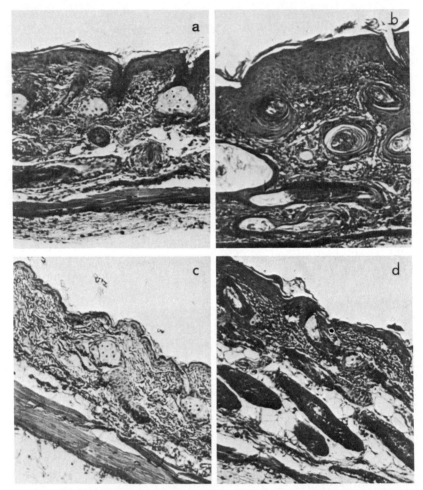

Fig. 3. Histology of skin of HRS/J, hr/hr and hr/+, mice administered TCDD. Six-week-old female HRS/J hr/hr and shaved hr/+ mice were administered acetone or a single application of TCDD (3 nmoles). Fourteen days after TCDD administration, epidermal hyperplasia, sebaceous cell metaplasia, and hyperkeratosis is observed in the skin of hr/hr mice, but not of hr/+ mice. (a) hr/hr, acetone; (b) hr/hr, TCDD; (c) hr/+, acetone; (d) hr/+, TCDD. Initial magnification 50×.

HRS/J mice of both phenotypes carry the Ah^b allele at the Ah locus, and have a high affinity receptor. The affinity and concentration of the receptor for TCDD measured in the liver of HRS/J hr/hr and hr/+ mice was the same (hr/hr, K_D = 0.30 ± .09 nM, n = 77 ± 20 fmol/mg cytosol protein; hr/+, K_D = 0.33 ± .07 nM, n = 62 ± 15 fmol/mg protein) (39).

Application of TCDD to the dorsal skin of HRS/J haired and hairless mice induces epidermal AHH activity with a similar time course and dose response relationship (Fig. 2). However, TCDD produces an obvious scaliness and extensive histologic changes in the skin of hr/hr mice, but not in $hr/+$ mice (Fig. 3). This response in hr/hr mice, which includes hyperplasia and hyperkeratosis of the interfollicular epidermis, squamous metaplasia of the sebaceous glands, and hyperkeratosis within the dermal cysts is time and dose dependent.

ROLE OF THE Ah LOCUS IN DETERMINING THE HISTOLOGICAL RESPONSE (39)

A series of halogenated aromatic hydrocarbons were examined for their potency to produce a histologic response of moderate severity and for their binding affinity for the hepatic cytosol receptor (Table I). There is an obvious correspondence between those compounds which evoke epidermal changes and those which bind to the receptor, but a quantitative comparison between the cumulative dose *in vivo* and the equilibrium dissociation constant for receptor binding in vitro is not possible.

We examined the epidermal response produced by 3,3′,4,4′,5,5′-hexabromobiphenyl, a congener of TCDD (depicted in Table I) in hairless C57BL/6J (B6) and DBA/2J (D2) mice and their cross and backcross (Fig. 4). The B6 hairless mice (Ah^b/Ah^b, hr/hr) administered this hexabromobiphenyl isomer developed a moderate epidermal hyperplasia, hyperkeratosis, and sebaceous metaplasia, whereas the skin of treated D2 (Ah^d/Ah^d, hr/hr) hairless mice appeared similar to that of untreated control mice. The epidermal response in D2B6F$_1$ (Ah^b/Ah^d, hr/hr) was intermediate to that of the two parental phenotypes. Application of 3,3′,4,4′,5,5′-hexabromobiphenyl to the hairless progeny of the backcross D2 (Ah^d/Ah^d, $hr/+$) × D2B6F$_1$ (Ah^b/Ah^d, hr/hr), which were phenotyped as heterozygous at the Ah locus (Ah^b/Ah^d) produced a skin lesion of moderate severity, whereas the skin of hairless offspring phenotyped as homozygous (Ah^d/Ah^d) were unaffected by this congener.

A sufficiently high dose of TCDD (total dose 3.6 nmol over 4 weeks) produced a maximal histologic lesion in the skin of both B6 and D2 hairless mice. Haired B6, D2, and B6D2F$_1$ mice, which were shaven and then treated with TCDD, do not develop any histologic changes in the skin.

To summarize, the proliferative/metaplastic response produced by TCDD and congeners in the skin of hairless (hr/hr) mice is dependent on the interaction of two genetic loci, Ah and hr. The epidermal re-

TABLE I
Halogenated Aromatic Hydrocarbons: Capacity to Produce Epidermal Hyperplasia–Metaplasia in HRS/J hr/hr Mice and Binding Affinity for Hepatic Cytosol Receptor

Structure	Epidermal hyperplasia–metaplasia (total nmol/mouse for 2+ response)[a]	Receptor affinity $K_D(nM)$[a]	Structure	Epidermal hyperplasia–metaplasia (total nmol/mouse for 2+ response)[a]	Receptor affinity $K_D(nM)$[b]
2,3,7,8-tetrachlorodibenzo-p-dioxin	0.36	0.27	1,3,7,8-tetrachlorodibenzo-p-dioxin	Inactive (360)	Inactive (27)
2,3,3',4-tetrachlorodibenzofuran analog	1.76	0.57	2,8-dichlorodibenzo-p-dioxin	Inactive (360)	Inactive (54)
2,3,7,8-tetrachlorodibenzofuran	1.2	0.73	2-chlorodibenzofuran	Inactive (2400)	Inactive (540)
3,3',4,4'-tetrachloroazoxybenzene	356	0.93	3,3',5,5'-tetrachloroazoxybenzene	Inactive (356)	Inactive (540)

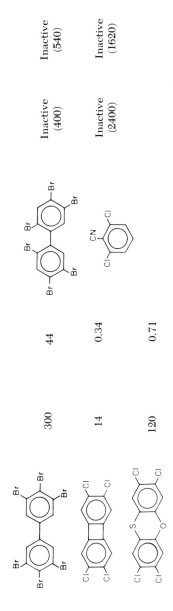

	300	44	Inactive (400)	Inactive (540)
	14	0.34	Inactive (2400)	Inactive (1620)
	120	0.71		

[a] Compounds were dissolved in acetone and applied to the dorsal skin of HRS/J *hr/hr* mice 2 or 3 times per week for 4 weeks. All compounds were tested on at least three mice; active compounds were tested at at least two concentrations which differed by 10-fold. Skin samples were fixed, sectioned, and stained with hematoxylin and eosin. Each compound was scored as the lowest dose which produced a moderate response (2+). For inactive compounds, the highest cumulative dose tested in nmol/mouse is shown in parentheses.

[b] The binding affinity of each compound for the mouse liver cytosol receptor was estimated by its capacity to compete with [³H]TCDD for specific binding sites in an ammonium sulfate-precipitated fraction of liver cytosol from C57BL/6J mice (5). For inactive compounds the highest concentration tested in nmol/liter is shown in parentheses.

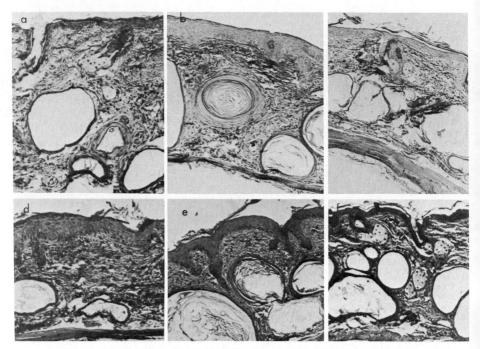

Fig. 4. Histology of skin of hairless C57BL/6J, DBA/2J, D2B6F$_1$, and D2B6F$_1$ × D2 backcross mice administered 3,3′,4,4′,5,5′-hexabromobiphenyl. Hairless (b) C57BL/6J (Ah^b/Ah^b, hr/hr), (c) DBA/2J (Ah^d/Ah^d, hr/hr), and (d) D2B6F$_1$ (Ah^b/Ah^d, hr/hr) were administered 3,3′,4,4′,5,5′-hexabromobiphenyl to the dorsal skin 3 times per week for 9 weeks (total dose, 0.6 μmol/mouse). The hairless progeny of the backcross DBA/2J (Ah^d/Ah^d, $hr/+$) × D2B6F$_1$ (Ah^b/Ah^d, hr/hr) were phenotyped for the Ah locus as (e) heterozygous (Ah^b/Ah^d, hr/hr) or (f) homozygous (Ah^d/Ah^d, hr/hr) and then administered 3,3′,4,4′,5,5′-hexabromobiphenyl to the dorsal skin twice a week for 12 weeks (total dose, 2.4 μmol/mouse). Skin of control mice of each type was similar to that of (a) the backcross (Ah^b/Ah^b, hr/hr) treated only with acetone. Initial magnification 50 ×.

sponse is mediated by the cytosol receptor as indicated by the correspondence of the structure–activity relationship for the skin lesion and for receptor binding, and segregation of the response with the Ah locus. The histologic response is produced in the epidermis of HRS/J hr/hr mice, but not in their haired ($hr/+$) congenic littermates. Both HRS/J hr/hr and $hr/+$ mice (1) possess the Ah^b allele, (2) have a hepatic cytosol receptor with a similar affinity and concentration and, (3) respond to TCDD with induction of epidermal AHH activity. However, TCDD evokes additional responses in the skin of hr/hr mice. These observations are depicted in a model in Fig. 5. In the skin of $hr/+$ or ($+/+$ mice), TCDD and other halogenated aromatic hydrocar-

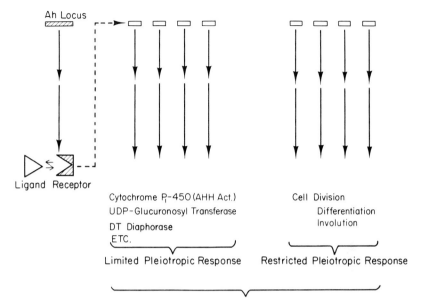

Fig. 5. A model for the interaction of the *Ah* and *hr* loci in the pleiotropic response of HRS/J mouse skin to the halogenated aromatic hydrocarbons. The *Ah* locus determines the cytosol receptor which binds halogenated aromatic hydrocarbons and mediates the ensuing gene expression. In the epidermis of *hr/+* mice, there is a limited pleiotropic response, consisting primarily of the induction of enzymes related to drug metabolism (shown on the left of the diagram). In the skin of *hr/hr* mice, an additional battery of genes is expressed which controls cell division and differentiation (shown on the right of the diagram).

bons are recognized by the cytosol receptor and induce a *limited* pleiotropic response; however, in *hr/hr* mice an additional battery of genes is expressed, which results in the proliferation/differentiation response. This additional battery is restricted in the skin of *hr/+* or *+/+* mice. Thus *sensitivity* of the epidermal response is determined by the *Ah* locus, whereas the *extent* of the response is determined by the *hr* locus.

TUMOR PROMOTION BY TCDD IN HRS/J MICE (40)

In light of the stringent conditions necessary for halogenated aromatic hydrocarbons to produce a hyperplastic/metaplastic response in mouse skin, it is perhaps not surprising that in previous studies, using CD-1 (25) and Swiss Webster mice (26) (which are wild-type, +/+, at the *hr* locus), TCDD was found to be inactive as a tumor promoter in a

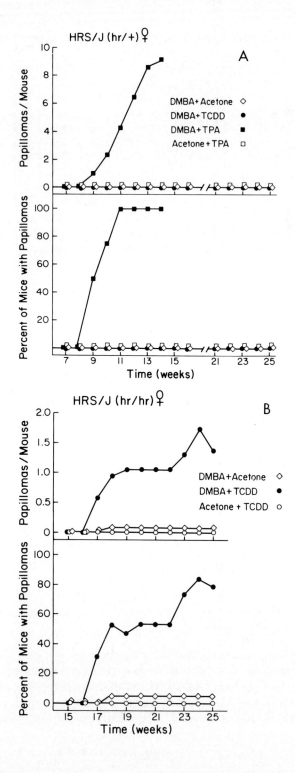

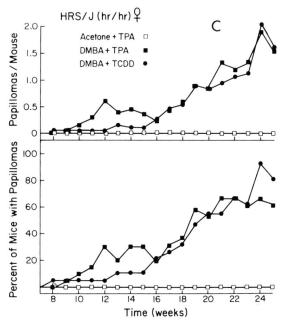

Fig. 6. Tumor promotion by TCDD and TPA in HRS/J $hr/+$ and hr/hr mice. (A) Shaved, HRS/J $hr/+$ were administered a single topical application of DMBA (0.2 µmol) dissolved in acetone or the solvent alone, and then treated twice weekly with TPA (2 µg/mouse), TCDD (50 ng/mouse for 5 weeks, then 20 ng/mouse) or the solvent, acetone. (B) HRS/J hr/hr were treated with the identical regimen as Fig. 6A. (C) HRS/J hr/hr mice received two applications of DMBA (0.2 µmol) 1 week apart or the solvent, acetone, and were then administered twice weekly applications of TPA (2 µg/mouse) or TCDD (20 ng/mouse). All groups in Fig. 6A-C consisted of 20 female mice, 6–8 weeks of age at the start of the experiment. The survival rate was ≥80% in all groups.

two-stage model of skin tumorigenesis. We were prompted to re-examine the potential of TCDD and congeners to act as tumor promoters using HRS/J hr/hr and $hr/+$ mice.

HRS/J ($hr/+$) haired mice developed skin papillomas when topically administered a single dose of 0.2 µmoles of 7,12-dimethylbenzanthracene (DMBA) to the dorsal skin, followed by twice weekly applications of the tumor promotor 12-O-tetradecanoylphorbol 13-acetate (TPA) 2 µg/mouse (Fig. 6A). Administration of DMBA alone, or repeated applications of TPA without prior initiation produced no tumors. Compared to other mouse strains, HRS/J ($hr/+$) mice are relatively sensitive to this regimen of DMBA/TPA: by 11 weeks of promotion, 100% of the animals developed papillomas and by 14 weeks, the tumor multiplicity was 9.2 papillomas/mouse. Following initiation with DMBA, repeated topical administration of TCDD (50 ng/mouse, twice weekly for 5 weeks, and then 20 ng/mouse twice weekly) failed to produce any papillomas over the 25-week course of the experiment.

In contrast, HRS/J *hr/hr* mice, initiated with DMBA and administered the same regimen of TCDD, did develop papillomas (Fig. 6B). After 25 weeks of TCDD promotion, 79% (15 of 19) of surviving mice had one or more papillomas, with an average multiplicity of 1.4 tumors/mouse. The time course and tumor multiplicity in *hr/hr* mice produced by twice weekly application of TCDD (20 ng/mouse) and TPA (2 µg/mouse) were very similar (Fig. 6C). Thus, TPA promotes tumor formation in the skin of both HRS/J haired and hairless mice, whereas TCDD is effective only in *hr/hr* mice.

A further comparison of TPA and TCDD is shown in Table II. HRS/J hairless mice were initiated with N-methyl-N'-nitro-N-nitrosoguanidine (MNNG) and twice weekly treated topically with various

TABLE II
Comparison of TCDD and TPA as Tumor Promoters in HRS/J Hairless Mice[a]

			At 20 weeks	
Initiation	Promotion		Tumor incidence (surviving mice with tumors/surviving mice, %)	Tumor multiplicity (average number of papillomas/surviving mice)
1. MNNG	Acetone		1/19	5	0.05
2. MNNG	TCDD	3.75 ng[b]	11/20	55	0.7
3. MNNG	TCDD	7.5 ng	13/17	77	1.5
4. MNNG	TCDD	15 ng	10/10	100	4.0
5. MNNG	TCDD	30 ng	15/19	79	1.6
6. Acetone	TCDD	30 ng	0/19	0	0
7. MNNG	TPA	1 µg	5/19	26	0.4
8. MNNG	TPA	3 µg	13/18	72	1.6

[a] HRS/J hairless female mice, 7 weeks of age, were administered a single dose of MNNG (5 µmol) in acetone or the solvent alone applied to the dorsal skin and then were treated twice weekly with various doses of TCDD or TPA (in 50 µl of acetone) for 20 weeks. There were initially 20 mice per group and the survival was 85% or more in all groups except group 4 (TCDD 15 ng). Ten animals died in this group after 15 weeks of promotion from dehydration coincident with being changed to a new type of cage. These deaths did not alter the tumor incidence or tumor multiplicity, that is at 15 weeks versus 16 weeks.

[b] Dose of compound/mouse given twice weekly.

doses of TCDD or TPA. TCDD produced a dose-related increase in the incidence and multiplicity of skin papillomas at doses of 3.75, 7.5, and 15 ng mouse, but at 30 ng/mouse TCDD produced more toxicity and was less effective. Comparable tumor promotion was produced by twice weekly administration of TCDD at 7.5 to 15.0 ng/mouse (1.2 to 2.3 × 10^{-9} mol/kg) and TPA at 3 µg/mouse (2.4 × 10^{-7} mol/kg). TCDD appears to be approximately 100 times more potent than TPA on a molar basis, but one cannot formally compare the relative potencies of two compounds unless they act by a similar mechanism.

We examined the capacity of several halogenated aromatic hydrocarbon congeners and commercial mixtures of halobiphenyl isomers to promote tumor formation in the skin of HRS/J hairless mice (Table III). 2,3,7,8-Tetrachlorodibenzofuran and 3,4,5,3',4',5'-hexabromobiphenyl, both agonists for the cytosol receptor, which produce epidermal hyperplasia and hyperkeratosis in hr/hr mice (shown in Table I), are effective tumor promoters. 2,7-Dichlorodibenzo-p-dioxin and 2,4,-5,2',4',5'-hexabromobiphenyl, congeners which neither bind to the receptor, nor produce the characteristic biochemical and toxic effects of TCDD (shown in Table I), are inactive as tumor promoters at the dose tested. Aroclor 1254, a commercial mixture of chlorobiphenyl isomers, at 1 mg/mouse twice weekly slightly enhanced the incidence of skin tumors in MNNG initiated mice, but this effect was not statistically significant. In contrast, a similar regimen of the commercial bromobiphenyl mixture Firemaster-FF-1 proved an effective tumor promoter but also quite toxic.

A few aspects of these experiments require further comment.

1. *Local and systemic effects.* Topical application of TPA produces an acute inflammation and hyperplasia of the epidermis in both hr/hr and $hr/+$ mice, with little systemic toxicity. Topical application of TCDD produces no gross or histologic changes in the skin of $hr/+$ mice; hr/hr mice develop a scaly, shiny, and taut appearance to the skin; histologically one observes epidermal hyperplasia, hyperkeratosis, and squamous metaplasia of the sebaceous glands (Fig. 3). TCDD produces no acute inflammatory response in epidermis. TCDD (and other halogenated aromatic hydrocarbons) produce dose-related systemic toxicity in both hr/hr and $hr/+$ mice: reduced weight gain, loss of subcutaneous and abdominal fat, involution of the lymphoid organs, and hepatomegaly and parenchymal cell changes.

2. *Tumor multiplicity.* The incidence and multiplicity of tumors produced by initiation and TPA promotion is less in HRS/J hr/hr mice than $hr/+$ mice as noted by Giovanella *et al.* (41) and seen in Fig. 6A

TABLE III
Tumor Promotion by Halogenated Aromatic Hydrocarbons in HRS/J Hairless Mice[a]

	Initiation	Promotion	Dose (dose/mouse, 2 times/week)	n	Tumor incidence (surviving mice with tumors/surviving mice, %)		Tumor multiplicity (average number of papillomas/surviving mice)
1	MNNG	Acetone		26	0/23	0	0
2	Acetone	TCDD	50 ng, 5 weeks, then 20 ng	26	0/18	0	0
3	MNNG	TCDD	50 ng, 5 weeks, then 20 ng	26	16/19	84	1.6
4	MNNG	2,3,7,8-Tetrachlorodibenzofuran	1.0 µg	20	19/19	100	4.9
5	Acetone	2,3,7,8-Tetrachlorodibenzofuran	1.0 µg	20	1/20	5	0.05
6	MNNG	3,4,5,3',4',5'-Hexabromobiphenyl	20 µg	20	12/20	60	1.5
7	Acetone	3,4,5,3',4',5'-Hexabromobiphenyl	20 µg	20	0/20	0	0
8	MNNG	2,4,5,2',4',5'-Hexabromobiphenyl	20 µg	20	0/20	0	0
9	Acetone	2,4,5,2',4',5'-Hexabromobiphenyl	20 µg	26	0/22	0	0
10	MNNG	2,7-Dichlorodibenzo-p-dioxin	20 µg	20	0/19	0	0
11	Acetone	2,7-Dichlorodibenzo-p-dioxin	20 µg	20	0/20	0	0
12	MNNG	Aroclor 1254	1 mg	20	4/19	21	0.3
13	Acetone	Aroclor 1254	1 mg	20	0/19	0	0
14	MNNG	Firemaster FF-1	2 mg, 5 weeks, then 1 mg	26	9/15	60	2.0
15	Acetone	Firemaster FF-1	2 mg, 5 weeks, then 1 mg	20	1/16	6	0.13

[a] HRS/J hairless female mice, 8 weeks of age, received a single administration of MNNG (5 µmol) in acetone, or solvent alone applied to the skin, and then were treated topically twice weekly with test compound for 20 weeks. Treatment with Firemaster FF-1 produced a significant mortality, and the dose was reduced after 5 weeks. The appreciable mortality in other groups (2, 3, and 6) occurred largely in the first 6 weeks of the experiment, resulted primarily from dehydration and did not appear to be a result of the

and 6C. Thus, the relatively low tumor multiplicity produced by TCDD and congeners in *hr/hr* mice is a constraint of the animal model and is not a reflection of the poor promoting efficacy of these compounds.

3. *Survival rate in hr/hr mice.* Although TCDD (and other halogenated aromatic hydrocarbons) produce considerable toxicity, dosage regimens were used (Fig. 6B and C) which permit ≥80% survival. HRS/J harbor a murine leukemia virus, and the incidence of lymphatic leukemia is greater in *hr/hr* mice (45% by 10 months of age) than *hr/+* mice (1% by 10 months of age). In all the experiments with *hr/hr* mice, 0-15% of the animals develop leukemia, and this incidence appeared to be independent of the chemical treatment of the animals.

TCDD and related halogenated aromatic hydrocarbons comprise a class of potent tumor promoters which appear to act by a mechanism that is distinct from that of TPA and its congeners. The epidermal hyperplasia and most probably the tumor promotion (as suggested by the structure-activity relationship in Table III) produced by halogenated aromatic hydrocarbons in *hr/hr* mice is mediated through the cytosol receptor. In contrast the promoting activity of TPA and TPA-like agonists appear to act through another mechanism: (1) they bind to a membrane receptor, (2) produce an inflammatory response in mouse skin, and (3) promote tumor formation in haired and hairless mice.

Two aspects of this model of tumor promotion in the skin of hairless mice may prove especially useful. First, the comparison of the pleiotropic responses produced by TCDD in the skin of congenic HRS/J *hr/hr* and *hr/+* mice may permit identification of the additional battery of genes expressed in *hr/hr* skin which are in some manner associated with tumor promotion. Second, it may be determined if any of the biochemical effects evoked by TPA and its congeners and implicated in tumor promotion by these compounds are also common to the response produced by halogenated aromatic hydrocarbons.

ACKNOWLEDGMENTS

We wish to thank David Palen, Mary Folz-Erbs, Jane Weeks, and Laurie Brumblay for their technical assistance. This study was supported in part by grants from the National Institute of Environmental Health Sciences ES 01884 and a National Cancer Institute Center Grant CA-07175. A.P. is a Burroughs Wellcome Scholar in Toxicology.

REFERENCES

1. Poland, A., Greenlee, W. F., and Kende, A. S. (1979) *Ann. N.Y. Acad. Sci.* **320**, 214–230.
2. Goldstein, J. A. (1980) *In* "Halogenated Biphneyls, Terphenyls, Naphthalenes, Dibenzodioxins and Related Products" (R. Kimbrough, ed.), pp. 151–190. Elsevier/North-Holland Biomedical Press, Amsterdam.
3. Poland, A., and Knutson, J. C. (1982) *Annu. Rev. Pharmacol.* **22**, 517–554.
4. Poland, A., and Kende, A. S. (1977) *Cold Spring Harbor Conf. Cell Proliferation* **4**, 847–867.
5. Poland, A., Glover, E., and Kende, A. S. (1976) *J. Biol. Chem.* **251**, 4936–4946.
6. Greenlee, W. F., and Poland, A. (1979) *J. Biol. Chem.* **254**, 9814–9821.
7. Okey, A. B., Bondy, G. P., Mason, M. E., Nebert, D. W., Forster-Gibson, C. J., Muncan, J., and Dufresne, M. J. (1980) *J. Biol. Chem.* **255**, 11418–11422.
8. Nebert, D. W., and Gielen, J. E. (1972) *Fed. Proc., Fed. Am. Soc. Exp. Biol.* **31**, 1315–1327.
9. Nebert, D. W., Goujon, F. M., and Gielen, J. E. (1972) *Nature (London), New Biol.* **236**, 107–110.
10. Taylor, B. A. (1971) *Life Sci.* **10**, 1127–1134.
11. Poland, A., Glover, E., Robinson, J. R., and Nebert, D. W. (1974) *J. Biol. Chem.* **249**, 5599–5606.
12. Schwetz, B. A., Norris, J. M., Sparschu, G. L., Rowe, V. K., Gehring, P. J., Emerson, J. L., and Gerbig, C. G. (1973) *Environ. Health Perspect.* **5**, 87–99.
13. Kimbrough, R. D. (1974) *CRC Crit. Rev. Toxicol.* **2**, 445–489.
14. McConnell, E. E. (1980) *In* "Halogenated Biphenyls, Terphenyls, Naphthalenes, Dibenzodioxins, and Related Products (R. Kimbrough, ed.), pp. 109–150. Elsevier/North-Holland Biomedical Press, Amsterdam.
15. McConnell, E. E., Moore, J. A., Haseman, J. K., and Harris, M. W. (1978) *Toxicol. Appl. Pharmacol.* **44**, 335–356.
16. Poland, A., and Glover, E. (1980) *Mol. Pharmacol.* **17**, 86–94.
17. Jones, K. G., and Sweeney, G. D. (1980) *Toxicol. Appl. Pharmacol.* **53**, 42–49.
18. Huff, J. E., Moore, J. A., Saracci, R., and Tomatis, L. (1980) *Environ. Health Perspect.* **36**, 221–240.
19. National Research Council of Canada (1981) "Polychlorinated Dibenzo-p-dioxins: Criteria for Their Effects of Man and His Environment," NRCC Publ. No. 18574. Natl. Res. Counc. Can., Ottawa.
20. Kociba, R. J., Keyes, D. G., Beyer, J. E., Carreon, R. M., Wade, C. E., Dittenber, D. A., Kalnins, R. P., Frawson, L. E., Park, C. N., Barnard, S. D., Hummel, R. A., and Humiston, C. G. (1978) *Toxicol. Appl. Pharmacol.* **46**, 279–303.
21. Wasson, J. S., Huft, J. E., and Loprieno, N. (1977/78) *Mutat. Res.* **47**, 141–160.
22. Geiger, L. E., and Neal, R. A. (1981) *Toxicol. Appl. Pharmacol.* **59**, 125–129.
23. Poland, A., and Glover, E. (1979) *Cancer Res.* **39**, 3341–3344.
24. Pitot, H. C., Goldsworthy, T., Campbell, H. A., and Poland, A. (1980) *Cancer Res.* **40**, 3616–3620.
25. Berry, D. L., DiGiovanni, J., Juchau, M. R., Bracken, W. M., Gleason, G. L., and Slaga, T. J. (1978) *Res. Commun. Chem. Pathol. Pharmacol.* **20**, 101–107.
26. National Toxicology Program Technical Report 201 (1982) "Carcinogenesis Bioassay of 2,3,7,8-Tetrachlorodibenzo-p-dioxin in Swiss Webster Mice (Dermal Study)," NTP 80–32 NIH Publ. No. 82-1757. U.S. Dept. Health, Education and Welfare.

27. Bradlaw, J. A., and Casterline, J. L. (1979) *J. Assoc. Off. Anal. Chem.* **62**, 904–916.
28. Niwa, A., Kumaki, K., and Nebert, D. W. (1975) *Mol. Pharmacol.* **11**, 399–408.
29. Knutson, J. C., and Poland, A. (1980) *Toxicol. Appl. Pharmacol.* **54**, 377–383.
30. Kimmig, J., and Schultz, K. H. (1957) *Naturwissenschaften* **44**, 337–338 (in German).
31. Allen, J. R., Barsotti, D. A., Van Miller, J. P., Abrahamson, L. H., and Lalich, J. J. (1977) *Food Cosmet. Toxicol.* **15**, 401–410.
32. Jones, E. L., and Krizek, H. A. (1962) *J. Invest. Dermatol.* **39**, 511–517.
33. Grieg, J. B., Jones, G., Butler, W. H., and Barnes, J. M. (1973) *Food Cosmet. Toxicol.* **11**, 585–595.
34. Inagami, K., Kaga, T., Kikuchi, M., Hashimoto, M., Takahashi, H., and Wada, K. (1969) *Fukuoka Igaku Zasshi* **60**, 548–553 (in Japanese).
35. Crew, F. A. E., and Mirskaia, L. (1931) *J. Genet.* **25**, 17–28.
36. Grüneberg, H. (1952) *Bibliogr. Genet.* **15**, 102–115.
37. Green, M. C. (1966) In "Biology of the Laboratory Mouse" (E. L. Green, ed.), pp. 87–150. McGraw-Hill, New York.
38. Womack, J. E., Davisson, M. T., Eicher, E., and Kendall, D. A. (1977) *Biochem. Genet.* **15**, 347–365.
39. Knutson, J. C., and Poland, A. (1982) *Cell* **30**, 225–234.
40. Poland, A., Palen, D., and Glover, E. (1982) *Nature (London)* **300**, 271–279.
41. Giovanella, B. C., Liegel, J., and Heidelberger, C. (1970) *Cancer Res.* **30**, 2590–2597.

PART III

GROWTH FACTORS AND RECEPTORS

Transforming Growth Factors Produced by Viral-Transformed and Human Tumor Cells

GEORGE J. TODARO, HANS MARQUARDT,
DANIEL R. TWARDZIK, FRED H. REYNOLDS, JR., AND
JOHN R. STEPHENSON

Laboratory of Viral Carcinogenesis
National Cancer Institute-FCRF
National Institutes of Health
Frederick, Maryland

INTRODUCTION

The isolation of retroviruses with acute transforming function has occurred with increasing frequency over the past few years. Such viruses represent genetic recombinants between host cellular sequences (oncogenes) and nontransforming type-C virus structural genes (Fischinger, 1980; Klein, 1982). Viruses of this nature transform cells in culture and induce neoplasms of a variety of histological classes *in vivo*. Although the number of independent retrovirus isolates is high, the total number of unique "oncogenes" so far represented in such viruses is only 13 or 14 (Coffin et al., 1981). Several of these are represented as multiple virus isolates of the same species or, in several instances, have even originated in different species (Weinberg, 1982). By comparison to transforming sequences identified within the DNA of *in vitro* propagated human tumor cells, one oncogene, c-*has*, has been implicated in the induction of human bladder carcinomas (Der et al., 1982; Parada et al., 1982; Santos et al., 1982), whereas a second gene, c-*kis*, appears to be associated with carcinomas of the lung (Der et al., 1982). Other cellular homologues of viral oncogenes including c-*myc* (Dalla-Favera *et al.*, 1982), c-*fes* (Dalla-Favera *et al.*, 1982; Heisterkamp *et al.*, 1982), c-*sis* (Swan *et al.*, 1982), and c-*abl* (Heisterkamp *et al.*, 1982) have been mapped on chromosomes involved in translocations frequently associated with human lymphoid neoplasms. Finally, when linked to efficient viral promoter sequences, the human cellular homologue of v-*has* directly transforms cells in culture (Chang *et al.*, 1982). On the basis of these considerations, transforming retroviruses provide an important potential model system for studies of the molecular basis of human cancers.

Upon functional analysis, transforming gene products of several viral oncogenes have been found to exhibit tyrosine-specific protein kinase activity. These include v-*src*, v-*fps*, v-*ros*, v-*fes*, and v-*abl* (Stephenson, 1980; Klein, 1982). With the exception of v-*has*, which encodes a threonine-specific protein kinase (Shih *et al.*, 1980), enzymatic activities have not been found for transforming proteins encoded by other retrovirus isolates. Despite efforts to identify cellular substrates for viral oncogene-encoded protein kinases and attempts to ascribe enzymatic function to the other viral-encoded transforming proteins, very little is currently known regarding mechanisms of transformation by these viruses. In this chapter, the role of low-molecular-weight growth factors in transformation that is mediated by viral oncogenes and the role of such factors in spontaneously arising human tumors are considered.

EGF BINDING BY RETROVIRUS-TRANSFORMED AND HUMAN TUMOR CELLS

Among the phenotypic properties characteristic of retrovirus transformed cells is a reduction in available epidermal growth factor (EGF) receptors. This decrease in receptor number, as measured by a reduction of capacity for binding ^{125}I-labeled EGF, occurs within a few days of viral infection and is observed with each of the retrovirus isolates analyzed to date. These include most of the available mammalian retroviruses with transforming activity.

The retrovirus isolates that have been studied in greatest detail with respect to reduced EGF binding, include Moloney MSV (v-*mos*) (Todaro *et al.*, 1976), Abelson MuLV (v-*abl*) (Blomberg *et al.*, 1980), and the Gardner (v-*fes*), Snyder-Theilen (v-*fes*), and McDonough (v-*fms*) strains of feline sarcoma virus (FeSV) (Todaro *et al.*, 1976; Reynolds *et al.*, 1981b) (Table I). The availability of well characterized morphological revertants and transformation defective (td) mutants in these virus

TABLE I
Production of TGFs by Viral Transformed and Human Tumor Cells

Cell line	[^{125}I]EGF (cpm) bound/10^6 cells[a]	TGF release into culture medium[b]	
		ng Equivalent of EGF/liter medium	Soft agar colonies/ng equivalent of EGF
FRE 3A			
Control	3,630	0.1	—
ST FeSV	<200	200	173
G FeSV	210	60	150
McD FeSV	280	10	130
Abelson MuLV	<200	70	185
Human tumor cells			
A204	11,400	0.1	—
A673	<200	50	207
A431	146,000	0.1	—
A549	23,400	0.1	—
9812	<200	25	170

[a] Cells were incubated for 4 hr at 4°C in HEPES-buffered (pH 7.4) DMEM containing 0.1% ovalbumin and 2 ng of ^{125}I-labeled EGF and washed 3 times in serum-free DMEM. Total ^{125}I-labeled EGF bound per 10^6 cells was measured in triplicate.

[b] Mean values from three separate determinations represent competition with ^{125}I-labeled EGF for EGF membrane receptors on formaldehyde-fixed A431 cells as described previously (Twardzik et al., 1982). The number of soft agar colonies represents the average number of colonies containing a minimum of 20 NRK cells per eight random low-power fields scored 7 days after seeding with growth factor.

systems has provided important genetic controls to establish the correlation between reduced EGF binding and expression of transformation. For instance, Abelson MuLV- and Snyder-Theilen FeSV-transformed cells are subject to low rates of morphologic reversion to a nontransformed phenotype (Sacks et al., 1979; Reynolds et al., 1981c). This phenomenon is due to hypermethylation of the proviral DNA (Groffen et al., 1983) and is associated with restoration of EGF binding to levels characteristic of nontransformed control cells (Blomberg et al., 1980). Similarly, in Abelson MuLV (Blomberg et al., 1980) and Snyder-Theilen FeSV (Reynolds et al., 1981c) td mutant-infected clones in which viral-encoded transforming proteins, although functionally inactive, are expressed at levels comparable to those of wild-type (wt) viral-transformed cells, EGF binding occurs to an equally high level as with control cells (Table II).

There is increasing evidence that reduced EGF binding is specifically associated with retrovirus oncogene mediated transformation,

TABLE II
Association of rTGF Production with Abelson MuLV Tyrosine Specific Protein Kinase Activity[a]

Cell line	Growth in soft agar (%)	P120$^{gag-abl}$ expression	Protein kinase	rTGF Production	
				ng Equivalent of EGF/liter medium	Soft agar colonies/ng equivalent of EGF
Control FRE 3A	0.01	0.02	0.01	0.1	—
wt Abelson MuLV[a]	16	0.9	0.7	70	185
td Abelson MuLV[a]	0.01	1.1	0.01	0.1	—

[a] The isolation and characterization of wild-type (wt) and transformation-defective (td) Abelson MuLV-infected cloned FRE 3A Fisher rat cells have been previously described (Reynolds et al., 1980a). Levels of P120$^{gag-abl}$ expressed in µg per mg cell protein were estimated by competition immunoassay for their p12 structural components. Quantitation of polyprotein associated protein kinase activity expressed as pmol ^{32}P-incorporated per immunoprecipitate was measured as described previously (Blomberg et al., 1980). TGF production was measured as described in Table I.

rather than representing a more generalized phenotypic property of transformed cells (Todaro et al., 1976; Todaro and De Larco, 1978). For instance, both SV40 and polyoma virus-transformed cells bind EGF at control levels. Moreover, of the chemically transformed mouse cells and spontaneously arising human tumor cells examined to date, some bind ^{125}I-labeled EGF to only a very low extent (Table I). The majority, however, bind EGF relatively efficiently. Thus as a model system, retroviral transformation may be relevant only to those naturally occurring human tumors characterized by reduced EGF binding (Todaro et al., 1981).

TRANSFORMING GROWTH FACTOR PRODUCTION

As one explanation for the reduction of EGF binding associated with retrovirus transformation, it was initially hypothesized that a critical step in expression of the transformed phenotype may involve synthesis of low-molecular-weight growth factors capable of competing with EGF for binding to its membrane receptor. In accordance with this suggestion, the production of growth factors by cells with available receptors would thus lead to nonregulated proliferative growth.

Tumor cells, especially those that lack available EGF membrane receptors, have been shown to produce low-molecular-weight proteins, designated transforming growth factors (TGFs), which, although related to EGF, exhibit the unique property of morphologically transforming cells in culture (Todaro et al., 1980; Ozanne et al., 1980).

In general, production of TGFs by transformed cells parallels loss of EGF binding. For instance, only the few spontaneous human tumor cell lines characterized by low levels of available EGF receptors have been found positive for TGF production (Todaro et al., 1980). Similarly, all of the retrovirus-transformed cell lines studied to date have been found to produce TGF, a finding consistent with their reduced EGF binding. Quantitatively, however, differences are observed between cells transformed by different retrovirus isolates. For instance, as shown in Table II, we have generally found induction of around 10-fold higher levels of TGF production by retroviruses encoding tyrosine-specific protein kinases than by other virus groups (Twardzik et al., 1983). The reason for this difference is as yet unknown. In both morphologic revertants of Abelson MuLV-transformed cells and in cells infected with Abelson MuLV td mutants, TGF production is undetectable, which is consistent with the high levels of EGF binding by such cells (Twardzik et al., 1982). Similarly, TGF production by mouse cells infected by a temperature-sensitive mutant of Moloney MSV is greatly reduced at the nonpermissive temperature (De Larco et al., 1981).

TGFs are acid- and heat-stable proteins that specifically bind to EGF receptors, morphologically transform rat and human fibroblasts in monolayer culture, and confer upon anchorage dependent cells the ability to grow in soft agar (De Larco and Todaro, 1980; Todaro et al., 1980). TGF-induced transformation is fully reversible; removal of TGF from the culture medium results in reversion of the cell phenotypes to a nontransformed morphology, and cells in soft agar colonies induced by TGF, if selected and plated as monolayer cultures in the absence of TGF, grow as normal contact-inhibited monolayers. Antisera directed against EGF do not crossreact with TGFs, even of the same species, as measured by either radioimmunoassay or immunoprecipitation analysis. Although TGFs are detected as a series of distinct size classes ranging from 6,000 to 20,000 M_r (De Larco and Todaro, 1978), the form that has been studied in greatest detail and that which is primarily considered below has a molecular weight of approximately 6,000. Whether the higher molecular-weight forms represent precursors to, or aggregates of, the 6,000 M_r TGF is not as yet resolved.

PURIFICATION OF RAT, MOUSE, AND HUMAN TGFs

TGFs from culture medium of Abelson MuLV- (Twardzik et al., 1982) and Snyder-Theilen FeSV-transformed rat embryo fibroblasts, (Marquardt et al., 1983) and from the human metastatic melanoma cell line A2058 (Marquardt and Todaro, 1982) have been purified to homogeneity. Cell monolayers were grown to 90–95% confluency and incubated overnight in serum-free medium; culture fluids were harvested at the sequential 12-24-hr time points. Conditioned medium was concentrated by use of a hallow fiber device, dialyzed against 0.1% acetic acid; the supernatant was clarified by centrifugation, concentrated by lyophilization, reconstituted in 1.0 M acetic acid, and analyzed by BioGel P-10 or P-100 gel permeation chromatography. Individual column fractions were assayed for competition with ^{125}I-labeled EGF for binding A431 cell-membrane receptors and for transformation of normal rat kidney (NRK) cells, as measured by ability to form progressively growing colonies in soft agar. As shown for Abelson MuLV-induced rTGF in Fig. 1A, major peaks of EGF competing activities were identified at apparent molecular weights of 10,500 and 6,700. In each case, these were well separated from the bulk of the protein that eluted in the exclusion volume of the column. Activities from both peaks morphologically transform cells in monolayer culture and support anchorage-independent growth of NRK cells in soft agar.

The lower molecular-weight TGF containing peaks from the Bio-Gel columns were pooled, reconstituted in 0.05% trifluoroacetic acid in water, and further purified on μBondapak C$_{18}$ columns. TGFs were eluted with a linear gradient of acetonitrile containing 0.045% trifluoroacetic acid. As shown in Fig. 1B, EGF competing and transforming activities co-eluted as single peaks at an acetonitrile concentration of 19.5%. None of the fractions contained one activity in the absence of the other. Analogous results were obtained upon analysis of TGFs from Snyder-Theilen FeSV-transformed rat and A2058 human tumor cells, with average recoveries of over 80% of the EGF

Fig. 1. Purification of rTGF from Abelson MuLV-transformed rat cells. (A) Cell monolayers were washed twice with serum-free medium; culture fluids were collected at three sequential 12-hr intervals, clarified by low-speed centrifugation, concentrated in a hollow fiber concentrator DC2, dialyzed against 1% acetic acid, and lyophilized. The lyophilized material from 6.0 liters of culture fluid was extracted according to a modification of the acid–ethanol procedure (Twardzik et al., 1982) and applied to a BioGel P100 column in 1.0 M acetic acid. Individual fractions (3.5 ml) were collected, and por-

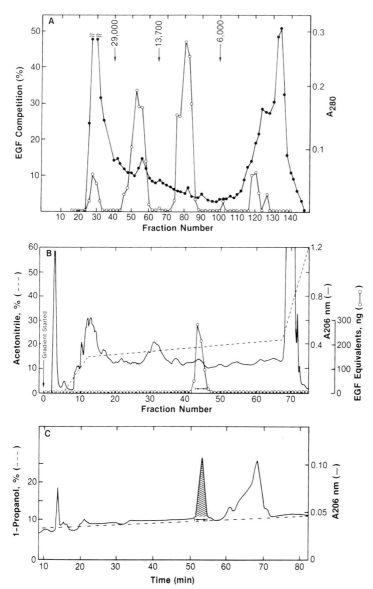

tions were tested for competition with ^{125}I-labeled EGF for binding to A431 cell membrane receptors (○) (De Larco and Todaro, 1978). (B) HPLC fractionation of pooled fractions from the 7000 M_r region of the BioGel P100 column on a μBondapak C_{18} support. Fractions (1.5 ml) were collected following elution using a linear gradient of acetonitrile containing 0.045% trifluoroacetic acid. Aliquots of the individual fractions were lyophilized and assayed for EGF competing activity (○). (C) Rechromatography of the TGF-containing fractions from column (B). Elution of 40 μg of protein was achieved with a linear gradient of 1-propanol containing 0.035% trifluoroacetic acid. The shaded area shows EGF competition.

competing activities from the pooled BioGel fractions. In contrast, upon similar analysis, mouse and human EGFs both eluted at an acetonitrile concentration of approximately 28%. These findings thus substantiate the physical association between EGF competing activity and induction of soft agar colony growth and further serve to distin-

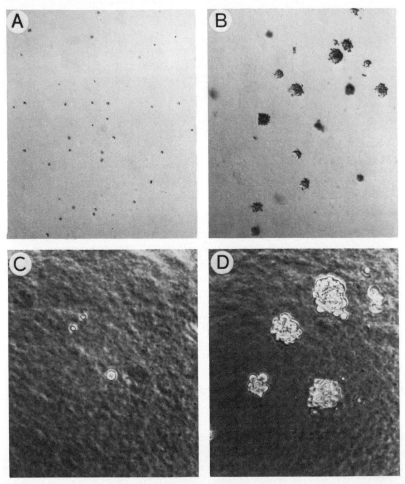

Fig. 2. Induction of growth of normal rat kidney fibroblasts in soft agar by Snyder-Theilen FeSV-transformed rat cell-derived TGF. Assays were performed according to previously described methods (De Larco and Todaro, 1978). Photomicrographs include: (A) untreated cells (×54); (B) cells treated for 7 days with 10 ng of Snyder-Theilen FeSV-induced rTGF (×54); (C) untreated cells (×460); and (D) typical normal rat kidney soft agar colonies formed 7 days posttreatment with Snyder-Theilen FeSV-induced rTGF (×460).

TABLE III
Purification of hTGFs from Conditioned Medium of the Human Melanoma Cell Line A2058[a]

Purification steps	Protein recovered (mg)	EGF competing activity recovered (units)	Relative specific activity (units/mg)	Degree of purification (-fold)	Recovery (%)
1. A2058-conditioned medium	1,020	4,525	4.4	1	100
2. Acid-soluble supernatant	837	4,299	5.1	1	95
3. Bio-Gel P-10					
pool P-10-A	29.7	2,077	70	16	45.9
pool P-10-B	14.5	2,033	140	32	44.9
4. μBondapak C$_{18}$ (acetonitrile)	0.202	1,628	8,059	1,823	36.0
5. μBondapak C$_{18}$ (1-propanol)	0.0015	1,476	984,000	223,636	32.6

[a] Purification of hTGF was performed according to methods described in detail previously (Marquardt and Todaro, 1982). Total protein was determined using bovine serum albumin as a standard. The quantitation of hTGF at step 5 was based on amino acid analysis. One EGF competing activity unit is defined as the amount of protein that inhibits the binding of [^{125}I]EGF to its receptor by 50%.

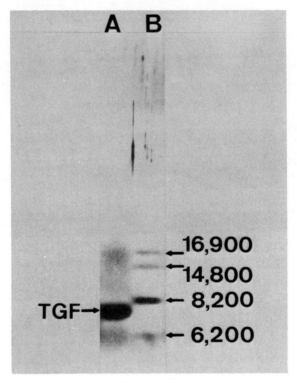

Fig. 3. SDS-PAGE analysis of rTGF isolated from Abelson MuLV-transformed rat cells. Purified rTGF (100 ng) was dissolved in 10 µl of sample buffer containing 1% SDS and 1% β-mercaptoethanol, incubated at 100°C for 2 min and applied to a 15–30% acrylamide gradient slab gel. After electrophoresis, gels were fixed in 50% methanol, 10% acetic acid for 2 hr, washed in 5% methanol, 7% acetic acid overnight, and stained with silver. The positions of marker polypeptides used to construct standard plots of log molecular weights versus mobility include horse heart myoglobin (16,900 M_r) and its cyanogen bromide peptides (14,800, 8,200, 6,200, and 2,500 M_r). The 2,500 M_r cyanogen bromide peptide was detected only with sufficient loading.

guish the TGF-related EGF competing activity from EGF itself. As a final purification step, TGF-containing fractions from the C_{18} µBondapak columns were pooled, lyophilized, reconstituted in 0.05% trifluoroacetic acid, and re-chromatographed over a second column using a linear gradient of 1-propanol containing 0.035% trifluoroacetic acid as the mobile phase modifier. With all four TGF preparations, EGF competing activities co-purified with distinct absorbance peaks and were effectively separated from contaminating UV-absorbing materials. The elution profile of TGF isolated from Abelson MuLV-transformed Fisher rat cells is shown in Fig. 1C.

To assess the final extent of purification achieved by the above scheme, each of the TGF preparations was analyzed by SDS-polyacrylamide gel electrophoresis and visualized by silver staining, following the final HPLC fractionation step. Single major polypeptide bands with apparent molecular weights of around 7,400 (relative to standards) were observed in each case (Fig. 3). The overall extents of purification and recoveries of TGF from the human melanoma cell line AZ058 are summarized in Table III. The first stage at which appreciable purification was achieved was the P100 chromatography step, which resulted in approximately 16-fold increase in specific activity accompanied by a loss of around 50%. Most of the purification, however, was at the stage of the two HPLC fractionation steps, the first of which resulted in nearly 80-fold purification, and the second step resulted in an additional 100-fold purification with very little additional loss. It was only subsequent to the final HPLC purification step, however, that the TGFs were sufficiently purified so as to be identified by SDS-PAGE (Fig. 3). With all three TGFs, the final extents of purification were around 200,000-fold with overall recoveries of total EGF competing activity in the 30% range.

RAT AND HUMAN TGF AMINO ACID SEQUENCES

With the availability of highly purified preparations of TGFs of both rat and human origin, a more precise comparison of their structures was possible. The data obtained from analysis of the amino terminal 10 residues of all three TGF preparations are summarized in Table IV. The most important feature of this data is the striking conservation of sequences between TGFs from all three species. No detectable differences are apparent between TGFs from Snyder-Theilen FeSV- and Abelson MuLV-transformed rat cells. Moreover, rat TGF is indistinguishable from Moloney MSV-transformed mouse cell TGF and differs from the human derived factor only at one position within the first 10 residues. This extent of conservation between rat and human TGFs at the amino terminal end is quite high considering, for example, that mouse and human EGFs differ at a total of four positions within the first 10 residues.

It is also of interest to compare the sequence of rat TGFs to that previously reported for mouse EGF. Such a comparison is significant in view of the highly conserved nature of the TGFs and the lack to date of detectable differences between mouse and rat TGFs. The most striking similarity between these factors is the position of the six cystine residues within the amino terminal domain of the molecules. By

TABLE IV
Amino Terminal Sequences of Mouse,
Rat, and Human TGFs[a]

	1	5	10
mTGF	Val-Val-Ser-His-Phe-Asn-Lys-Pro-Asp		
rTGF	Val-Val-Ser-His-Phe-Asn-Lys-Cys-Pro-Asp		
hTGF	Val-Val-Ser-His-Phe-Asn-Asp-Pro-Asp		

[a] Semi-automated amino terminal Edman degradations were performed in an 890C Beckman sequenator. The thiazolinone derivatives of amino acids were converted to PTH-amino acids with 1 N HCl at 80°C for 10 min under nitrogen. The PTH derivatives at each cycle were separated by reverse-phase HPLC on a μBondapak phenylalkyl column. The details of this analysis and more extensive sequence data will be presented elsewhere (Marquardt et al., submitted for publication).

introduction of a single gap between the second and third cystines within each of the TGFs, all six cystines map at corresponding positions. The positions of these residues appear to be of functional significance, because despite extensive differences in these amino terminal sequences, mouse- and human-derived EGFs correspond completely with respect to the positions of their cystine residues. These findings suggest that EGF receptor binding, a property common to TGF and EGF, may be mediated through the amino terminal region of the molecules and involve the region of secondary structure resulting from the three disulfide bonds. Besides the six cystine residues, mouse EGF and rat TGF exhibit correspondence only at four additional positions within their first 30 residues. These include a proline at residue 7, serine at residue 9, tyrosine at residue 13, and glycine at residue 17. On the basis of the preliminary sequence analysis of the carboxy terminal half of TGF isolated from Snyder-Theilen FeSV-transformed rat cells, the molecule appears to share some, but only distant, sequence homology with EGF (Marquardt et al., submitted for publication).

TYROSINE PHOSPHORYLATION OF THE EGF RECEPTOR

As discussed above, the transforming gene products of several avian and mammalian retrovirus oncogenes exhibit tyrosine-specific protein kinase activity. These include among others, the v-*fes* gene common to the Snyder-Theilen and Gardner isolates of FeSV (Van de Ven et al., 1980; Barbacid et al., 1980; Reynolds et al., 1980b) and the v-*abl* gene of Abelson MuLV (Van de Ven et al., 1980; Witte et al., 1980). Similarly, Ushiro and Cohen (1980) demonstrated phosphorylation of the 160,000 M_r EGF receptor in response to EGF. On the basis of the similarities of the positions of the disulfide bonds within their amino terminal domains and the fact that TGF competes with EGF for receptor binding, it was of interest to test whether TGFs would also stimulate tyrosine phosphorylation of the EGF receptor. For this purpose, A431 cells were cultured for 1.0 min in medium containing either EGF or TGF. Extracts were prepared, incubated in [γ-^{32}P]ATP containing buffer for an additional 1.0 min, and analyzed by SDS-PAGE. As shown in Fig. 4, phosphorylation of the EGF receptor in response to both mTGF and hTGF under these conditions was comparable to that observed in response to EGF. Similarly, TGFs produced by Abelson MuLV-transformed rat cells have been found to induce phosphorylation of the EGF receptor (Twardzik et al., 1982). By two-dimensional phosphoamino acid analysis, the predominant ^{32}P-labeled amino acid residue was established in each case to be phosphotyrosine (Fig. 5).

To further test the specificity of TGF-induced phosphorylation of the EGF receptor, a synthetic peptide closely resembling the known site of tyrosine phosphorylation in the Rous sarcoma virus-transforming protein, pp60^{src}, was prepared (Pike et al., 1982a). This acceptor-site peptide is of particular interest because of its high degree of relatedness to acceptor peptides of other classes of retroviruses with tyrosine-specific protein kinase activity, including the Gardner and Snyder-Theilen strains of FeSV (v-*fes*) (Hampe et al., 1982), Fujinami sarcoma virus (v-*fps*) (Shibuya and Hanafusa, 1982), and Y-73 (v-*yes*) (Kitamura et al., 1982). Both EGF and partially purified preparations of TGF stimulated phosphorylation of the acceptor peptide by a membrane preparation of the EGF receptor (Pike et al., 1982b). The fact that the effects of the two growth factors were not additive suggests that TGF and EGF stimulate peptide phosphorylation through the same EGF receptor system. These findings thus demonstrate stimulation by TGFs of a protein kinase activity either closely associated with, or intrinsic to, the EGF membrane receptor, which phosphorylates tyrosine acceptor sites both within the membrane acceptor itself and in exogenous substrates.

By tryptic peptide analysis of the 160,000 M_r EGF receptor in A431 cells following either EGF- or TGF-induced phosphorylation, one major phosphorylated peptide has been identified (Reynolds et al., 1981a). The predominant phosphorylated residue is tyrosine. Al-

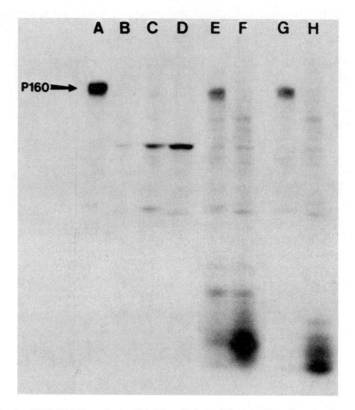

Fig. 4. SDS-PAGE analysis of A431 cellular substrates phosphorylated in response to either mEGF or mouse and human cell derived TGFs. A431 human tumor cells (A,C,E,G) and Fisher rat embryo cells (B,D,F,H) were grown to confluency; culture fluids were aspirated, and cells were rinsed with 10 mM sodium phosphate, pH 7.2, 200 mM NaCl, and 5 mM MgCl$_2$. Buffer (0.05 ml) containing 1.0 μg EGF (A,B), 200 ng partially purified hTGF (E,F), or mTGF (G,H) was added. Cells were incubated at 25°C for 1 min, disrupted by treatment for 1 min at 4°C in 0.05 ml of buffer consisting of 20 mM sodium phosphate, pH 7.2, 200 mM NaCl, 1% Triton X-100, and 5 mM MgCl$_2$, [γ-^{32}P]ATP (50 μCi) added, and incubation continued for 1.0 min. Reactions were terminated by addition of 0.05 ml of buffer consisting of 0.65 M Tris-HCl, pH 6.7, 1.0% SDS, 10% glycerol, 2.5% 2-mercaptoethanol, and 0.1% bromphenol blue; the reactions were heated for 2 min at 90°C, and analyzed by SDS-PAGE.

though phosphorylation of several minor peptides in tyrosine is also observed, this is somewhat variable depending upon reaction conditions. The map positions and relative intensities of phosphorylation of both major and minor acceptor sites in response to TGFs of mouse and human origins are analogous to those observed in response to EGF itself.

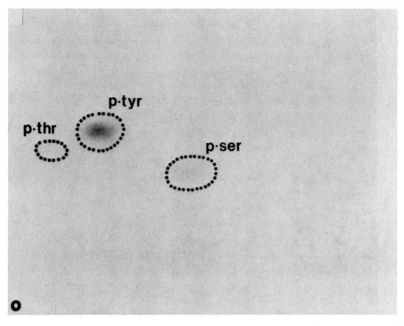

Fig. 5. Phosphoamino acid analysis of the 160,000 M_r EGF membrane receptor of A431 cells phosphorylated in response to EGF. *In vitro* phosphorylation was performed as described in the legend to Fig. 4, ^{32}P-labeled proteins were recovered from gel slices by incubation in TPCK–trypsin (50 µg per ml in 0.05 M ammonium bicarbonate, pH 8.0; Worthington) for 6 hr at 37°C. Supernatant fluids were filtered, lyophilized, washed by resuspension in H_2O, and relyophilized. Tryptic peptides were hydrolyzed in 6.0 M HCl at 100°C for 1 hr, and two-dimensional phosphoamino acid determination was performed as described elsewhere. The positions of unlabeled standards including phosphothreonine (p-thr), phosphotyrosine (p-tyr) and phosphoserine (p-ser) are indicated by dotted circles.

The similar receptor phosphorylation of A431 cell membrane receptors observed in response to TGFs as compared to EGF must be reconciled with differences in the phenotypic response of cells to these two classes of growth factors. In addition to their binding to, and phosphorylation of, the 160,000 M_r EGF membrane receptor, TGFs may specifically interact with another cell receptor. This model is supported by the recent demonstration of a 60,000 M_r NRK membrane receptor, which binds to highly purified preparations of ^{125}I-labeled mouse TGF but does not react with EGF (Massague *et al.*, 1982). Alternatively, interaction of growth factors with, and subsequent tyrosine phosphorylation of, the 160,000 M_r EGF receptor could be on a pathway resulting directly in expression of the transformed phenotype, or this interaction may simply provide a means of transporting the factors into the cell. Further studies are required to resolve these different possibilities.

CONCLUSIONS

As discussed above, the production of TGFs is a common property of cells transformed by retroviruses (De Larco and Todaro, 1978; Ozanne *et al.*, 1980; Twardzik *et al.*, 1982). Moreover, levels of TGF production appear generally to be much higher with retrovirus isolates whose transforming proteins exhibit tyrosine-specific protein kinase activity than by cells transformed by retroviruses lacking such activity (Twardzik *et al.*, 1983). Although a few of the chemically transformed rodent cell lines and spontaneous human tumors tested to date have also been found to produce TGF, others have not (Todaro *et al.*, 1980). Similarly, SV40 and polyoma virus-transformed cells do not exhibit the reduced levels of EGF binding characteristic of cells producing TGFs (Todaro *et al.*, 1976). The acquisition of cellular oncogenes by retroviruses may thus have selected for a subclass of sequences whose transforming activity is associated with TGF production. Alternatively, observed differences in TGF production may be only quantitative; TGF may be produced by many human tumor cells at levels too low for detection by presently available methods.

Although the weight of the available evidence strongly implicates TGFs in human tumors, many questions remain unanswered. For instance, it is not as yet known whether release of TGFs from the cells in which they are produced and their subsequent interaction with membrane receptors is necessary for transformation, or alternatively whether their transforming function is mediated entirely through intracellular pathway(s). The inability to detect production of the TGFs by many tumor cells could be explained by a model in which TGF-induced transformation could be entirely mediated through an intracellular pathway. From comparison to other growth factor systems, the former model, however, appears the most likely and certainly with respect to possible approaches to diagnosis and therapy would be the most desirable. It also remains to be resolved whether TGFs mediate their transforming function on their own or whether they act in concert with other defined cellular growth factors. For instance, Anzano *et al.* (1982) have recently described a factor that does not bind to the

EGF receptor, lacks transforming activity on its own, but which acts to enhance TGF-induced transformation and, in combination with EGF, induces growth of cells in soft agar.

Finally, the striking evolutionary conservation of TGFs between species as divergent as rat and human and their shared receptor recognition site with EGFs must be accounted for. These findings strongly argue that such factors have a normal host function. Even though the TGF associated transforming function is not necessarily mediated through the EGF receptor, the fact that binding of this receptor is a conserved property of TGFs indicates that this interaction has physiological significance. Studies of the role of TGFs in transformation should provide insight into the normal function(s) of both EGF and TGF.

REFERENCES

1. Anzano, M. A., Roberts, A. B., Meyers, C. A., Komoriya, A., Lamb, L. C., Smith, J. M., and Sporn, M. B. (1982) Cancer Res. **42**, 4776–4778.
2. Barbacid, M., Beemon, K., and Devare, S. G. (1980) Proc. Natl. Acad. Sci. U.S.A. **77**, 5158–5162.
3. Blomberg, J., Reynolds, F. H., Jr., Van de Ven, W. J. M., and Stephenson, J. R. (1980) Nature (London) **286**, 504–507.
4. Carpenter, G., King, L., and Cohen, S. (1979) J. Biol. Chem. **254**, 4884–4891.
5. Chang, E. H., Furth, M.E., Scolnick, E. M., and Lowy, D. R. (1982) Nature **297**, 479–483.
6. Coffin, J. M., Varmus, H. E., Bishop, J. M., Essex, M., Hardy, W. D. Jr., Martin, G. S., Rosenberg, N. E., Scolnick, E. M., Weinberg, R. A., and Vogt, P. K. (1981) J. Virol. **40**, 953–957.
7. Dalla-Favera, R., Franchini, G., Martinotti, S., Wong-Staal, F., Gallo, R. C., and Croce, C. M. (1982) Proc. Natl. Acad. Sci. U.S.A. **79**, 4714–4717.
8. De Larco, J. E., and Todaro, G. J. (1978) Proc. Natl. Acad. Sci. U.S.A. **75**, 4001–4005.
9. De Larco, J. E., and Todaro, G. J. (1980) J. Cell. Physiol. **102**, 267–277.
10. De Larco, J. E., Preston, Y. A., and Todaro, G. J. (1981) J. Cell. Physiol. **109**, 143–152.
11. Der, C. J., Krontiris, T. G., and Cooper, G. M. (1982) Proc. Natl. Acad. Sci. U.S.A. **79**, 3637–3640.
12. Fischinger, P. J. (1980) In "Molecular Biology of RNA Tumor Viruses" (J. R. Stephenson, ed.), pp. 163–198. Academic Press, New York.
13. Groffen, J., Heisterkamp, N., and Stephenson, J. R., (1983) Virology, in press.
14. Hampe, A., Laprevotte, I., Galibert, F., Fedele, L. A., and Sherr, C. J. (1982) Cell **30**, 775–785.
15. Heisterkamp, N., Groffen, J., Stephenson, J. R., Spurr, N. K., Goodfellow, P. N., Solomon, E., Carritt, B., and Bodmer, W. F. (1982) Nature (London) **299**, 747–749.
16. Kitamura, N., Kitamura, A., Toyoshima, K., Hirayama, Y., and Yoshida, M. (1982) Nature (London) **297**, 205–208.

17. Klein, G. (1982) "Advances in Viral Oncology," Raven Press, New York.
18. Marquardt, H., Hunkapiller, M. W., Hood, L. E., Twardzik, D. R., De Larco, J. E., Stephenson, J. R., and Todaro, G. J. (1983). Submitted for publication.
19. Marquardt, H., and Todaro, G. J. (1982) *J. Biol. Chem.* **257**, 5220–5225.
20. Massague, J., Czech, M. P., Iwata, K., De Larco, J. E., and Todaro, G. J. (1982) *Proc. Natl. Acad. Sci. U.S.A.*, **79**, 6822–6826.
21. Ozanne, B., Fulton, R. J., and Kaplan, P. L. (1980) *J. Cell. Physiol.* **105**, 163–180.
22. Parada, L. F., Tabin, C. J., Shih, C., and Weinberg, R. A. (1982) *Nature (London)* **297**, 474–478.
23. Pike, L. J., Gallo, B., Casnellie, J. E., Bornstein, P., and Krebs, E. G. (1982a) *Proc. Natl. Acad. Sci. U.S.A.* **79**, 1443–1447.
24. Pike, L. J., Marquardt, H., Todaro, G. J., Gallis, B., Casnellie, J. E., Bornstein, P., and Krebs, E. G. (1982b) *J. Biol. Chem.*, **257**, 14628–14631.
25. Reynolds, F. H., Jr., Van de Ven, W. J. M., and Stephenson, J. R. (1980a) *J. Virol.* **36**, 374–386.
26. Reynolds, F. H., Jr., Van de Ven, W. J. M., and Stephenson, J. R. (1980b) *J. Biol. Chem.* **255**, 11040–11047.
27. Reynolds, F. H., Jr., Todaro, G. J., Fryling, C., and Stephenson, J. R. (1981a) *Nature (London)* **292**, 259–262.
28. Reynolds, F. H., Jr., Van den Ven, W. J. M., Blomberg, J., and Stephenson, J. R. (1981b) *J. Virol.* **38**, 1084–1089.
29. Reynolds, F. H., Jr., Van de Ven, W. J. M., Blomberg, J., and Stephenson, J. R. (1981c) *J. Virol.* **37**, 643–653.
30. Sacks, T. L., Hershey, E. J., and Stephenson, J. R. (1979) *Virology* **97**, 231–240.
31. Santos, E., Tronick, S. R., Aaronson, S. A., Pulciani, S., and Barbacid, M. (1982) *Nature (London)* **298**, 343–347.
32. Savage, C. R., Jr., Hash, J. H., and Cohen, S. (1973) *J. Biol. Chem.* **248**, 7669–7672.
33. Shibuya, M., and Hanafusa, H. (1982) *Cell* **30**, 787–795.
34. Shih, T. Y., Papageorge, A. G., Stokes, P. E., Weeks, M. O., and Scolnick, E. M. (1980) *Nature (London)* **287**, 686–691.
35. Stephenson, J. R. (1980) "Molecular Biology of RNA Tumor Viruses." Academic Press, New York.
36. Swan, D. C., McBride, O. W., Robbins, K. C., Keithley, D. A., Reddy, E. P., and Aaronson, S. A. (1982) *Proc. Natl. Acad. Sci. U.S.A.* **79**, 4691–4695.
37. Todaro, G. J., De Larco, J. E., and Cohen, S. (1976) *Nature (London)* **264**, 26–31.
38. Todaro, G. J., and De Larco, J. E. (1978) *Cancer Res.* **38**, 4147–4154.
39. Todaro, G. J., Fryling, C., and De Larco, J. E. (1980) *Proc. Natl. Acad. Sci. U.S.A.* **77**, 5258–5262.
40. Todaro, G. J., Marquardt, H., De Larco, J. E., Reynolds, F. H., Jr., and Stephenson, J. R. (1981) *In* "Cellular Responses to Molecular Modulators" (W. Scott, R. Werner, and J. Schultz, eds.), pp. 183–204. Academic Press, New York.
41. Twardzik, D. R., Todaro, G. J., Marquardt, H., Reynolds, F. H., Jr., and Stephenson, J. R. (1982) *Science* **216**, 894–897.
42. Twardzik, D. R., Todaro, G. J., Reynolds, F. H., Jr., and Stephenson, J. R. (1983) *Virology*, **124**, 201–207.
43. Ushiro, H., and Cohen, S. (1980) *J. Biol. Chem.* **255**, 8363–8365.
44. Van de Ven, W. J. M., Reynolds, F. H., Jr., and Stephenson, J. R. (1980) *Virology* **101**, 185–197.
45. Weinberg, R. A. (1982) *Cell* **30**, 3–4.
46. Witte, O. N., Dasgupta, A., and Baltimore, D. (1980) *Nature (London)* **283**, 826–831.

Receptor-Mediated Endocytosis of Peptide Hormones and Macromolecules in Cultured Cells

MARK C. WILLINGHAM AND IRA H. PASTAN

Laboratory of Molecular Biology
National Cancer Institute
National Institutes of Health
Bethesda, Maryland

INTRODUCTION

The process of endocytosis of materials from the cell surface of animal cells occurs either by nonspecific (fluid-phase) uptake or by binding to components of the cell surface (adsorptive). For many physiologically important ligands, this binding occurs through interaction with specific, saturable receptors which exist as components of the plasma membrane. In the last decade, the events that occur after this binding have been more cleary elucidated for some ligand systems (1,2). In most of these ligand–receptor interactions, the complex is trapped in special surface structures called bristle-coated pits (3), from which the complex is rapidly internalized into the cell. The potential mechanisms by which this process occurs and the fate of the internalized components through the compartments within the cell will be the subject of this chapter.

THE MORPHOLOGIC STRUCTURES OF ENDOCYTOSIS ON THE CELL SURFACE

FLUID-PHASE ENDOCYTOSIS

In cell culture, there are certain distinct morphologic structures that exist on the cell surface that can mediate endocytosis (Fig. 1). In the

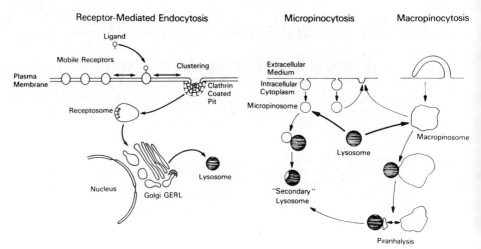

Fig. 1. Diagrammatic summary of the morphologic structures of endocytosis in cultured fibroblasts.

late 1920s, Lewis described the "drinking" of extracellular fluid by cultured macrophages, which he termed "pinocytosis" (4). This activity of these rapidly moving cells could be seen by light microscopy, and the description made by Lewis is still precise even by today's standards. In most cultured cells, however, this activity is in reality relatively slow and is only pronounced in those cell types that show rapid cell movement. This form of macroendocytosis is mediated by large folds in the cell surface called "ruffles," which fold back against adjacent areas of the plasma membrane with which they fuse. This fusion event creates a large membrane-limited bubble of extracellular medium within the cytoplasm called a macropinosome (5). With the advent of electron microscopy in the late 1940s and 1950s, other structures not visible by the light microscope became evident. Small invaginations in the cell surface were observed and called "caveolae" (6). These small (~800 Å diameter) flask-shaped invaginations are quite numerous on some cell types and have been reported to mediate the fluid-phase entry of smaller bubbles of medium called "micropinosomes," analogous to the larger macropinosomes formed by surface ruffles. Phagocytosis is a special form of endocytosis seen in certain specialized cell types (7) and will not be dealt with here. The main feature of these forms of endocytosis is that the materials which enter the cell in these pathways meet a similar fate: rapid fusion with mature lysosomes in the cell cytoplasm leading to destruction of the internalized molecules. In the case of cultured fibroblastic cells, the macropinosomal fusion with the lysosomal system can be visualized in living cells by light microscopy (5). The fusion process of macropinosome and lysosome does not occur by a simple single fusion event, but requires repeated interactions by directed motion of lysosomes, each time pinching off small pieces of the larger macropinosomal vesicle with which the lysosomes fuse. This process has been termed "piranhalysis" to note the frenzied activity of the lysosome, similar to a feeding frenzy in piranha fish (Fig. 1).

The directed motion of these lysosomes is another important feature of cellular activity, in that there is a special directed form of motion of intracellular organelles called "saltatory motion" (Latin-to dance) (8). This motion is restricted to certain organelles in the cell, and is most easily seen by light microscopy in living cells for mitochondria and lysosomes. The jumps, or saltations, of these organelles occur along tracks of microtubules in the cytoplasm and this motion can be interrupted by agents that interfere with microtubule function, such as colchicine (9). This motion is temperature- and energy-dependent and can be inhibited by concentrations of ATPase inhibitors that affect dynein-like but not actomyosin-like ATPases (10). The main feature of this motion, however, is its rapidity with many of the saltations visible without the use of time-lapse recording at rates of microns/second.

COATED PIT-MEDIATED ENDOCYTOSIS

Electron microscopy also revealed another previously unknown structure on the cell surface—the bristle-coated pit. In the 1950s images of these structures were published, but the exact nature and importance of these structures was not fully alluded to until 1964 when Roth and Porter described the presence of these structures on the surface of mosquito oocytes and their potential role in the concentrative uptake of macromolecules from the cell surface (3). In the early 1970s, it was shown that LDL (low density lipoproteins) could be found concentrated in coated regions of the cell surface just prior to endocytosis (11,12). Eventually, LDL could be found in lysosomes where it was degraded. Since that time, a number of other ligands for which specific receptors exist on the surface of cells were examined, both in intact tissue as well as in cultured cells. Evidence exists for the concentrative internalization through coated pits for α_2-macroglobulin(α_2M), insulin, triiodothyronine, epidermal growth factor

(EGF), asialoglycoproteins, a number of viruses, *Pseudomonas* toxin, transferrin, lysosomal enzymes (β-galactosidase), some lectins, maternal immunoglobulin, anti-IgM (in lymphoblastoid cells), yolk proteins, and in preliminary experiments, interferon. All of these ligands share a common feature of having highly specific cell surface receptors (reviewed in 1,2,11–15).

The coated pit is an indentation in the plasma membrane, generally 1200–1800 Å in dimension which has a latticework or cage of protein on the cytoplasmic face of the membrane. In most types of cultured cells, each cell has 500–1000 of these pits on its surface. The latticework is composed chiefly of a single protein called "clathrin" (13,16) which forms the structural basis for the lattice. Although the lattice can be manipulated *in vitro* to assemble and disassemble (17), we do not believe this process normally occurs in living cells (18–20). A second population of coated pits exists in the Golgi system and has a smaller size (~800 Å) than those at the plasma membrane. The function of these structures has not been clear, but recently their participation in intracellular traffic has been observed (21, see below). Both of these groups of coated structures contain clathrin, and antibodies to clathrin decorate both groups of coated elements when examined by electron microscopic immunocytochemistry (19). No other sites in the cultured cell types we have examined have substantial amounts of clathrin.

FLUORESCENCE EXPERIMENTS

One method of observing the entry of specific ligands is to label the ligand with a fluorescent tag and follow its distribution by fluorescence microscopy. Even if the size of an organelle that the ligand might associate with might be smaller than the refractile limit of resolution of light microscopy, fluorescence can still detect these objects because they emit light as point sources. This phenomenon is similar to our ability to "see" but not "resolve" stars in the night sky. Thus, one can label a ligand, such as $\alpha_2 M$, by producing a covalent derivative coupled to a fluorescent dye such as tetramethylrhodamine. Such a derivative can be shown to bind specifically and competably to its appropriate receptor on the surface of cells, in this case cultured fibroblasts (22). To follow the pathway of internalization binding is carried out at 4°C, where specific receptor binding occurs but not endocytosis (7). In this way most of the receptors on the cell surface can be labeled, and the endocytosis process can be synchronously started simply by raising the temperature to 37°C. Thus, very rapid endocytic events can be started and then halted by fixation to map precisely the time course of events.

With a rhodamine-labeled ligand such as α_2M, binding at 1°–4°C results in a diffuse labeling of the cell surface when observed by fluorescence. This image is actually due to a mixture of diffusely distributed occupied receptors and receptor–ligand complexes clustered in surface coated pits (2,23,24). This was only made clear, however, in electron microscopic experiments, because the complexes concentrated in coated pits seem to selectively decrease their fluorescence output. The exact reason for this decrease is not clear, but it does not occur when ligands are labeled indirectly with fluorescent labels such as antibodies (25,26). The labeling seen in the 4°C binding experiments can be shown to be specific by competition controls using excess unlabeled ligand (2,27).

When cells with surface-bound ligand are warmed to 37°C, the diffuse labeling of the surface changes within a few minutes to a punctate pattern of bright fluorescent dots which represent intracellular vesicles containing labeled ligand (2,28). These labeled organelles were shown to be intracellular vesicles since they moved by saltatory motion; this motion is known to be restricted to certain intracellular vesicles. These vesicles were termed "receptosomes" because of their participation in receptor-mediated endocytosis (28). More recent evidence has demonstrated the presence of receptor as well as ligand in these vesicles in some ligand systems (29,30). The movement of these receptosomes along microtubule tracks carries them back to the Golgi perinuclear region of the cell. It takes between 10 and 30 min for receptosomes to bring their ligand into the Golgi region at 37°C, depending on the cell type. Eventually, the ligand appears in large lysosomes, the rate of disappearance depending on the type of ligand. The results of light microscopic experiments cannot conclusively identify the structures involved, but they do give specific dynamic information about the process as it occurs in living cells. One result that could only be obtained from such experiments is the observation that receptosomes move as individual vesicles; this result could not be obtained from studies with fixed cells. To perform some of these fluorescence experiments, the amount of light emitted by a small number of labeled molecules must be greatly amplified. Often in such experiments the fluorescent signal is too small to be easily seen even by the dark-adapted eye. Our approach has been to use image intensification methods. Originally we used intensifier video cameras (22); more recently, we have also employed an electron acceleration intensifier (for a recent review of these methods, see 31).

ELECTRON MICROSCOPIC EXPERIMENTS

The organelles involved in this endocytosis process can be seen in greater detail by using ligands labeled for electron microscopic visual-

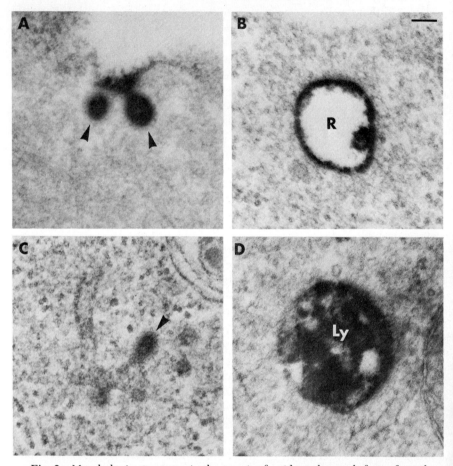

Fig. 2. Morphologic structures in the transit of epidermal growth factor from the cell surface to lysosomes. KB cells were incubated at 4°C with epidermal growth factor conjugated to horseradish peroxidase (EGF-HRP) (see text). Following washing, the cells were warmed to 37°C for 1 min (A), 5 min (B), 12 min (C), or 20 min (D) prior to fixation in glutaraldehyde. The peroxidase was detected by incubation with diaminobenzidine substrate, and the cells were embedded in Epon. EGF-HRP can be seen clustered in coated pits on the cell surface (arrowheads) in (A) followed by internalization into a receptosome (R) in (B). Following the delivery of the ligand to the Golgi system from the receptosome, EGF-HRP can be seen clustered in a coated pit (arrowhead) in the Golgi in (C) just prior to its appearance in lysosomes (Ly) as shown in (D). (All mags: ×90,000; bar = 0.1 μm; lead citrate counterstain)

ization (Fig. 2). These include the use of electron dense labels such as horseradish peroxidase, ferritin, colloidal gold, or the use of autoradiography. In a 4°C binding experiment, ligands such as α_2M can be found diffusely over the cell surface as well as concentrated in clathrin-coated pits (27). Other ligands such as certain viruses and EGF do not cluster in coated pits in the cold (21), in spite of the fact that EGF bound to its receptor has been found to be mobile in the plane of the plasma membrane at 4°C (32). LDL receptors, on the other hand, have been reported to cluster in coated pits even in an unoccupied state (11,12). A variety of drugs are capable of interfering with the clustering of occupied receptors into coated pits (1,2,24). The process of clustering appears to be able to occur in the cold and in cells treated with inhibitors of ATP synthesis. Endocytosis itself, the process of endocytic vesicle formation, requires elevated temperature and is blocked by inhibitors of ATP synthesis (7).

Numerous double-label light and electron microscopic experiments have clearly shown that any one coated pit and any one receptosome can mediate the entry of multiple ligands at the same time (1,2,23,25,33). This demonstrates one important point: that receptors which concentrate in the coated pit must have something in common to make them redistribute morphologically in the same structure. We imagine that this similarity might be reflected by a common sequence on the cytoplasmic face of the plasma membrane for these receptors, whereas the part of the receptor exposed on the exterior of the cell would contain ligand-specific sequences.

Although the coated pit serves as a nidus for the concentration of certain physiological receptor–ligand complexes, it also may serve a nonconcentrative function as the entry pathway for general membrane turnover and recycling. Considering that the time course of the endocytic event is extremely rapid (<20 sec at 37° for each coated pit) then a great deal of the cell surface lipid probably must flow through these structures to form receptosomes. This would result in a general turnover of cell surface lipids with a $T^{1/2}$ of 45–90 min. Calculations based on the amount of membrane delivered into each receptosome from each coated pit and the measured amounts of ligand internalized agree with this concept.

THE FORMATION AND STRUCTURE OF THE RECEPTOSOME

Electron microscopy has shown that the saltating vesicle derived from the coated pit is larger than the pit (2–4000 Å) and has no clathrin coat either morphologically or by immunocytochemistry. We termed this special vesicle the "receptosome." A number of properties are now reasonably clear about the receptosome: the membrane which it contains is most likely derived from the plasma membrane; the interior of the vesicle appears empty and does not seem to contain fluid derived from the culture medium; the ligand present in the coated pit is efficiently delivered to the receptosome; the receptor is also present with the ligand in at least two cases that have been examined; the receptosome often contains a single small inclusion vesicle (the receptosome therefore is initially a monovesicular body); the pH within the receptosome rapidly falls after its formation ($\cong$pH 5.0) (34); the vesicle fails to show detectable concentrations of lysosomal hydrolases cytochemically; the vesicle does not interact with or fuse with mature lysosomes in the cytoplasm. Following its formation, the receptosome moves by saltatory motion and appears to fuse with elements of the reticular portion of the Golgi system.

Receptosomes also have distinguishing morphologic characteristics. One is the presence of a fuzzy-appearing material on a flattened region at one side of the receptosome. This material is present on the cytosol face of the membrane. The nature of this material is not clear but it is seen on no other intracellular organelle.

The actual process by which receptosomes form has been difficult to study because it is very rapid. The original suggestion, which was based on the observation of coated vesicular images near the cell surface, was that coated pits pinch off to form isolated coated vesicles (13). Morphologic images of such vesicles are frequently seen in thin sections, but could result from a plane of section though a coated pit which did not pass through the neck of the pit. More recent experiments have suggested that the coated pits are stable structures that do not pinch off, but remain attached to the plasma membrane. One set of experiments has been performed by microinjection of antibodies to the coat protein, clathrin. In living cells this injected antibody reacts with the existing coated pits in the cell, decorating them specifically. If soluble unassembled clathrin or free-floating coated vesicles were present in the living cell, one would expect a precipitation reaction in the cytosol. This does indeed occur if the cell is later injected with soluble clathrin or isolated clathrin baskets from an exogenous source, but it does not occur in the living cell injected only with antibody. This suggests that free clathrin-coated vesicles probably do not exist in living fibroblasts (18,20).

Further evidence for the stable nature of coated pits comes from electron microscopic labeling experiments using impermeant external markers of the plasma membrane (20). Using ruthenium red and concanavalin A–peroxidase to label all structures in communication with the surface, every coated structure with a diameter of ~1400 Å that appeared to be a closed vesicle in thin section was found to be in communication with the cell surface at 4°C.

When warmed to 37°C, a significant number of these pits appeared to be isolated from the cell surface, but in the appropriate plane of section these pits were seen to have narrowed necks but not to be separated from the plasma membrane. These "cryptic" pits may be involved in the formation of receptosomes. When the cells were cooled back to 4°C, all of the pits reopened to the cell surface. These experiments suggest that these coated pits never actually pinched off as individual separate vesicles. Instead, we favor the idea that coated pits are stable elements of the plasma membrane and form receptosomes as an invagination of adjacent plasma membrane. Recent evidence using monovalent ionophores suggests that this process of vesicle formation may require the participation of an active H^+ pump in the coated pit membrane (35). This would be consistent with the observed energy and temperature dependence of vesicle formation and the rapid acidification of receptosomes after formation.

INTRACELLULAR FATE OF THE RECEPTOSOME

To follow the pathway of ligand entry, one needs a cell in which the entry and delivery to sequential compartments occurs in a synchronous wave. Further, the cell should have large numbers of receptors for a specific ligand for which an electron microscopic label is available. These requirements are satisfied by KB cells, a human carcinoma cell with $>10^5$ receptors per cell for EGF. Using a recently developed conjugate of EGF and horseradish peroxidase (EGF-HRP), we have followed the entry of EGF not only into the coated pit and receptosome, but also further into the cell (21,36). Between 3 and 10 min after warming cells to which EGH-HRP was bound at 4°C, the EGF-HRP is located exclusively in receptosomes. Between 10 and 13 min after entry, the receptosomes fuse with elements of the reticular Golgi system, the part of the Golgi that contains the second population of clathrin-coated pits in the cell (Fig. 2). The EGF-HRP can be followed into the lumen of these reticular elements and seen to concentrate again in the coated pits of the Golgi. Following this, the EGF-HRP can be seen in small lysosomes in the Golgi region no sooner than 15 min after entry. This strongly suggests that the coated elements of the Golgi function in a manner analogous to those at the plasma membrane in concentrating ligands prior to delivery to a specific compartment. In the case of those at the plasma membrane, it is the receptosome, for those in the Golgi it is lysosomes, both newly forming and mature. A similar concentration of ligand in coated pits of the Golgi has been seen for β-galactosidase in Chinese hamster ovary cells. In that case the binding to the cell surface, and presumably also in the Golgi system, is through a specific receptor that recognizes mannose-6-PO$_4$ residues in lysosomal enzymes. This mechanism could, therefore, be of physiological importance in the proper delivery of materials destined for lysosomes rather than exocytosis.

COMPARMENTALIZATION WITHIN THE GOLGI SYSTEM

Materials synthesized within the endoplasmic reticulum are either generally destined for lysosomal delivery (such as lysosomal enzymes) or for exocytosis to the outside of the cell. Very little is known specifically about the mechanism of delivery and addressing of materials for these separate destinations in cells, such as cultured fibroblasts, which do not form specific secretory granules. The finding that a lysosomal enzyme and a surface ligand such as EGF mix in the Golgi system for delivery to lysosomes suggests that such a morphologic pathway might also be used by the cell for materials synthesized in the endoplasmic reticulum (21). We have recently studied the exocytosis and processing of the G protein of vesicular stomatitis virus (VSV) as a model for the exocytic pathway (37). G protein is made in the endoplasmic reticulum, transported to the Golgi, and delivered to the cell surface in a highly synchronous fashion after infection by VSV. By electron microscopic immunocytochemistry, the location of G protein has been determined. No G protein could be detected in the coated pits of the Golgi, whereas large amounts could be found in the Golgi stacks, in the lumen of the endoplasmic reticulum, and on the plasma membrane. This is in keeping with the fact that little or no G protein is ever delivered to lysosomes. Further, it suggests that the coated regions of the Golgi may be organelles that direct materials strictly to lysosomes and do not participate in exocytosis. Thus, compartmentalization of materials destined for lysosomes may be determined by the presence of specific receptors for those materials that have the capacity to concentrate these ligands into coated regions of the Golgi.

PATHOLOGIC PROCESSES: TOXINS AND VIRUSES

A number of viruses and toxins have been suggested to take advantage of the coated pit endocytosis pathway to gain access to the cell. Many viruses have been found trapped in coated pits on the cell surface by electron microscopy, including both membrane-limited viruses (such as VSV and SFV) and non-membrane-limited viruses (such as adenovirus) (14,23,33,38). We have studied VSV and adenovirus and have observed their entry into cells in the coated pit–receptosome pathway. Analogous to studies with semliki forest virus (38), VSV appears to fuse with intracellular membranes, probably at the receptosome stage and probably in response to the low pH environment of the receptosome. Adenovirus enters the same pathway, but gains access to the cytosol directly as an intact virus particle apparently by lysis of the receptosome (39). It then migrates to the nuclear envelope where it selectively delivers its core into the nucleus, a most unusual mechanism. Preliminary experiments have shown that macromolecular materials delivered into the same receptosome in a double-label experiment also are delivered intact to the cytosol when they accompany an adenovirus viral particle (39).

Toxins, such as *Pseudomonas* exotoxin, also have been found to enter the coated pit–receptosome pathway (40). Whether the low pH environment favors their ability to cross the receptosome membrane remains to be established, but analogies to the viral systems are close. Both pathologic materials enter the same pathway and rapidly gain access to the cytosol where they have a specific site of action, a route that has yet to be clearly shown for physiologic ligands.

THE PHYSIOLOGICAL SIGNIFICANCE OF COATED PIT ENDOCYTOSIS

Since the coated pit system is found in virtually all functioning animal cells, its function must be of paramount importance. As yet, no one has isolated a mutant which lacks this system. The simplest reason may be that it is the main system by which cells recycle and turnover in a controlled way the components of their surface. Without such turnover, the lack of specific surface functions may rapidly lead to cell death. Certainly, for the concentrative ligand systems, much of the action of the endocytosis system may be to recycle specific receptors or to remove ligands from the surface where they exert their physiologic functions. As yet, no clearly proven examples exist for the role of this entry system in the physiologic function of such a ligand. One potential candidate for such a ligand system may be interferon (15). Another is the appearance of triiodothyronine in nuclear fractions that is dependent on the endocytosis of T_3 through the receptosome pathway (41). In studies on the site of action and the binding parameters of specific ligands, the existence of this very rapid and specific endocytosis system on the cell surface must be taken into account.

SUMMARY

Almost all mammalian cells contain a special morphologic system for selecting and internalizing specific molecules from their environment. The elements of this system include: bristle-coated pits on the plasma membrane, a special intracellular transport organelle called the "receptosome," the reticular membranous portion of the Golgi, the bristle-coated pits in the Golgi system, and lysosomes. In a larger sense, the exocytic elements of the Golgi system and the endoplasmic reticulum also interact with these other components to carry materials either derived from endocytosis or from intracellular synthesis to their proper destination. This traffic of intracellular materials is coordinated and directed by the Golgi, much of it probably directed by receptor-dependent interactions.

REFERENCES

1. Pastan, I. H., and Willingham, M. C. (1981) *Annu. Rev. Physiol.* **43**, 239–250.
2. Pastan, I. H., and Willingham, M. C. (1981) *Science* **214**, 504–509.
3. Roth, T. F., and Porter, K. R. (1964) *J. Cell Biol.* **20**, 313–332.
4. Lewis, W. H. (1931) *Bull. Johns Hopkins Hosp.* **49**, 14–36.
5. Willingham, M. C., and Yamada, S. S. (1978) *J. Cell Biol.* **78**, 480–487.
6. Bruns, R. R., and Palade, G. E. (1968) *J. Cell Biol.* **37**, 277–280.
7. Silverstein, S. C., Steinman, R. M., and Cohn, Z. A. (1977) *Annu. Rev. Biochem.* **46**, 669–722.
8. Rebhun, L. I. (1971) *Int. Rev. Cytol.* **32**, 93–137.
9. Wehland, J., and Willingham, M. C. (1983) (submitted).
10. Forman, D. S. (1981) *J. Cell. Biol.* **91**, 414a.
11. Anderson, R. G. W., Brown, M. S., and Goldstein, J. (1977) *Cell* **10**, 351–364.
12. Goldstein, J. L., Anderson, R. G. W., and Brown, M. S. (1979) *Nature (London)* **279**, 679–685.
13. Pearse, B. M. F., and Bretscher, M. S. (1981) *Annu. Rev. Biochem.* **50**, 85–101.

14. Dales, S. (1973) *Bacteriol. Rev.* **37**, 103–135.
15. Zoon, K. C., Arnheiter, H., ZurNedden, D., FitzGerald, D. J. P., and Willingham, M. C. (1983) (submitted).
16. Pearse, B. M. F. (1976) *Proc. Natl. Acad. Sci. U.S.A.* **73**, 1255–1259.
17. Keen, J. H., Willingham, M. C., and Pastan, I. H. (1979) *Cell* **16**, 303–312.
18. Wehland, J., Willingham, M. C., Dickson, R. B., and Pastan, I. H. (1981) *Cell* **25**, 105–119.
19. Willingham, M. C., Keen, J. H., and Pastan, I. H. (1981) *Exp. Cell Res.* **132**, 329–338.
20. Willingham, M. C., Rutherford, A. V., Gallo, M. G., Wehland, J., Dickson, R. B., Schlegel, C. R., and Pastan, I. H. (1981) *J. Histochem. Cytochem.* **29**, 1003–1013.
21. Willingham, M. C., and Pastan, I. H. (1982) *J. Cell Biol.* **94**, 207–212.
22. Willingham, M. C., and Pastan, I. H. (1978) *Cell* **13**, 501–507.
23. Willingham, M. C., Haigler, H. T., Dickson, R. B., and Pastan, I. H. (1981) *In* "International Cell Biology 1980–1981" (H. G. Schweiger, ed.), pp. 613–621. Springer-Verlag, Berlin and New York.
24. Dickson, R. B., Schlegel, C. R., Willingham, M. C., and Pastan, I. H. (1982) *Exp. Cell Res.* **140**, 215–226.
25. Via, D. P., Willingham, M. C., Pastan, I. H., Gotto, A. M., and Smith, L. C. (1982) *Exp. Cell Res.* **141**, 15–22.
26. Anderson, R. G. W., Goldstein, J. L., and Brown, M. S. (1980) *J. Recept. Res.* **1**, 17–39.
27. Willingham, M. C., Maxfield, F. R., and Pastan, I. H. (1979) *J. Cell Biol.* **82**, 614–625.
28. Willingham, M. C., and Pastan, I. H. (1980) *Cell* **21**, 67–77.
29. Willingham, M. C., Pastan, I. H., and Sahagian, G. G. (1983) *J. Histochem. Cytochem.* **31**, 1–11.
30. Dickson, R. B., Willingham, M. C., and Pastan, I. H. (1983) In preparation.
31. Willingham, M. C., and Pastan, I. H. (1983) *In* "Methods in Enzymology" (in press).
32. Hillman, G. M., and Schlessinger, J. (1982) *Biochemistry* **21**, 1667–1692.
33. Dickson, R. B., Willingham, M. C., and Pastan, I. H. (1981) *J. Cell Biol.* **89**, 29–34.
34. Tycko, B., and Maxfield, F. R. (1982) *Cell* **28**, 643–651.
35. Dickson, R. B., Schlegel, C. R., Willingham, M. C., and Pastan, I. H. (1982) *Exp. Cell Res.* **142**, 127–140.
36. Willingham, M. C., Haigler, H. T., FitzGerald, D. J. P., Gallo, M. G., Rutherford, A. V., and Pastan, I. H. (1983) *Exp. Cell Res.* (in press).
37. Wehland, J., Willingham, M. C., Gallo, M. G., and Pastan, I. H. (1982) *Cell* **23**, 831–841.
38. Helenius, A., Kartenbeck, J., Simons, K., and Fries, E. (1980) *J. Cell Biol.* **84**, 404–420.
39. FitzGerald, D. J. P., Padmanabhan, R., Pastan, I. H., and Willingham, M. C. (1983) *Cell* **32**, 607–617.
40. FitzGerald, D., Morris, R. E., and Saelinger, C. B. (1980) *Cell* **21**, 867–873.
41. Horiuchi, R., Cheng, S.-y., Willingham, M. C., and Pastan, I. H. (1982) *J. Biol. Chem.* **257**, 3139–3144.

Thyroid Dependence of Neoplastic Transformation

DUANE L. GUERNSEY
Department of Physiology and Biophysics
University of Iowa College of Medicine
Iowa City, Iowa

CARMIA BOREK
Radiological Research Laboratory
Department of Radiology and Department of Pathology
College of Physicians and Surgeons
Columbia University
New York, New York

PAUL B. FISHER
Department of Microbiology
College of Physicians and Surgeons
Columbia University
New York, New York

ISIDORE S. EDELMAN
Department of Biochemistry
College of Physicians and Surgeons
Columbia University
New York, New York

INTRODUCTION

Hormones are known to play an important role as modifiers of tumor development: Experiments on endocrine gland ablation and hormone administration to animals have implicated many humoral agents, including thyroid hormone, as modulators of neoplasia (1,2). Interpretation of these findings, however, is complex in that the results may vary depending on a variety of factors including: (1) the mechanism and degree of thyroid ablation, (2) the dose of administered thyroid hormones [thyroxine (T_4) and triiodothyronine (T_3)], (3) accompanying alterations in hypothalamic and pituitary secretions, (4) the time sequence of thyroid modification in relation to the carcinogenic event or tumor transplant, and (5) differences in thyroid hormone effects on cell metabolism and biochemistry in various specific cell types.

HISTORICAL BACKGROUND

A relationship between breast cancer and thyroid status was reported as early as 1890 (3). Recently, however, clinical studies have reported conflicting data, in that some studies found growth of mammary tumors suppressed (4,5) and others enhanced (6,7) in hypothyroid patients. Indeed, the relationship between breast cancer in women and thyroid disease has prompted a controversy concerning the safety of thyroid medication in women (8,9). The difficulty in interpreting these clinical studies is the unresolved question of whether the increased incidence of breast cancer in these patients is a consequence of the hypothyroidism or the thyroid hormone therapy.

Laboratory investigations in rodents have also directed attention to the role of thyroid hormones in mammary tumorigenesis. Thus, several reports noted that hypothyroidism decreased the incidence of experimentally induced mammary tumors (10–13). Recently, Goodman et al. (14) found that experimental hypothyroidism in rats suppressed the growth of carcinogen-induced mammary tumors and that administration of thyroid hormone, sufficient to maintain the euthyroid state restored tumor incidence to the control rate. It is important to note that in the experiments by Goodman et al. (14) the animals were challenged with the carcinogen *after* thyroid status was altered. Previous studies of thyroid effects on mammary tumors using a similar protocol (10,12,15,16) are consistent with the results of Goodman et al. (14).

In 1952 and 1953, Bielschowsky and Hall (17,18) reported that surgical thyroidectomy completely eliminated the induction of liver tumors by 2-aminofluorene and 2-acetylaminofluorene. In the 1960s, Goodall (19,20) published a series of papers confirming these results and found that induction of hepatomas by fluroenylamines were thyroid hormone dependent. Goodall (20) also noted that thyroidectomy had no effect on progression of tumor foci but suppressed the initiating action of the carcinogen. In his studies, however, some extrahepatic tumors were not thyroid dependent.

The role of the thyroid in the biology of hepatic tumors has been of renewed recent interest, as well. Mishkin and colleagues (21) reported that experimental hypothyroidism (propylthiouracil-induced or radio-ablation) retarded the growth of transplanted Morris 44 hepatomas, inhibited metastatic growth, and increased the survival of rats bearing hepatomas. Short *et al.* (22) also demonstrated that surgical thyroidectomy inhibits the growth of transplanted Morris hepatoma 777 and 5123tc and that the inhibition is reversed by thyroid hormone administration.

The effects of thyroid status on several other induced tumor systems have also been reported. Gillman and Gilbert (23,24) found that the thyroid gland was not essential for the initiation of estrogen-induced pituitary adenoma, although the growth and size of the tumor were thyroid-dependent. In contrast the thyroid was necessary for initiation and promotion of trypan blue-induced reticulosarcoma (23,24). In recent studies in well-defined syngeneic mouse sarcoma systems, hypothyroidism suppressed growth and metastases, whereas thyroid hormone treatment enhanced both tumor growth and metastatic potential (25).

Thus, it has been clearly demonstrated that thyroid status effects the induction of certain tumors, the growth of transplanted tumors, the metastatic potential of tumors, and the survival of animals bearing tumors. In some systems, thyroid status is a determinant in the initiation of tumor development and in others the promotion/progression of tumor growth. Not all tumors, however, exhibit a marked dependence on thyroid status.

THYROID HORMONE MODULATION OF NEOPLASTIC TRANSFORMATION IN CELL CULTURE

Our studies addressed the question of whether thyroid hormone was involved in the pathogenesis of neoplasia *in vitro*, as had previously been reported *in vivo*. The use of cell culture systems, of course, has the advantage of well-characterized stages of initiation and promotion in oncogenesis. We chose early-passage primary hamster embryo cells (HE) and C3H/10T1/2 clone 8-mouse embryo cells for our studies: X-ray induced transformation has been well-characterized in HE cells in earlier studies (26–29) and analyzed further in C3H/10T1/2 cells more recently (30–32).

In our studies on thyroidal modulation of X-ray induced transformation, the HE cells were scored with a clonal assay, as described previously (26–28) and the C3H/10T1/2 cells as described by Reznikoff

TABLE I

The Effect of Thyroid Hormone on X-Ray Irradiation-Induced Cell Transformation *in Vitro* of C3H/10T1/2 Mouse Cells and Hamster Embryo Cells in Culture[a]

Cells	Serum treatment/ (T_3 condition)	X-rays (Gy)	Total surviving colonies	No. of colonies transformed	Transformation frequency
C3H/10T1/2	Untreated FBS	0	23,313	0	0
	Untreated FBS	3	27,737	24	8.65×10^{-4}
	Resin-treated FBS/($-T_3 - T_4$)	0	23,154	0	0
	Resin-treated FBS/($-T_3 - T_4$)	3	33,747	0	0
	Resin-treated FBS + T_3/($+T_3$)	0	15,884	0	0
	Resin-treated FBS + T_3/($+T_3$)	3	21,040	16	7.60×10^{-4}
HE	Untreated FBS	0	2,800	0	0
	Untreated FBS	2.2	4,800	16	3.33×10^{-3}
	Resin-treated FBS/($-T_3 - T_4$)	0	2,400	0	0
	Resin-treated FBS/($-T_3 - T_4$)	2.2	1,500	0	0
	Resin-treated FBS + T_3/($+T_3$)	0	2,600	0	0
	Resin-treated FBS + T_3/($+T_3$)	2.2	3,600	15	4.17×10^{-3}

[a] Values given are totals of three separate experiments with C3H/10T1/2 cells and two separate experiments with hamster embryo cells. Stock 10^{-3} M T_3 in 50% *n*-propanol was diluted in media with 10% resin-treated fetal bovine serum (FBS) to give a final concentration of 10^{-7} M T_3 (designated $+T_3$). Media without thyroid hormone was prepared with 10% resin-treated FBS and an amount of diluent equal to that added in the $+T_3$ media. Stock cultures of C3H/10T1/2 cells were maintained in Eagle's basal medium +10% heat-inactivated FBS; hamster embryo cells in Dulbecco's modified Eagle's medium + 10% FBS. Both cultures contained penicillin (50 U/ml) and streptomycin (50 μg/ml). All cells were maintained at 37°C with 5% CO_2 in air throughout the experiment. One week before seeding, stock cultures were placed in the experimental culture media, with and without T_3 (as described in the text) and maintained in these conditions for the duration of the experiment. Twenty-four hours after seeding, the cells were irradiated with X rays at room temperature (2.2 Gy for HE and 3 Gy for C3H/10T1/2 cells), at a dose rate of 0.322 Gy/min, thereafter receiving weekly media changes. After an appropriate incubation period (6 weeks for C3H/10T1/2 and 2 weeks for HE), the cells were fixed and stained with Giemsa and scored for transformation. Both type II and III foci were scored in C3H/10T1/2 experiments (35).

TABLE II
Time-Dependence of Thyroid Hormone Modulation of X-Ray-Induced Neoplastic Transformation in Vitro[a]

Pretreatment	Medium				Transformed foci/ surviving cells	Transformation frequency (foci/surviving cells)
	At 12 hr before X-ray	At X-ray	At 12 hr after X-ray	At 24 hr after X-ray and thereafter for 6 weeks		
+T3	+T3	+T3[b]	+T3	+T3	0/27,168[c]	—
−T3	−T3	−T3[b]	−T3	−T3	0/27,160[c]	—
+T3	+T3	+T3	+T3	+T3	4/21,746[c]	1.8×10^{-4}
−T3	−T3	−T3	−T3	−T3	0/35,760[c]	—
+T3	+T3	−T3	−T3	−T3	1/9,450	1.1×10^{-4}
+T3	+T3	+T3	+T3	−T3	6/12,600	4.7×10^{-4}
−T3	+T3	+T3	+T3	+T3	5/11,210	4.5×10^{-4}
−T3	−T3	+T3	+T3	+T3	2/23,570[c]	8.5×10^{-5}
−T3	−T3	−T3	+T3	+T3	0/12,750	—
−T3	−T3	−T3	−T3	+T3	0/16,480[c]	—

[a] T3 was at 0.1 μM when present. X-irradiation was at 3 grays (C3H/10T1/2 cells).
[b] No X-irradiation.
[c] Composite of two separate experiments (36).

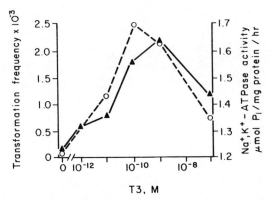

Fig. 1. Effect of varying concentrations of T_3 on transformation frequency (○) and Na^+,K^+-ATPase activity (▲). For transformation experiments, cells were pretreated in various doses of T_3 for 1 week prior to irradiation (4 grays) and maintained in the same conditions for the remainder of the experiment (36).

et al. (33). Endogenous thyroid hormones (T_3/T_4) were removed from fetal bovine serum (FBS) by adsorption to AG-1x10 resin, as described by Samuels et al. (34).

In both C3H/10T1/2 and HE cells, no spontaneous transformants were detected in cultures that were not exposed to X-rays during the 7 weeks of analysis (Table I) (35). In unmodified FBS, a single exposure to 2–3 Grays yielded transformation frequencies of 10^{-3} in both populations. In resin-treated FBS, however, no transformants were detected at these X-ray exposures. Enrichment of the resin-treated FBS with only triiodothyronine (T_3) restored the transformation frequency to the control rates, implying an absolute dependence on thyroid hormone.

To distinguish between possible early effects (initiation) and late effects during the period between challenge with X-rays and the appearance of foci, the time course of the response to T_3 was also investigated (36). Addition of T_3 to the media 12 hr before irradiation gave the maximum yield of transformants (Table II). In contrast, supplementation with T_3, 12 or 24 h after irradiation failed to restore sensitivity to the challenge. In view of the well-documented latent period (8 to 12 hr) in the inducing action of T_3 and the relatively brief (i.e., days) relaxation times on removal of the hormone (34,37), our results imply a crucial dependence on thyroid hormone during initiation and virtually complete independence during the period between initiation and expression as a focus.

The time-course studies raised the possibility of T_3 induction of a host cell protein that is an obligatory participant in initiation. Accordingly, the dependence of x-ray induced transformation of C3H/10T1/2 cells on T_3 concentration was analyzed (36). As noted in the preceding experiments no transformants were obtained in the resin-treated FBS (Fig. 1). Progressive enrichment with T_3 gave a maximum yield at 10^{-10} M (Fig. 1); the threshold concentration was 10^{-12} M. These values are within the range of physiological concentrations of thyroid hormone in the blood of intact mammals. Further trials revealed that the dependence on thyroid hormone was not a result of differences in survival or growth characteristics of the cells maintained in the presence or absence of thyroid hormone.

The fact that pretreatment with the hormone for at least 12 hr prior to irradiation is necessary for the expression of transformation is consistent with the characteristic lag period of thyroid hormone induction of protein synthesis (37,38). Earlier studies noted that T_3 regulated the synthesis of rat renal Na,K-ATPase with a latent period of 12 hr (39,40). Thus, Na,K-ATPase was used as an index of the inducing action of T_3 and simultaneous measurements were made of the T_3 concentration-dependence of Na/K,ATPase and neoplastic transformation in the C3H/10T1/2 cells (Fig. 1). The close similarity in these responses, including the fall-off at concentrations in excess of 10^{-9} M supports the inference that thyroid hormone induces the synthesis of a host protein(s) that is necessary for the initiation of transformation.

Further evidence for the involvement of protein synthesis was ad-

TABLE III
Effect of T3, rT3, and T3 + Cycloheximide on X-Ray-Induced Neoplastic Transformation *in Vitro*[a]

Pretreatment	Medium 12 hr before X-ray	Medium 24 hr after X-ray	Transformed foci/ surviving cells	Transformation frequency
−T3	+T3	−T3[d]	0/8,820	—
−T3	+T3	−T3	14/15,040	9.3×10^{-4}
−T3	−T3	−T3	0/15,130	—
−T3	+rT3	−T3	0/16,005	—
−T3	+T3[b]	−T3	7/12,354	5.7×10^{-4}
−T3	+T3[c]	−T3	0/13,200	—

[a] T3 was at 1 nM when present; rT3, reverse T3 at 1 nM. Irradiation was 4 grays. From ref. 36.
[b] Plus cycloheximide at 50 ng/ml.
[c] Plus cycloheximide at 100 ng/ml.
[d] No X-irradiation.

TABLE IV
Total Number of Surviving Cells and Transformed
Colonies from All X-Ray Transformation
Experiments ± Triiodothyronine $(T_3)^a$

	$+T_3$	$-T_3$
Total surviving cells	149,231	129,200
Total transformed colonies	129	0
Transformation frequency	8.6×10^{-4}	0

[a] Summary of all experiments with C3H/10T1/2 cells.

duced in that cells exposed to the biologically inactive isomer of T_3, reverse T_3 (rT_3), during the critical time period (i.e., 12 hr before to 24 hr after irradiation) were not transformed by X-rays (Table III) (36). Additionally, when cycloheximide, a potent inhibitor of protein synthesis was added and removed concurrently with T_3, transformation frequency was suppressed by 40% at an inhibitor concentration of 50 ng/ml and by 100% at 100 ng/ml (Table III). These concentrations of cycloheximide inhibited 20% and 50% of protein synthesis, respectively.

The absolute dependence of X-ray induced transformation on T_3 in the C3H/10T1/2 cell line is documented in Table IV. In media containing T_3 (all doses) X-irradiation induced neoplastic transformation in 129 of 149,231 surviving cells. In media depleted of thyroid hormones, there has been a total of 129,200 surviving cells and not one transformed focus was detected. These results include all of the experiments completed to date.

THYROID HORMONE MODULATION OF ADENOVIRUS TRANSFORMATION OF CELLS IN CULTURE

Analysis of the multifactorial nature of carcinogenesis has been aided by the development of cell culture systems which respond to the combined actions of chemical carcinogens, tumor promoters, and transforming viruses (41). Accordingly, we made use of a cloned population of Fischer rat embryo cells (CREF), which is transformed 150-fold more efficiently than secondary rat embryo cells (42) to evaluate the effects of T_3 on transformation induced by a temperature-sensitive DNA-minus mutant of type 5 adenovirus (H5ts125). In media devoid

TABLE V
Effect of T_3 on H5ts125 Transformation of CREF Cells[a]

Experiment number	$-T_3$	$+T_3$	T_3 removed 72-hr postinfection
1	18 ± 5	52 ± 5	48 ± 11
2	21 ± 6	60 ± 8	56 ± 5

[a] Cells were infected with 3 PFU/cell of H5ts125 at 36°C. Three hours postinfection the cells were subcultured and replated at 39.5°C. Seventy-two hours postinfection, cultures were shifted to low Ca^{2+} media; the media was changed 2 times per week, and assays were scored 21 to 28 days postinfection. Cultures were preconditioned in media $\pm T_3$ for 1 week. The media were depleted of T_3 by treatment of FBS with Aglx10 resin ($-T_3$) and repleted by addition of T_3 to the resin-treated media to a final concentration of 10^{-9} M. Values given are the number of foci per 10^5 infected cells, mean ± SD from 7 to 10 plates per condition for each experiment. (Unpublished observations of P. B. Fisher, D. L. Guernsey, I. B. Weinstein, and I. S. Edelman.)

of thyroid hormone for the entire transformation assay (including 1 week preconditioning), there was a threefold reduction in H5ts125 transformation of CREF cells compared to cultures incubated in the presence of T_3 (Table V) (P. B. Fisher, D. L. Guernsey, I. B. Weinstein, and J. S. Edelman, unpublished data). No significant reductions in transformation frequencies were observed, however, if T_3 was removed 72 hr post infection. The reduction in H5ts125 transformation of CREF cells when grown in the absence of T_3 cannot be ascribed simply to an inhibitory effect on cell growth because both normal and Ad5 transformed cells grew at similar rates and did not differ in saturation densities when grown in the absence or presence of T_3. Although the underlying mechanism(s) involved in T_3 modulation of H5ts125 transformation of CREF cells is not known, T_3 may alter transformation by (a) regulating viral uptake, (b) affecting viral DNA integration into the cell genome, or (c) modifying the early expression of viral and/or cellular genes required for establishing the transformed state. Studies are now in progress to distinguish between these alternative mechanisms.

As previously discussed, in addition to its effects on the initiation of chemically induced transformation *in vivo*, thyroid hormone also affects the growth of transplantable tumors and the metastatic potential of tumors. Although other possibilities have not been excluded, these findings suggest that thyroid hormone might exert a direct effect on

TABLE VI
Effect of T_3 on Anchorage-Independent Growth in H5ts 125 Transformed CREF Clones[a]

CREF Clone	Experiment number	$-T_3$	$+T_3$
wt-3A	1	8.6 ± 1.3	11.6 ± 0.8
	2	9.4 ± 0.8	15.5 ± 0.7
ts-7E	1	5.3 ± 0.4	12.0 ± 0.8
	2	6.4 ± 0.8	13.7 ± 1.1

[a] wt-3A is a wild-type Ad5-transformed CREF clone and ts-7E is an H5ts125 transformed CREF clone. Cells were preconditioned for 1 week in a low Ca^{2+} medium $\pm T_3$. At the time of the agar assay cells were seeded in agar in low Ca^{2+} medium $\pm T_3$. 5×10^3 cells were seeded in agar, and the numbers represent the agar cloning efficiency (%) ± SD for 4 plates per experimental condition. (Unpublished observations of P. B. Fisher, D. L. Guernsey, I. B. Weinstein, and I. S. Edelman.)

expression of the transformed phenotype of tumor cells. One of the best *in vitro* indicators of *in vivo* tumorigenic potential of transformed fibroblast and epithelial cells is anchorage-independent growth, i.e., growth in agar, agarose, or methylcellulose (43). We, therefore, tested the effect of T_3 on anchorage-independence in two clones of adenovirus transformed CREF cells—wt-3A (transformed by wild-type 5 adenovirus) and ts-7E (transformed by H5ts125) (44). Both transformed clones displayed higher agar cloning efficiencies when T_3 was present during both the preconditioning and growth in agar stages (Table VI). The two clones differed, however, in their sensitivity to T_3. The ts-7E clone was more sensitive to T_3 modulation of growth in agar and exhibited approximately a twofold higher cloning efficiency when T_3 was present. In addition, in both clones the final size of the colonies in agar was always greater if T_3 was incorporated in the agar assay.

T_3 NUCLEAR RECEPTORS IN NORMAL AND TRANSFORMED CELLS

A substantial body of evidence indicates that many of the biological responses to thyroid hormone are initiated by T_3-nuclear-receptor interaction and subsequent regulation of RNA and protein synthesis

TABLE VII
The T_3 Binding Capacities (N_{max}) of Isolated Nuclei From Normal and Transformed C3H/10T1/2 Cells and CREF Cells[a]

Cell type	N_{max} (fmol T_3 bound/μg DNA)
C3H/10T1/2	0.80 ± 0.11
X-Ray transformed C3H/10T1/2	0.51 ± 0.03[b]
CREF	0.65[c]
wt-3A transformed CREF clone	0.15
ts-7E transformed CREF clone	0.12

[a] Unpublished observations of D. L. Guernsey, P. B. Fisher, and I. S. Edelman.
[b] $P < 0.05$.
[c] Values are the mean of 2 separate experiments.

(37,38). It was of interest, therefore, to assess nuclear binding of T_3 in the various normal and transformed cell lines used in the previously described experiments. The results indicate that prior to transformation the C3H/10T1/2 and CREF cells contain a reasonable complement of high-affinity T_3 binding sites, or similar magnitude (Table VII). Transformation of the C3H/10T1/2 cells was attended by a modest decline in the abundance of these binding sites (35%). This effect was even greater in the transformed CREF clones (i.e., ~80% decrease). If these binding sites represent authentic receptors, the residual population (20%) was sufficient to mediate the T_3-dependent, anchorage-independent growth in agar, noted above. It should be noted that there were no significant differences in the apparent K_d's for T_3 in any of the cell populations (before or after transformation).

The existence of high affinity nuclear T_3 binding sites in pretransformed C3H/10T1/2 and CREF cells raises the distinct possibility of the participation of this pathway in the dependence of these cells on thyroid hormone for initiation of transformation. In the case of X-ray transformation of C3H/10T1/2 cells, the dependence is absolute, whereas in the case of adenovirus-induced transformation of CREF cells the dependence is relative. The basis for these differences is currently under study.

CONCLUSION

Previous studies demonstrating that experimental hypothyroidism prevents the appearance of carcinogen induced tumors in rodents suggested that the hormone was necessary for the initiation of tumorigenesis. Our results support this inference. Hypothyroid C3H/10T1/2 cells and primary hamster embryo cells are insensitive to the initiation of transformation induced by X-rays indicating that thyroid status at the time of irradiation is critical to the response. Additionally, CREF cells rendered hypothyroid at the time of adenovirus infection are less susceptable to transformation. The mechanism by which thyroid hormone modulates the initiation stages of carcinogenesis is unknown. The hormone could be modulating genetic events, such as DNA repair, mutagenesis, recombination, or epigenetic pathways by modifying susceptability to initiation of transformation by oncogenic agents. In any case, it appears that T_3 acts via induction of protein synthesis in the target cells.

Recent studies indicate that transplanted tumors grow less and exhibit a decreased metastatic potential when placed in hypothyroid animals and that this can be reversed by thyroid hormone administration to the animal. These reports suggest that thyroid status may also influence the promotion/progression stages of oncogenesis. Our data showing that T_3 modifies the anchorage-independent growth of neoplastic cells supports the suggestions of a thyroid effect on the expression of the transformed phenotype in tumor cells.

REFERENCES

1. Kellen, J. P., and Hilf, R., eds. (1979) "Influences of Hormones in Tumor Development," Vol. 1. CRC Press, Boca Raton, Florida.
2. Kellen, J. A., and Hilf, R., eds. (1979) "Influences of Hormones in Tumor Development," Vol. 2. CRC Press, Boca Raton, Florida.
3. Beatson, G. T. (1890) *Lancet* **2**, 162–164.
4. Humphrey, L. J., and Swerdlow, M. (1964) *Cancer (Brussels)* **17**, 1170–1173.
5. Backwinkel, K., and Jackson, F. S. (1964) *Cancer (Brussels)* **17**, 1174–1177.
6. Itoh, K., and Maruchi, N. (1975) *Lancet* **2**, 1119–1121.
7. Ellerker, A. G. (1956) *Med. Press* **235**, 280–285.
8. Kopdi, C. C., and Wolfe, J. N. (1976) *JAMA, J. Am. Med. Assoc.* **236**, 1124–1127.
9. Gorman, C. A., Becker, D. B., Greenspan, F. S., Levy, R. P., Oppenheimer, J. H., Riulin, R. S., Robbins, J., and Vanderlaan, W. P. (1977) *JAMA, J. Am. Med. Assoc.* **237**, 1459–1463.
10. Jull, J. W., and Huggins, C. (1960) *Nature (London)* **188**, 73–75.
11. Helfenstein, J. E., Young, S., and Currie, A. R. (1962) *Nature (London)* **191**, 1108–1109.

12. Jabara, A. G., and Maritz, J. S. (1973) *Br. J. Cancer* **28**, 161–166.
13. Grice, O. D., Fairchild, S., and Thomas, C. G. (1967) *Cancer Res.* **8**, 23 (abstr.).
14. Goodman, A. D., Hoekstra, P., and Marsh, P. S. (1980) *Cancer Res.* **40**, 2336–2342.
15. Newman, W. C., and Moon, R. C. (1968) *Cancer Res.* **28**, 864–868.
16. Dubnik, C. S., Morris, H. P., and Dalton, A. J. (1950) *J. Natl. Cancer Inst. (U.S.)* **10**, 815–839.
17. Bielschowsky, F., and Hall, W. H. (1952) *Proc. Univ. Otaga Med. Sch.* **30**, 26.
18. Bielschowsky, F., and Hall, W. H. (1953) *Br. J. Cancer* **7**, 358–366.
19. Goodall, C. M. (1966) *Cancer Res.* **26**, 1880–1885.
20. Goodall, C. M. (1968) *N. Z. Med. J.* **67**, 32–43.
21. Mishkin, S. Y., Pollack, R., Yalovsky, M. A., Morris, H. P., and Mishkin, S. (1981) *Cancer Res.* **41**, 3040–3043.
22. Short, J., Klein, K., Kibert, L., and Ove, P. (1980) *Cancer Res.* **40**, 2417–2422.
23. Gillman, J., and Gilbert, C. (1955) *Nature (London)* **175**, 724–725.
24. Gillman, J., Gilbert, C., and Spence, I. (1955) *Experientia* **11**, 157–158.
25. Kumar, M. S., Chiang, T., and Deodhor, S. D. (1979) *Cancer Res.* **39**, 3515–3518.
26. Borek, C., and Sachs, L. (1966) *Nature (London)* **210**, 276–278.
27. Borek, C., and Sachs, L. (1967) *Proc. Natl. Acad. Sci. U.S.A.* **57**, 1522–1527.
28. Borek, C., and Hall, E. J. (1973) *Nature (London)* **243**, 450–453.
29. Borek, C., and Hall, E. J. (1974) *Nature (London)* **252**, 499–501.
30. Terzaghi, M., and Little, J. B. (1976) *Cancer Res.* **36**, 1367–1374.
31. Harr, A., and Elkind, M. M. (1979) *Cancer Res.* **39**, 123–130.
32. Borek, C., Miller, P., Pain, C., and Troll, W. (1979) *Proc. Natl. Acad. Sci. U.S.A.* **76**, 1800–1803.
33. Reznikoff, C. A., Bertran, J. S., Brankow, D. W., and Heidelberger, C. (1973) *Cancer Res.* **33**, 3239–3249.
34. Samuels, H. H., Stanley, F., and Casanova, J. (1979) *Endocrinology* **105**, 80–85.
35. Guernsey, D. L., Ong, A., and Borek, C. (1980) *Nature (London)* **288**, 591–592.
36. Guernsey, D. L., Borek, C., and Edelman, I. S. (1981) *Proc. Natl. Acad. Sci. U.S.A.* **78**, 5708–5711.
37. Oppenheimer, J. H. (1979) *Science* **203**, 971–979.
38. Tata, J. R. (1970) In "Biochemical Actions of Hormones (G. Litwack, ed.), Vol. 1, pp. 89–133. Academic Press, New York.
39. Lo, C.-S., and Edelman, I. S. (1976) *J. Biol. Chem.* **251**, 7834–7840.
40. Lo, C.-S., and Lo, T. N. (1979) *Am. J. Physiol.* **326**, F9-F13.
41. Fisher, P. B., Mufson, R. A., Weinstein, I. B., and Little J. C. (1981) *Carcinogenesis* **2**, 183–187.
42. Fisher, P. B., Weinstein, I. B., Eisenberg, D., and Ginsberg, H. S. (1978) *Proc. Natl. Acad. Sci. U.S.A.* **75**, 2311–2314.
43. Fisher, P. B., and Weinstein, I. B. (1980) *IARC Sci. Publ.* **27**, 113–131.
44. Fisher, P. B., Babiss, L. E., Weinstein, I. B., and Ginsberg, H. S. (1982) *Proc. Natl. Acad. Sci. U.S.A.* **79**, 3527–3531.

PART IV

VIRAL ONCOGENES

A Novel Retrovirus, Adult T-Cell Leukemia Virus (ATLV): Characterization and Relation to Malignancy

YORIO HINUMA
Institute for Virus Research
Kyoto University,
Kyoto, Japan

INTRODUCTION

In 1977, Takatsuki *et al.* (1) first described in Japan a probable new disease entity, adult T-cell leukemia (ATL). In 1980, a nationwide survey of leukemia and lymphoma in Japan by the T and B cell Malignancy Study Group (consisting of 48 investigators) (2) confirmed Takatsuki's observation and further proposed that this disease could also be called adult T cell leukemia/lymphoma (ATLL) because of its leukemic lymphomatous nature. A characteristic hematologic finding in the disease was polymorphic nuclear deformation of leukemic or lymphoma cells having T cell markers (Fig. 1). Another striking feature of this malignancy was the clustering of patients in southwestern Japan. This unique geographical clustering of cases of Japanese T cell malignancy (ATL) was analogous to that of African B cell malignancy (Burkitt lymphoma). This prompted investigators to suspect its possible viral etiology.

Attempts were made to find a putative virus causally related to ATL. As a result of studies started in October 1980, a retrovirus was demonstrated in a T cell line originating from leukemic cells of a patient with ATL (3), and subsequent virological and serological studies suggested that the virus was etiologically related to ATL. This retrovirus was named adult T cell leukemia virus (ATLV) (4). The initial studies were described in a previous paper (5).

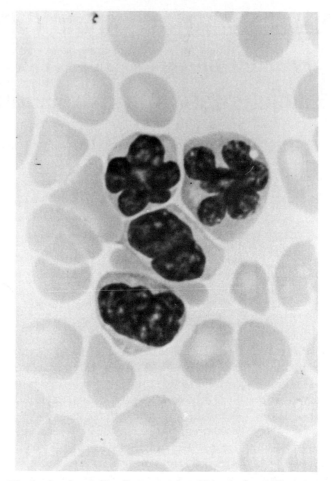

Fig. 1. Leukemic T cells in peripheral blood of an ATL patient.

This chapter reports the general properties of ATLV and the possible etiological relation to ATLV to ATL.

PROPERTIES OF ATLV

MORPHOLOGICAL OBSERVATIONS

Particles of ATLV seem to correspond to type C, but not B or D, of known retroviruses, judging from their appearance by electron microscopy in thin sections of ATL cell lines (3,6). However, ATLV appeared to differ from known mammalian retroviruses in two points. First, budding and completely electron-lucent forms of ATLV were extremely rare (3,6). This fact was unlikely to be due to an extremely low level of virus production in the cells because numerous virus particles were seen in extracellular spaces of cultures of the MT-1 line after induction by 5'-iodo-2-deoxyuridine (IUdR) (3) and in the MT-2 line (6). Second, the size of virus particles was extremely variable, ranging from about 50 to 150 nm in diameter (7). The reason for this is unknown, but suggested that the mechanism of maturation and/or release of ATLV differed from that of known retroviruses.

Type C virus particles similar to those in established ATL cell lines have also been found in short-term cultures of peripheral leukocytes from ATL patients and healthy adults having antibodies to ATLV (8–11).

BIOCHEMICAL PROPERTIES OF ATLV

ATLV was first characterized by biochemical techniques in two ATL-associated human T-cell lines—MT-1 and MT-2 (4). ATLV produced in cultures of the MT-2 cell line was examined after sucrose-density gradient centrifugation. ATLV in fractions with densities of 1.152–1.155 g/cm^3 showed abundant type C virus particles, reverse transcriptase activity that was insensitive to actinomycin D, RNA labeled with [^{3}H]uridine and specific proteins with molecular weights of 11,000, 14,000, 17,000, 24,000, and 45,000. Furthermore, complementary DNA (cDNA) prepared by an endogenous reaction with detergent-treated virions hybridized to 35 S RNA and 26 S component containing poly(A), which was inducible by IUdR treatment of the MT-1 cell line. In the MT-2 cell line, 35 S RNA was also detected, but several other smaller RNAs were observed as strong bands. All these data fit the requirements for retrovirus. However, for some unknown reason we have been unsuccessful in detecting 35 S RNA in purified virions produced by MT-2 cells. The integrated form of ATLV proviral DNA was detected in both MT-1 and MT-2 cell lines.

ATLV-ASSOCIATED ANTIGEN(S) (ATLA)

ATLA was first detected in a small fraction (1 to 5%) of acetone-fixed cells of an ATL cell line (MT-1) by indirect immunofluorescence using natural human antibodies in sera from either ATL patients or healthy adults in ATL-endemic areas (3). It was demonstrated that 100% of the cells in the MT-2 cell line were ATLA-positive (6). However, ATLA has not been detected in any other human cell lines unrelated to ATLV so far tested (3). The fluorescence was somewhat granular and often seen as large aggregates in the cytoplasm, but not the nucleus. However, it is noticeable that ATLA may also involve the antigen(s) associated with the cell membrane, because ATLA antibody-positive sera could react with the surface of living cells of ATLA-positive cell lines (12). The membrane antigen was named ATLMA. Most multinuclear giant cells in ATLV-positive cell lines were ATLA-positive, probably reflecting the formation of multinuclear cells by fusion of cells having virus-antigen(s) in the cell surface. ATLA has been found not only in established long-term ATL cell lines, but also in short-term cultures of either fresh leukemic cells from ATL patients (8) or fresh peripheral lymphocytes from healthy ATLV-carriers (8–11). ATLA has not been found in fresh uncultured ATL leukemic cells (8), suggesting that the formation of ATLV or ATLA is suppressed *in vivo*, but activated during *in vitro* cultivation.

Accumulated evidence suggests that ATLA could be related to the type C virus particles detected in ATL cell lines or short-term cultures of peripheral leukocytes of ATL patients or ATLA antibody-positive healthy adults, because (1) IUdR caused remarkable induction of ATLA; (2) the proportion of ATLA-positive cells and of cells with type C virus were roughly parallel; and (3) ATLA was localized in the cytoplasm. Furthermore, the following experiments (4,13) provide information on the ATLA specific for ATLV: Sera from patients with ATL or healthy adults, with titers of anti-ATLA predetermined by indirect immunofluorescence, were analyzed by immunoprecipitation followed by SDS-polyacrylamide gel electrophoresis. For this, an ATLV producer cell line (MT-2) was labeled with [^{35}S]methionine. All anti-ALTA positive sera, but no anti-ALTA negative sera tested, specifically reacted with four polypeptides with molecular weights of 76,000, 53,000, 36,000 and 24,000. Furthermore, enrichment of three polypeptides was observed on reaction with anti-ATLA positive sera. In control experiments with the ATLA-negative human T-cell lines (Molt-4 and HPB-ALL) none of these seven polypeptides were precipitated by reaction with anti-ATLA positive sera (3). All anti-ATLA-positive, but not -negative sera, tested were shown to react with a polypeptide with a molecular weight of 24,000 of ATLV purified by sucrose density gradient centrifugation (4,13).

MONOCLONAL ANTIBODIES TO ATLV

Mouse monoclonal antibodies against ATLA were prepared by the hybridoma technique (14). The ATLA specificity of the antibodies was determined by indirect immunofluorescence assay. Among several monoclonal antibodies, two ATLV-specific antibodies (GIN-2 and GIN-14) were studied. GIN-2 and GIN-14 antibodies both reacted with cytoplasm of acetone-fixed ATLA-positive cells, such as MT-1 and MT-2, but not with the membranes of living cells. The features of the antigens seen by immunofluorescence with these two antibodies were clearly distinguishable: with GIN-2 antibody, the fluorescence was localized as a large mass in the cytoplasm, whereas with GIN-14 the fluorescence was more diffusely distributed in the entire cytoplasm. Analysis of polypeptides of MT-2 cells labeled metabolically with [^{35}S]methionine by immunoprecipitation followed by SDS-polyacrylamide gel electrophoresis showed that GIN-2 preferentially precipitated p28 and p20 with slight precipitation of p19, whereas GIN-14 precipitated only p28. However, no polypeptides reacting with either one of the two monoclonal antibodies were detected in [^{35}S]methionine-labeled virions purified by sucrose density-gradient centrifugation. These results strongly suggest that both GIN-2 and GIN-14 monoclonal antibodies react with components of an ALTA complex and that the antigenic determinants or antigens of ATLA recognized by these antibodies are distinct.

These analyses of ATLA, which were first detected with natural human antibodies and then by the monoclonal antibody technique, are leading to definite determination and characterization of ATLV specific antigens.

BIOLOGIC ACTIVITIES OF ATLV

So far, transformation of human cells by cell-free ATLV has been unsuccessful (15). However, transformation of human lymphocytes by co-cultivation with lethally X-ray irradiated MT-2 cells was consistently observed (15,16). Leukocytes from either cord blood or adult peripheral bloods were target cells for transformation. The cells transformed by this co-cultivation method were not only T-cells but also frequently null cells (15). All transformed cell lines expressed ATLA. It was reported that peripheral blood lymphocytes of a Japanese monkey were transformed by co-cultivation with irradiated MT-2 cells (17).

Although co-cultivation with MT-2 cells consistently resulted in transformation of normal cells, two other ATLV-carrying cell lines (MT-1 and MT-4) did show this activity. The reason why these two cell lines could not transform cells is unknown, but it is possible that the MT-2 strain, but not the MT-1 or MT-4 strain of ATLV, has a transforming capacity.

It was demonstrated that cell-free MT-2 ATLV could induce formation of ATLA in peripheral lymphocytes of normal adults (15). This may mean that cell-free ATLV can infect and generate viral synthesis in cells but not induce this transformation.

ATLV-CARRYING CELL LINES

Table I lists ATLV-carrying cell lines. Two cell lines MT-1 and Ha originated directly from leukemic cells of ATL patients. Their growth is TCGF-independent. MT-2 and MT-4 were cell lines derived from co-cultures of leukemic cells with cord lymphocytes, and they appeared to originate from cord T-cells, but possibly transformed by ATLV in leukemic cells of individual ATL patients. Three other ATL leukemic cell lines (Su, Ni and Ya) were still TCGF-dependent for growth. We obtained several cell lines by transformation of peripheral lymphocytes from healthy adults or cord lymphocytes by co-cultivation with lethally irradiated MT-2 cells, and all of them should have

TABLE I
ATLV-Carrying Cell Lines

Line	Origin	TCGF[a] requirement	Surface marker	ATLA (+) cells (%)	Type C virus particles[a]	Reference
MT-1	ATL	−	T	5	+	3, 18
MT-2	ATL + CBL[b]	−	T	>95	+	6, 19
MT-4	ATL + CBL	−	T	>95	+	20
YAM	MT-2 + PBL[c]	−	T	>95	+	15
AIB	MT-2 + CBL	−	non-T, non-B	>95	ND[d]	15
Ha	ATL	−	T	40	ND	21
Su	ATL	+	T	20	+	21
Ni	ATL	+	T	20	ND	21
Ya	ATL	+	T	40	ND	21
Te-Sl	Healthy	−	T	>95	ND	11
Mo-Sl	Healthy	+	T	>95	+	11
Ke-Sl	Healthy	+	T	>95	+	11
As-Sl	Healthy	+	T	>95	ND	11
Su-Sl	Healthy	+	T	>95	ND	11

[a] Detected by electron microscopy.
[b] Co-cultivation of leukemic cells with normal cord blood lymphocytes (CBL).
[c] Co-cultivation of X-ray irradiated MT-2 cells with peripheral blood lymphocytes from healthy adults (PBL).
[d] Not done.

the genome of ATLV in MT-2 cells. YAM and AIB are representatives. It is noticeable that AIB cells have no T or B cell surface markers and that such non-T, non-B cell lines were frequently derived from cord lymphocytes co-cultured with MT-2 cells (15). We have continuous cultures of ATLV-carrying T-cell lines derived from healthy ATLV-carrier adults. One of them (Te-Sl) became able to grow without TCGF.

RELATION OF ATLV TO ATL

DETECTION OF ATLV IN FRESH LEUKEMIC CELLS OF ATL PATIENTS

ATLV has been demonstrated in fresh ATL leukemic cells by immunological, morphological, and biochemical procedures. As mentioned already, ATLA was detected in short-term mass cultures of peripheral blood mononuclear cells from 6 patients with ATL (8). In two of these cultures, type-C virus particles were also detected. Furthermore, proviral DNA was detected with the probe cDNA of ATLV in peripheral leukocytes of all five ATLV patients tested (4). *In vitro* cultivation of fresh cells was required for detection of ATLA or type C virus particles, but not for detection of provirus DNA.

Only a single band of proviral DNA was obtained with such DNAs of fresh lymphocytes from ATL patients and its size varied in different patients (4). These findings suggest that the cellular site of integration of ATLV provirus differs, thus implying that ATLV is not a widely destributed endogenous virus, but is acquired exogeneously. Provirus DNA was not detected in leukocytes from the parents of an ATL patient (22). This may indicate that ATLV is not vertically transmitted at a genetic level.

HEALTHY CARRIERS OF ATLV

Persistence of ATLV or ATLA-carrying cells could be detected not only in ATL patients but also in healthy adults having anti-ATLA (8,11). ATLA-carrying T cells could be demonstrated by short-term mass culture or by clonal culture using TCGF. Many anti-ATLA-positive adults are found among healthy residents in ATL-endemic areas (3,23), indicating that most of them were healthy carriers of ATLV. The natural routes of ATLV infection are not yet clear, although accumulated data on familial clustering of anti-ATLA-positive individuals

(24) suggest perinatal infection from mother to child or transmission by contact between husband and wife. The transmission of ATLV by blood transfusion from ATLV-positive persons to ATLV-negative persons is also highly possible (25), and this possibility seems important in medical practice.

NATURAL HUMAN ANTIBODY TO ATLV

A nation-wide seroepidemiologic survey of ATLV, detected as anti-ATLA, was made in Japan (23). Sera from 15 different areas were screened for antibody. High incidences (6 to 37%) of antibody-positive donors were found in seven regions, one in northern Japan, and the others in southwestern regions. These areas correspond to ATL-endemic areas. Examination of sera from healthy donors of 6- to 80-years-old in ATL-endemic areas showed an age-dependent increase of seropositive donors with a maximum of about 30% at 40 years of age. Anti-ATLA was found in all but two of 142 patients with ATL. Anti-ATLA positive patients with ATL were mainly found in ATLV-endemic areas, with only a few in ATL-nonendemic areas. It seems note worthy that areas in which the incidence of anti-ATLA positive residents is high may be ATLV-endemic areas, because anti-ATLA positive individuals, including both ATL patients and healthy donors, had ATLA-bearing or ATLV-carrying T cells in their peripheral blood (8,11). In other words, it was now clear that most, if not all, anti-ATLA positive individuals are ATLV-carriers. However, it is still uncertain whether there are any ATLV-carriers among anti-ATLA-negative individuals in ATLV-endemic areas.

There is no serologic evidence for an etiological relation of ATLV with any malignant diseases other than ATL or ATLL except cutaneous T cell lymphoma (CTCL) including mycosis fungoides and Sézary syndrome (4,23,25). Table II shows the frequencies of anti-ATLA positive cases among patients having ATL (or ATLL) or CTCL. Of the patients with ATL, most, if not all, had anti-ATLA in both ATL endemic and nonendemic areas. However, of the patients with CTCL, none of the 12 patients in ATL-nonendemic areas had anti-ATLA. This finding suggests that ATLV is not associated with CTCL. It is noteworthy that six of nine patients with CTCL in ATL-endemic areas had anti-ATLA, possibly because it is sometimes difficult to distinguish CTCL from ATL or ATLL in cutaneous affections, the cytological features of the two diseases being very similar, especially without long-term observation: distinction of CTCL from ATL on the basis of clinical data only is not always possible.

TABLE II
Incidence of Anti-ATLA Positive Sera in Adult Patients with ATL and
CTCL in ATL-Endemic and ATL-Nonendemic Areas of Japan

Disease	Positive cases/cases tested	
	ATL endemic area	ATL nonendemic area
ATL[a]	140/142	15/15
CTCL[b]		
Mycosis fungoides	5/7	0/10
Sézary syndrome	1/2	0/2

[a] From the data of Hinuma *et al.* (23).
[b] From the data of Hinuma *et al.* (23) and Shimoyama *et al.* (25).

This fact and the high rate (about 25%) of antibody-positive adults in ATLV-endemic areas (4,23) may explain the reported high incidence of anti-ATLA-positive patients with CTCL in these areas. Thus ATLV may not in fact be related to CTCL, although Poiesz *et al.* (26,27) considered that their retrovirus (HTLV) was closely related to CTCL, because it was isolated from patients with CTCL. Very recently, Robert-Guroff *et al.* (28) reported that the sera of six of seven patients with Japanese ATL had antibodies to HTLV isolated from American patients with CTCL. However, they could not understand why antibodies to HTLV was isolated, but very frequently in Japanese patients with ATL. As a possible explanation, they suggested that the antibody-positive American cases of CTCL might be of the same type as these of Japanese ATL. This problem must be examined by direct comparative studies not only on isolated Japanese ATLV and American HTLV, but also on Japanese and American CTCL and Japanese ATL.

CONCLUDING REMARKS

A new human retrovirus has been isolated and characterized from leukemic cells of patients with ATL and named ATLV. ATLV was also detected in peripheral blood cells of healthy adults having antibody to ATLV. ATLV (virus)-endemic areas correspond to ATL (disease)-endemic areas and are located mainly in southwestern Japan. Provirus DNA of ATLV could be detected in leukemic cells of ATL, but not in other human cells unrelated to ATL. These facts strongly suggest that ATLV is not an endogenous retrovirus, but is transmitted exogeneously in some restricted areas. Detailed studies on the genes and proteins of ATLV are needed not only for clarifying ATL-leukemogenesis but also in understanding more about the nature of all human tumors.

ACKNOWLEDGMENTS

This work was supported by Grants-in-Aid for Cancer Research from the Ministry of Education, Science, and Culture and from the Ministry of Health and Welfare of Japan.

REFERENCES

1. Takatsuki, K., Uchiyama, T., Sagawa, K., and Yodoi, J. (1977) In "Topics in Hematology" (S. Seno, F. Takaku, and S. Irino, eds.), pp. 73–77. Excerpta Medica, Amsterdam.
2. The T- and B-cell Malignancy Study Group (1981) Jpn. J. Clin. Oncol. 11, 15–38.
3. Hinuma, Y., Nagata, K., Hanaoka, M., Nakai, M., Matsumoto, T., Kinoshita, K., Shirakawa, S., and Miyoshi, I. (1981) Proc. Natl. Acad. Sci. U.S.A. 78, 6476–6480.
4. Yoshida, M., Miyoshi, I., and Hinuma, Y. (1982) Proc. Natl. Acad. Sci. U.S.A. 79, 2031–2035.
5. Hinuma, Y. (1981) Annu. Rep. Inst. Virus Res., Kyoto Univ. 24, 1–8.
6. Miyoshi, I., Kubonishi, I., Yoshimoto, S., Akagi, T., Ohtsuki, Y., Shiraishi, Y., Nagata, K., and Hinuma, Y. (1981) Nature (London) 296, 770–771.
7. Nakai, M., personal communication (February 2, 1982).
8. Hinuma, Y., Gotoh, Y., Sugamura, K., Nagata, K., Goto, T., Nakai, M., Kamada, N., Matsumoto, T., and Kinoshita, K. (1982) Gann 73, 341–344.
9. Miyoshi, I., Taguchi, H., Fujita, M., Niiya, K., Kitagawa, T., Ohtsuki, Y., and Akagi, T. (1982) Gann 73, 339–340.
10. Miyoshi, I., Fujishita, M., Taguchi, H., Ohtsuki, Y., Akagi, T., Morimoto, Y., and Nagasaki, A. (1982) Lancet 1, 683–684.
11. Gotoh, Y., Sugamura, K., and Hinuma, Y. (1982) Proc. Natl. Acad. Sci. U.S.A. 79, 4780–4782.
12. Chosa, T., Yamamoto, N., and Hinuma, Y. (1983) Microbiol. Immunol. (submitted for publication).
13. Yamamoto, N., and Hinuma, Y. (1982) Int. J. Cancer 30, 289–293.
14. Tanaka, Y., Koyanagi, Y., Chosa, T., Yamamoto, N., and Hinuma, Y. (1982) Gann (submitted for publication).
15. Yamamoto, N., Koyanagi, Y., Okada, M., Kannagi, M., and Hinuma, Y. (1982) Science 217, 737–739.
16. Miyoshi, I., Kubonishi, I., Yoshimoto, S., and Shiraishi, Y. (1981) Gann 72, 978–981.
17. Miyoshi, I., Taguchi, H., Fujishita, M., Yoshimoto, S., Kubonishi, I., Ohtsuki, Y., Shiraishi, Y., and Akagi, T. (1982) Lancet 1, 1016–1017.
18. Miyoshi, I., Kubonishi, I., Sumida, M., Hiraki, S., Tsubota, T., Kimura, I., Miyamoto, K., and Sato, I. (1980) Gann 71, 155–156.
19. Miyoshi, I., Kubonishi, I., Yoshimoto, S., and Shiraishi, Y. (1981) Gann 72, 978–981.
20. Miyoshi, I., personal communication (May 25, 1981).

21. Gotoh, Y., and Hinuma, Y., unpublished data (May 5, 1981).
22. Yoshida, M., personal communication (March 5, 1982).
23. Hinuma, Y., Komoda, H., Chosa, T., Kondo, T., Kohakura, M., Takenaka, T., Kikuchi, M., Ichimaru, M., Yunoki, K., Sato, I., Matsuo, R., Takiuchi, Y., Uchino, H., and Hanaoka, M. (1982) *Int. J. Cancer* **29**, 631–635.
24. Tajima, K., Tominaga, S., Suchi, T., Kawagoe, T., Komoda, H., Oda, T., Fujita, K. and Hinuma, Y. (1982) *Gann* **73**, 893–901.
25. Shimoyama, M., Minato, K., Tobinai, K., Horikoshi, N., Ikuba, T., Deura, K., Nagatani, T., Ozaki, Y., Inada, N., Komoda, H., and Hinuma, Y. (1982)*Jpn. J. Clin. Oncol.* **12**, 109–116.
26. Poiesz, B. J., Ruscetti, F. W., Gazdar, A. F., Bunn, P. A., Minna, J. D., and Gallo, R. C. (1980) *Proc. Natl. Acad. Sci. U.S.A.* **77**, 7415–7419.
27. Poiesz, B. J., Ruscetti, F. W., Reitz, M. S., Kalyanaraman, V. S., and Gallo, R. C. (1981) *Nature (London)* **294**, 268–271.
28. Robert-Guroff, M., Nakao, Y., Notake, K., Ito, Y., Sliski, A., and Gallo, R. C. (1982) *Science* **215**, 975–978.

The v-*myc* and c-*myc* Genes: Roles in Oncogenesis

WILLIAM S. HAYWARD

Sloan-Kettering Institute for Cancer Research
New York, New York

INTRODUCTION

Retroviruses can be classified into two groups: those that contain oncogenes and those that do not (for reviews, see 1–3). Members of the first group (acute, or rapidly transforming retroviruses) induce neoplastic disease in infected animals within a few weeks after infection and cause rapid transformation of target cells in tissue culture. These viruses contain oncogenes (v-*onc* genes) that were derived from normal cellular genes (proto-*onc* or c-*onc* genes) by recombination. More than 15 different v-*onc* genes (and corresponding c-*onc* genes) have been identified thus far (4). Viruses of the second group (slowly transforming retroviruses), which lack oncogenes, induce neoplastic disease in animals only after a long latent period (4–12 months) and do not cause transformation of tissue culture cells at detectable frequency.

The *myc* gene has been implicated in malignancies induced by both classes of retroviruses. Four independently isolated acute retroviruses (MC29, MH-2, OK-10, and CMII) have been shown to carry the v-*myc* gene in their genomes (5,6). These viruses induce neoplasms of hematopoietic tissues and carcinomas (see Table I). Avian leukosis virus (ALV), which lacks an oncogene, induces primarily B-cell lymphomas, but also occasionally erythroleukemias, nephroblastomas, and sarcomas (Table I). Induction of B-cell lymphomas results from activation of the cellular homolog (c-*myc*) of the v-*myc* gene (7). Transcriptional activation of c-*myc* is caused by rare integrations of the ALV provirus adjacent to this host gene, which place c-*myc* under the control of viral regulatory sequences (3,7–13).

TABLE I
Oncogenic Spectra of Selected Avian Retroviruses

Virus[a]	Oncogene	Neoplasm	Latent period
ALV	—	B-cell lymphoma, erythroleukemia, nephroblastoma, fibrosarcoma	4–12 months
MC29	myc	Myeloid leukemia, B-cell lymphoma, carcinoma, fibrosarcoma	3–6 weeks
RSV	src	Fibrosarcoma	1–2 weeks
AEV	erb (A,B)	Erythroblastosis, fibrosarcoma	1–2 weeks

[a] ALV (avian leukosis virus); MC29 (myelocytomatosis virus-29); RSV (Rous sarcoma virus); AEV (avian erythroblastosis virus)

THE v-myc GENE

The v-myc gene was first identified in MC29 virus (5,6). This virus was presumably derived by recombination between ALV and the cellular c-myc gene. During the recombination process, portions of the ALV genome (pol and portions of gag and env) were replaced by myc gene sequences (see Fig. 1). MC29 is thus defective, requiring an ALV helper virus for replication.

MC29 was initially isolated from a bird that exhibited myelocytomatosis. Infection with this virus, however, can result in a number of different types of neoplasms, including myeloid leukemias, carcinomas, and (under certain conditions) sarcomas (1,5,14). Recently, bursal lymphomas have also been observed at fairly high frequency in MC29-infected birds (15). Tumors induced by MC29 arise somewhat more slowly (3–6 weeks) (15) than those induced by many other acute

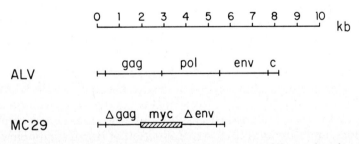

Fig. 1. Genetic maps of ALV and MC29. Genes required for viral replication are gag (coding for viral structural proteins), pol (reverse transcriptase), and env (envelope glycoprotein). Noncoding sequences are located at the extreme 5' and 3' ends of the viral genomes. The cell-derived myc insert of MC29 has replaced portions of the gag and env genes and all of pol.

viruses (see, for example, RSV and AEV, Table I). This latent period, however, is much shorter than that associated with viruses such as ALV, which lack oncogenes (4–12 months).

THE c-*myc* GENE

The c-*myc* gene, from which v-*myc* was derived, is present in the genomes of all vertebrates (6,16). Although little is known about its function in the normal cell, the fact that this gene (and other c-*onc* genes) is highly conserved throughout evolution suggests that it is essential for normal development of the organism. Several observations point to a possible role in hematopoietic cell growth control or differentiation. (1) Although c-*myc* is expressed at low levels in most tissues, it is expressed at high levels in hematopoietic tissues (bursa, thymus, spleen, bone marrow) of young birds (up to 4 weeks after hatching) (15; C.-K. Shih, unpublished) and in isolated B and T cells (17). (2) c-*myc* is expressed at elevated levels in human tissue culture cells representing early stages of B-cell differentiation (18). (3) c-*myc* is expressed at elevated levels in a human promyelocytic leukemia

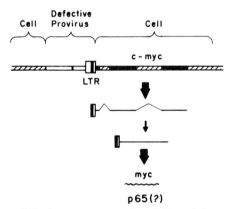

Fig. 2. Activation of the host c-*myc* gene by ALV. A defective provirus is shown integrated upstream from the c-*myc* gene and in the same transcriptional orientation. This juxtaposition of viral and cellular sequences is found in a majority of ALV-induced lymphomas (see text). The coding sequences of c-*myc* (solid bars) are located in two exons (0.8–0.9 kb each), separated by an intron of approximately 1 kb (13,21,22). The putative c-*myc* gene product (p65) has not yet been isolated. Its molecular weight (55-65K) is estimated from the apparent coding capacity of the *myc* gene. Analyses of the tumor-specific DNA and RNA sequences (7,8) indicate that the *myc* protein in ALV-induced lymphomas does not contain any virus-encoded peptides.

cell line, HL-60 (19). Treatment of HL-60 cells with retinoic acid or other compounds induces terminal differentiation and a simultaneous shut-off of c-*myc* (and c-*myb*) expression (19,20).

The c-*myc* gene differs somewhat in structure from v-*myc* (13,21,22). Whereas the v-*myc* coding sequences are contiguous (Fig. 1), the coding sequences of c-*myc* are contained in two exons, separated by an intron of approximately 1 kb (see Fig. 2). Presumably the intervening sequences were lost during the recombination event that generated MC29 virus (or during subsequent replication of the virus). The absence of introns is a common feature of all v-*onc* genes.

ACTIVATION OF c-*myc* BY ALV

Previously, we presented evidence that ALV induces B-cell lymphomas by activating the host c-*myc* gene (7). Activation results from insertion of viral regulatory sequences adjacent to the gene (7–13; see Fig. 2). These studies demonstrated that (1) in nearly all ALV-induced lymphomas, a provirus is integrated adjacent to a single cellular gene, c-*myc*; (2) levels of c-*myc* mRNA are elevated 30– to 100-fold in the lymphomas, as compared to equivalent normal tissues; (3) in most tumors, the *myc*-specific sequences are present in RNA transcripts that contain approximately 100 nucleotides of viral information derived from the long terminal repeat (LTR) of the integrated provirus.

In the vast majority of B-cell lymphomas, the provirus is integrated upstream from the c-*myc* gene and in the same transcriptional orientation (3,7–10,13) (see Fig. 2). Transcription initiates within the LTR— in most cases the 3' LTR—and reads into the adjacent cellular sequences. Integrations appear to be clustered in several discrete regions located within a 2-kb region immediately upstream from the c-*myc* coding sequences (3,8,10,23). Payne *et al.* (12), however, have identified one tumor in which integration was downstream from c-*myc*, and several tumors in which integration was upstream, but in the opposite transcriptional orientation. Because the viral promoter could not be utilized for c-*myc* activation in these tumors, it seems likely that initiation occurs on a cellular promoter. "Enhancer" sequences within the viral LTR may exert some positive regulatory influence over transcription in a manner analogous to that demonstrated for the 72-bp repeat of SV40 (24–26). This mechanism, however, is apparently less efficient at activating c-*myc* because more than 90% of tumors analyzed thus far contain proviruses integrated such that transcription could initiate on the viral promoter.

Interestingly, most, and perhaps all, of the proviruses integrated next to c-*myc* are defective (8–10,23). In most cases, the 5' LTR, plus additional coding information, is deleted. This observation suggests that defectiveness may play an essential role in c-*myc* activation. One explanation is that efficient transcription from the 3' LTR can occur only when active transcription from the 5' LTR is abolished. Because retroviral transcription normally proceeds from the 5' LTR into, and beyond, the initiation site on the 3' LTR (27), efficient utilization of the 3' promoter might be blocked.

More than 15 different c-*onc* genes have been identified thus far (4), and it is probable that more will be identified. Thus, it was surprising that a single gene, c-*myc*, was involved in such a high proportion of ALV-induced lymphomas. Although it is possible that sequences upstream from c-*myc* are preferred sites for ALV integration, this explanation seems unlikely. No specificity for ALV integration has been demonstrated (28). Furthermore, it is clear that ALV integration near c-*myc* can occur at more than a single site (3,7–10,12,13,23). It seems more likely, therefore, that involvement of c-*myc* in B-cell lymphoma relates to some property of the target cell. Two possibilities can be considered: (1) Only certain target cells respond to the c-*myc* product in a way that leads to neoplastic transformation; (2) integration may occur preferentially in a transcriptionally active region, due to conformational changes in the chromatin. As mentioned above, c-*myc* is expressed at elevated levels in early bursal cells. Relevant to this question is the recent observation that a different c-*onc* gene, c-*erb*, appears to be activated in ALV-induced erythroleukemias (H.-J. Kung, personal communication). Furthermore, neither c-*myc* nor c-*erb* are involved in ALV-induced sarcomas, nephroblastomas, or carcinomas (M. C. Simon; H. Varmus; personal communications). Integration appears to be specific in these tumors, but the cellular genes involved have not been identified.

Cooper and Neiman (11) have identified a second cellular gene that appears to play a role in ALV-induced lymphomagenesis. These authors have proposed a multistep model in which c-*myc* activation is an early event. A change at a second genetic locus would be involved in some later stage of tumor development.

CONCLUSIONS

The common feature of transformation by acute retroviruses (e.g., MC29) and slowly transforming retroviruses (ALV) is the activation of a c-*onc* gene. Although the mechanisms are different, in both cases a cellular gene has been placed under the control of viral regulatory sequences and is thus expressed constitutively at high levels. The inappropriate expression of a normal cellular gene (a c-*onc* gene) is apparently sufficient to induce aberrant growth. In some acute viruses, additional alterations have occurred within the coding regions of the oncogenes. These may also contribute to the oncogenic potential of the v-*onc* genes, but the possible influence of these changes has not been evaluated. Experiments in which a viral LTR was linked to the c-*mos* (29,30) or the c-*ras* (31) genes have demonstrated that transcriptional activation alone is sufficient to induce transformation in these cases. These *in vitro* constructs are analogous to the structures observed *in vivo* in ALV-induced tumors.

We have previously proposed that c-*onc* gene activation might be a common pathway for both viral and nonviral cancers (3,7). Specific chromosomal rearrangements have long been known to be associated with certain types of human neoplasms (for reviews, see 32,33). If such a rearrangement occurred within or near a c-*onc* gene, the activity of this gene could be drastically altered. For example, a translocation or transposition might place the coding sequences of a c-*onc* gene in juxtaposition with a transcriptionally active regulatory element from another cellular gene (7,33). Alternatively, point mutations or deletions within regulatory or coding sequences might alter the expression of the gene, or the properties of the gene product. Several groups (34,35) have recently shown that DNA sequences from human carcinomas, identified by their ability to transform NIH 3T3 cells, are related to the *ras* family of c-*onc* genes. In this case, activation of the transforming potential of the cellular gene (as assayed by transfection of NIH 3T3 cells) appears to be related to a single base change in the coding region of c-*ras* (R. Weinberg, personal communication). It seems likely, however, that examples of transcriptional activation of c-*onc* genes, analogous to that in ALV-induced lymphomas, will also be found among human cancers of nonviral origin.

ACKNOWLEDGMENTS

The author would like to thank Lauren O'Connor for help in preparation of the manuscript, and the many colleagues who have contributed to the studies cited in this short review. Work from this laboratory was supported by NIH grant CA-34502 from the National Cancer Institute and by the Flora E. Griffin Fund.

REFERENCES

1. Graf, T., and Beug, H. (1978) *Biochim. Biophys. Acta* **516**, 269–299.
2. Bishop, J. M. (1981) *Cell* **23**, 5–6.
3. Hayward, W. S., Neel, B. G., and Astrin, S. M. (1982) *Adv. Viral Oncol.* **1**, 207–233.
4. Coffin, J. M., Varmus, H. E., Bishop, J. M., Essex, M., Hardy, W. D., Martin, G. S., Rosenberg, N. E., Scolnick, E. M., Weinberg, R. A., and Vogt, P. K. (1981) *J. Virol.* **49**, 953–957.
5. Hayman, M. J. (1981) *J. Gen. Virol.* **52**, 1–14.
6. Roussel, M., Saule, S., Lagrou, C., Rommens, C., Beug, H., Graf, T., and Stehelin, D. (1979) *Nature (London)* **281**, 452–455.
7. Hayward, W. S., Neel, B. G., and Astrin, S. M. (1981) *Nature (London)* **290**, 475–480.
8. Neel, B. G., Hayward, W. S., Robinson, H. L., Fang, J., and Astrin, S. M. (1981) *Cell* **23**, 323–334.
9. Payne, G. S., Courtneidge, S. A., Crittenden, L. B., Fadly, A. M., Bishop, J. M., and Varmus, H. E. (1981) *Cell* **23**, 311–322.
10. Fung, Y.-K., Fadley, A. M., Crittenden, L. B., and Kung, H.-J. (1981) *Proc. Natl. Acad. Sci. U.S.A.* **78**, 3418–3422.
11. Cooper, G. M., and Neiman, P. E. (1981) *Nature (London)* **292**, 857–858.
12. Payne, G. S., Bishop, J. M., and Varmus, H. E. (1982) *Nature (London)* **295**, 209–214.
13. Neel, B. G., Gasic, G. P., Rogler,C. E., Skalka, A. M., Ju, G., Hishinuma, F., Papas, T., Astrin, S. M., and Hayward, W. S. (1982) *J. Virol.* **44**, 158–166.
14. Moscovici, C., and Gazzolo, L. (1982) *Adv. Viral Oncol.* **1**, 83–106.
15. Hayward, W. S., Shih, C.-K., and Moscovici, C. (1983) "Cetus-UCLA Symposium on Tumor Viruses and Differentiation." Alan R. Liss, Inc., New York, (in press).
16. Sheiness, D. K., Hughes, S., Varmus, H. E., Stubblefield, E., and Bishop, J. M. (1980) *Virology* **105**, 415–424.
17. Gonda, T. J., Sheiness, D. K., and Bishop, J. M. (1982) *Mol. Cell. Biol.* **2**, 617–624.
18. Eva, A., Robbins, K. C., Andersen, P. R., Srinivasan, A., Tronick, S. R., Reddy, E. P., Ellmore, R. C., and Aaronson, S. A. (1982) *Nature (London)* **295**, 116–119.
19. Westin, E. H., Wong-Staal, F., Gelmann, E. P., Dalla Favera, R., Papas, T. S., Lautenberger, J. A., Eva, A., Reddy, E. P., Tronick, S. R., Aaronson, S. A., and Gallo, R. C. (1982) *Proc. Natl. Acad. Sci. U.S.A.* **79**, 2490–2494.
20. Westin, E. H., Gallo, R. C., Arya, S. K., Eva, A., Sonza, L. M., Baluda, M. A., Aaronson, S. A., and Wong-Staal, F. (1982) *Proc. Natl. Acad. Sci. U.S.A.* **79**, 2194–2198.
21. Robins, T., Bister, K., Garon, C., Papas, T., and Duesberg, P. (1982) *J. Virol.* **41**, 635–642.
22. Vennstrom, B., Sheiness, D., Zabielski, J., and Bishop, J. M. (1982) *J. Virol.* **42**, 773–779.
23. Rovigatti, U. G., Rogler, C. E., Neel, B. G., Hayward, W. S., and Astrin, S. M. (1982) "Fourth Annual Bristol-Myers Symposium on Cancer Research." Academic Press, New York, pp 319–330.
24. Benoist, C., and Chambon, P. (1981) *Nature (London)* **290**, 304–310.
25. Gruss, P., Dhar, R., and Khoury, G. (1981) *Proc. Natl. Acad. Sci. U.S.A.* **78**, 943–947.
26. Moreau, P., Hen, R., Wasylyk, B., Everett, R., Gaub, M. P., and Chambon, P. (1981) *Nucleic Acids Res.* **9**, 6047–6068.

27. Hayward, W. S., and Neel, B. G. (1981) *Curr. Top. Microbiol. Immunol.* **91**, 217–276.
28. Temin, H. M. (1980) *Cell* **21**, 599–600.
29. Oskarsson, M., McClements, W. L., Blair, D. G., Maizel, J. V., and Vande Woude,— G. F. (1980) *Science* **207**, 1222–1224.
30. Blair, D. G., Oskarsson, M., Wood, T. G., McClements, W. L., Fischinger, P. J., and Vande Woude, G. F. (1981) *Science* **212**, 941–943.
31. DeFeo, D., Gonda, M. A., Young, H. A., Chang, E. H., Lowy, D. R., Scolnick, E. M., and Ellis, R. W. (1981) *Proc. Natl. Acad. Sci. U.S.A.* **78**, 3328–3332.
32. Rowley, J. D. (1980) *Annu. Rev. Genet.* **14**, 17–39.
33. Klein, G. (1981) *Nature (London)* **294**, 313–318.
34. Der, C. J., Krontiris, T. G., and Cooper, G. M. (1982) *Proc. Natl. Acad. Sci. U.S.A.* **79**, 3637–3640.
35. Parada, L. F., Tabin, C. J., Shih, C., and Weinberg, R. A. (1982) *Nature (London)* **297**, 474–478.

Biological Properties of the Human DNA Sequence Homologous to the *mos* Transforming Gene of Moloney Sarcoma Virus

ALESIA WOODWORTH, MARIANNE OSKARSSON,
DONALD G. BLAIR, MARY LOU McGEADY,
MICHAEL TAINSKY AND GEORGE F. VANDE WOUDE

Laboratory of Molecular Oncology
National Cancer Institute
National Institutes of Health
Bethesda, Maryland

INTRODUCTION

We have previously described the molecular cloning of a 9 kilobase pair (kb) *Bam*HI fragment from human placental DNA that contains a sequence homologous to the transforming gene v-*mos* of Moloney murine sarcoma virus (Mo-MSV) (1). The DNA sequence of the homologous region of the human DNA (termed hu*mos*) was resolved and compared to that of the mouse cellular homologue of v-*mos* (termed mu*mos*) (2). The hu*mos* gene contained an open reading frame of 346 codons that was aligned with the equivalent mu*mos* DNA sequence by the introduction of gaps of 15 and 3 bases in mu*mos* and a single gap of 9 bases in hu*mos*. The aligned coding sequences were 77% homologous and terminated at equivalent opal codons. The hu*mos* open reading frame initiates at an ATG found internally in the mu*mos* coding sequence. From this ATG, the polypeptides predicted from the hu*mos* and mu*mos* DNA sequences were 75% homologous. Four regions near the middle of the polypeptide chain, ranging from 19 to 26 consecutive amino acids, were conserved.

We have previously shown that mu*mos* is inactive in a biological DNA transfection assay, but could be activated when the long terminal repeat (LTR) of the Mo-MSV was covalently linked to it (3). These newly constructed LTR-mu*mos* hybrid clones transformed cells as efficiently as cloned subgenomic Mo-MSV fragments containing both v-*mos* and LTR sequences (4). However, in spite of the homology between mu*mos* and hu*mos*, we were unable to transform mouse cells with transfected hu*mos* DNA fragments or with hybrid recombinants containing hu*mos* and retroviral LTR sequences (1). In this chapter, we will show that a hybrid recombinant between the polypeptide coding regions of hu*mos* and v-*mos* of Mo-MSV can efficiently transform mouse cells in the DNA transfection assay. This suggests that differences in the hu*mos* and mu*mos* open reading frames can be responsible for the inability of the former to transform mouse cells. Sequence differences between hu*mos* and mu*mos* results in 84 amino acid changes (1), but as few as seven may prevent mouse cell transformation by the former.

RESULTS

The direct DNA sequence comparison between the cellular *mos* sequences of human and mouse (1) is shown in Fig. 1. The DNA sequence is presented in triplets for the *mos* open reading frames. Both open reading frames converge at a common ATG (position 241) and terminate at equivalent opal codons (position 1279). There are 94 third position changes that result in only two amino acid changes (positions 432 and 1263), suggesting that selective pressure for conservation of the *mos* gene is at the protein level. Therefore, it would be reasonable to expect that the hu*mos* gene could be activated to transform mouse NIH 3T3 cells in a manner analogous to activation of the LTR mu*mos* transforming potential (4,5,6). To test this, five unique recombinants were generated between hu*mos* and portions of Mo-MSV (Fig. 2). The first four recombinants were analogous to the LTR mu*mos* constructs previously shown to have biological activity in the mouse transfection assay (4,5,6). These recombinants were gen-

---

frame (2) and the surrounding region of homology with hu*mos* are underscored. The restriction enzymes sites above the line are in the hu*mos* sequence; below the line they are in mu*mos*. The *Pst*1 and *Sac*II sites of hu*mos* are not present in p*Vh*, whereas the *Ava*II sites of mu*mos* is missing and the *Sac*I site of hu*mos* is present. Therefore, the crossover occurs between positions 478 and 523. A plus or minus after the enzyme indicates the presence or absence of this restriction site in the p*Vh* recombinant.

```
c-mos
(human)   1  CGGAAGGGAAATGCTTTCATCTGAAAGGGATAGCTGTGCTTCATTCCGGTTTCTCCCTCCATCTGATAAAAACTCTTGCT
(mouse)                                                                          CCTTGTGGAG

         81  GAGTGACAGCACAGATGTAGCTCATTTGGAACAAGTGAAGGAAAAGGAGAAAAGGGATGAGGTGGAGCGAAGGAGTAGTC
             TAGTGATAGCACAGATGTGGCT         GGTTTTGAGAATCAAGGAAGAAGGGAAAGGAACTGGGATTGAAGGCAGCAATC
                         ---

        161  AGTCATGTTTCCAAAGTCCCGCGGTTTCCCCTAGTCTCTTCATTCACTCCAGCGGCCCTGGTGTCCCCCTGCAAAGTGCG
             TGCCATGCTGCCAAACTTCCCTGGCTGTTCCTAATCATTTCTCCCTAGTGTCTCATGTGACTGTCCCATCTGAGGGTGTA

        241  ATG CCC TCG CCC CTG GCC CTA CGC CCC TAC CTC CGG AGC GAG TTT TCC CCA TCG GTG GAC
                 T       T   A AG     G TT  G        CT C T     C G   G

        301  GCG CGG CCC TGC AGC AGT CCC TCA GAG CTA CCT         GCG AAG CTG CTT CTG GGG
                 T       T           T   T TG  T   GCC  G(AGG AAG GCA) G         C T C

                                    |      Bgl I     |
        352  GCC ACT CTT CCT CGG GCC CCG CGG CTG CCG CGC CGG CTG GCC TGG TGC TCC ATT GAC TGG
             A  C              T         C G A     A                        T         A
                     Pst I-
        412  GAG CAG GTG TGC TTG CTG CAG AGG CTG GGA GCT GGA GGG TTT GGC TCG GTG TAC AAG GCG
                 A       A T C  A   T               C T                            T  A   C

                 Sac II-
        472  ACT TAC CGC GGT GTT CCT GTG GCC ATA AAG CAA GTG AAC AAG TGC ACC AAG AAC CGA CTA
                     A                   C           A                           G   T   GT
                                                                             Ava II-
        532  GCA TCT CGG CAG AGT TTC TGG GCT GAG CTC AAC GTA GCA AGG CTG CGC CAC GAT AAC ATC
                 C A                         A   G   AT     A    A                 C       A

        592  GTG CGC GTG GTG GCT GCC AGC ACG CGC ACG CCC GCA GGG TCC AAT AGC CTA GGG ACC ATC
                 T G                                     A   AC      C               T       A

        652  ATC ATG GAG TTC GGT GGC AAC GTC ACT TTA CAC CAA GTC ATC TAT GGC GCC GCC GGC CAC
                         T G         G       C                               C T   A C   TCA

        712  CCT GAG(GGG GAC GCA GGG GAG)CCT CAC TGC CGC(ACT)GGA GGA CAG TTA AGT TTG GGA AAG
                 G                         T A T     A     A   ACG                 G

        772  TGT CTC AAG TAC TCA CTA GAT GTT GTG AAC GGC CTG CTC TTC CTC CAC TCG CAA AGC ATT
                 C       T C             T                       T       T               A

        832  GTG CAC TTG GAC CTG AAG CCC GCG AAC ATC TTG ATC AGT GAG CAG GAT GTC TGT AAA ATT
                 T                           A       T                   A  C   T       G C

        892  AGT GAC TTC GGT TGC TCT GAG AAG TTG GAA GAT CTG CTG TGC TTC CAG ACA CCC TCT TAC
                         C       C C       C   CG  T       G           CGG     G G T   C   C

                                                  Sac I+
        952  CCT CTA GGA GGC ACA TAC ACC CAC CGC GCC CCG GAG CTC CTG AAA GGA GAG GGC GTG ACG
                 AC A       G       G       AA  T           A                       ATT  CC

       1012  CCT AAA GCC GAC ATT TAT TCC TTT GCC ATC ACT CTC TGG CAA ATG ACT ACC AAG CAG GCG
                 C   T       C   C T       GA       C   G       C           G G       T

       1072  CCG TAT TCG GGG GAG CGG CAG CAC ATA CTG TAC GCG GTG GCC TAC GAC CTG CGC CCG
                 T   C   C   C   A CT     T   GG  A   T       A          A T           C

       1132  TCC CTC TCC GCT GCC GTC TTC GAG GAC TCG CTC CCC GGG CAG CGC CTT GGG GAC GTC ATC
             A   G G A   GA  G   G           ACC C       C   G AT     A A     A CA  A       A

       1192  CAG CGC TGC TGG AGA CCC AGC GCG GCG CAG AGG CCG AGC GCG CGG CTG CTT TTG GTG GAT
                 A               GAG G   C       C CT             T T   A GAA   C CAA  AG   C

       1252  CTC ACC TCT TTG AAA GCT GAA CTC GGC TGA CTGAAAACTTGGTCAAGATAAG
             AG  G           C CG   GG   C       A       CTCCATCGAGCCGATGTAGAGA
```

Fig. 1. Comparison of hu*mos* DNA sequence (1) to the mu*mos* sequence of Van Beveren *et al.* (2). The complete nucleotide sequence of the hu*mos* through position 1251 is given. In the putative coding sequence of the hu*mos* gene (residues 241 to 1281), the DNA sequence is represented as triplets corresponding to the codons. The mu*mos* sequence (2) is represented as an uninterrupted sequence in the hu*mos* noncoding regions and is given in the hu*mos* coding sequence only where the corresponding bases differ. Where spaces were inserted to align the two sequences in the hu*mos* coding region, the bases which have no homologues to the other sequences are enclosed by parentheses. In the 5′ noncoding regions, the first ATG in the mu*mos* open reading

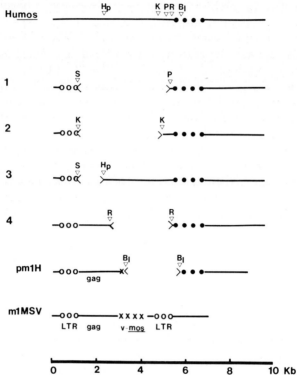

Fig. 2. Mo-MSV-hu*mos* recombinants. The top and bottom line drawings represent the parental cloned DNA fragments of hu*mos* (1) and Mo-MSV (7,8) that were used to construct the hybrid molecules shown in between. The symbols represent: ●, humos; ○, LTR sequences; X, v-*mos*; Hp, *Hpa*I; K, *Kpn*I; P, *Pvu*II; R, *Eco*RI; Bl, *Bgl*I; S, *Sma*I. Recombinants 1 thru 4 represent the LTR of Mo-mSV placed at variable distances 5' to the hu*mos* sequence. pm1H is a hybrid gene consisting of the entire 5' region of m1Mo-MSV (7,8) fused at a common *Bgl*I site in *mos* (position 382, Fig. 1) to 3' sequences from hu*mos*. With the exception of m1Mo-MSV all the molecules tested in the biological DNA transfection assay (3) were inactive for transformation.

erated by forming hybrid molecules between portions of the Mo-MSV LTR and unique restriction sites upstream from the beginning of the hu*mos* sequence (Fig. 2). The first three were constructed by covalently linking the LTR at the *Sma*I or *Kpn*I restriction sites to either the *Pvu*II, *Kpn*I, or *Hpa*I restriction sites (1). The fourth construct utilized an *Eco*R1 site outside of the LTR sequence (7) which was covalently linked to an *Eco*R1 site about 200 base pairs upstream from the beginning of the hu*mos* open reading frame. None of these recombinants were active in the biological transfection assay. Moreover, pm1H (Fig. 2), consisting of the entire 5' region of Mo-MSV and 3' sequences from hu*mos* fused at a common *Bgl*I in *mos* Fig. 1 position 382) was also activated in the transformation assay. Analogous mu*mos* recombinants transformed mouse NIH3T3 cells efficiently (3,4,5).

To test if differences in hu*mos* 3' to position 382 (Fig. 1) were responsible for its inability to transform mouse cells, we generated *mos* gene hybrids in *E. coli* utilyzing the miniplasmid, πVX recombination system (developed by B. Seed, personal communication). The scheme for generating these recombinants is shown in Figure 3. The insert from a cloned integrated Mo-MSV provirus: (8,9) was recloned into the charon 4A vector (λHT-1), whereas a 2.5-kb fragment containing the entire hu*mos* gene (1) was cloned into the miniplasmid πVX termed π*mos*-hu. This miniplasmid recombinant, possessing the supF tRNA gene, was used to transform the sup⁰ *Escherichia coli* strain W3110 p3. These transformed *E. coli* cells were then infected with 10^6 plaque-forming units (pfu) of λHT-1, and after lysis, recombinant phage designated λHTh, possessing the miniplasmid recombined at the homologous *mos* sequence were selected on the sup⁰ W3110 p3 *E.coli* strain. By this process, greater than 10^4 recombinants could be isolated and have the general structure shown in Fig. 3 (λHTh). Each recombinant has the miniplasmid π*mos*-hu inserted in a permuted fashion into the homologous v-*mos* region thereby generating two LTR-*mos*-gene domains. From left to right in λHTh, the first *mos* gene domain consists of the 5' portion of Mo-MSV (from the 5' LTR through the *mos* sequence) with the 5' portion of the *mos* gene derived from v-*mos* and the 3' portion derived from hu*mos*. We refer to this domain as V*h*. The V*h* domain is separated from a 3' *mos* gene domain by πVX vector sequences. The 3' *mos* gene region begins with hu*mos* sequences and terminates with v-*mos* sequences. This *mos* gene contains the reciprocol hu*mos*/v-*mos* elements that the 5' V*h* domain possesses. For example, two *Sac*I sites uniquely present only in hu*mos* divide into three domains (π*mos*-hu; Fig. 3). In λHTh (Fig. 3), both *Sac*I sites have segregated to the V*h* *mos* domain. The 3' *mos* gene is followed by the Mo-MSV 3' LTR. This *mos* gene-LTR region is referred to as *h*V. Both the V*h* and *h*V regions were separately subcloned into the plasmid pBR325 vector and are referred to as pV*h* and p*h*V (Fig. 3).

We have shown that v-*mos* recombinants containing either a 5' or 3' LTR element will efficiently transform mouse NIH 3T3 cells in DNA transfection assays (10). The analogous LTR hu*mos*/v-*mos* recombinants pV*h* and p*h*V (Fig. 3) were each tested in the NIH3T3

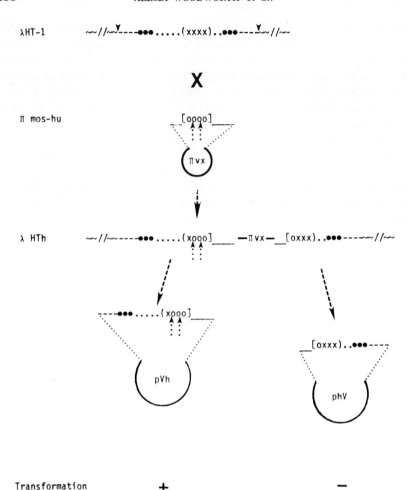

Transformation   +   −

Fig. 3. A transformation active v-*mos*/hu*mos* recombinant. Recombinants betwen the v-*mos* region of Mo-MSV and hu*mos* were generated in *E.coli* by the miniplasmid πVX system of B. Seed (personal communication). The *Eco*R1 cell genome fragment containing the integrated provirus of HT1MSV (7,8) was inserted into Ch 4a (λHT-1). A 2.5 kb *Xba*I fragment of hu*mos* containing the entire *mos* gene was cloned into πVX (refered to as π*mos*-hu) and used to transform the sup⁰ *E.coli* strain W3110 p3 according to B. Seed (personal communication). These cells were infected with λHT-1 (10⁶ pfu/10⁸ cells), and phages recovered after cell lysis were plated onto the sup⁰ strain to select for recombinants. We estimate that 10⁴ to 10⁵ recombinants containing the sup F miniplasmid in the *mos* gene (γHTh is one type) were generated from infection of W3110 cells containing π*mos*-hu. Each portion of λHTh, containing the 5' (V*h*) and 3' (*h*V) LTR-*mos* domain, was subcloned into pBR322 to yield pV*h* and p*h*V. These subclones were tested in the NIH3T3 transfection–transformation assay (9). A plus or minus indicates the biological transforming activity. In λHT-1: ∼, phage lambda se-

transfection–transformation assay. Only the pVh recombinant was biologically active (Fig. 3). By restriction enzyme analyses, we determined the locus of the v-mos/humos recombination. This occured between nucleotide positions 478 and 523 (Fig. 1) and indicates that the coding sequences in humos, 5' to position 523, may prevent this gene from transforming mouse cells. Rechavi et al. (11) have recently shown that the activation of mumos can occur when the open-reading frame of the mos gene is completely substituted 5' to the position equivalent to position 355 in humos (Fig. 1). This suggests that this N-terminal portion of mumos is not essential for transformation (11). From position to the recombination point in Vh (position 523, Fig. 1), there are only seven differences between humos and mumos. Specifically, they are as shown in the following tabulation:

Position (Fig. 1)	mumos	humos
358	Pro	Leu
373	Gly	Arg
397	Phe	Cys
427	Met	Leu
430	His	Gln
442	Ser	Ala
478	His	Arg

The transformation by the pVh mos hybrid plasmid shows that the C-terminal two-thirds of the humos coding sequence (position 478-1282) is functionally active as part of the mos transforming gene. It is not possible to conclude on the basis of these results whether the humos gene has oncogenic potential in human cells. It may fail to interact with the mouse cell components in the same way as the mumos gene product. However, we have begun to pinpoint the locus in the humos sequence that prevents mouse cell transformation. In the absence of functional assays for the transforming protein (12) or conditional mutants in the mos gene, these analyses can begin to reveal important domains of the mos protein.

quences; ●●●, LTR; ..., Moloney leukemia virus derived sequences in Mo-MSV; (XXXX), v-mos; ▼, 12.5 kb EcoR1 fragment (7,8) inserted in Ch 4a. In π mos-hu, the solid line represents the πVX vector sequence; [OOOO], humos; ▲, two SacI sites that divide the humos gene into three domains. Therefore, in γHTh (XOOO] and [OXXX) depict the hybrid viral (V)/human (h) mos genes.

REFERENCES

1. Watson, R., Oskarsson, M., and Vande Woude, G. F. (1982) *Proc. Natl. Acad. Sci. U.S.A.* **79**, 4078–4082.
2. Van Beveren, D., van Straaten, F., Galleshaw, J. A., and Verma, I. M. (1981) *Cell* **27**, 97–108.
3. Blair, D. G., Oskarsson, M. K., Wood, T. G., McClements, W. L., Fischinger, P. J., and Vande Woude, G. F. (1981) *Science* **212**, 941–943.
4. Blair, D. G., McClements, W., Oskarsson, M., Fischinger, P., and Vande Woude, G. F. (1980) *Proc. Natl. Acad. Sci. U.S.A.* **77**, 3504–3508.
5. Oskarsson, M., McClements, W., Blair, D. G., Maizel, J. V., and Vande Woude, G. F. (1980) *Science* **207**, 1222–1224.
6. McClements, W. L., Dhar, R., Blair, D. G., Enquist, L., Oskarsson, M., and Vande Woude, G. F. (1981) *Cold Spring Harbor Symp. on Quant. Biol.* **45**, 699–705.
7. McClements, W. L., Enquist, L. W., Oskarsson, M., Sullivan, M., and Vande Woude, G. F. (1980) *J. Virology* **53**, 488–497.
8. Vande Woude, G. F., Oskarsson, M., Enquist, L. W., Nomura, S., Sullivan, M., and Fischinger, P. J. (1979) *Proc. Natl. Acad. Sci., U.S.A.* **76**, 4464–4468.
9. Vande Woude, G. F., Oskarsson, M., McClements, W. L., Enquist, L. W., Blair, D. G., Fischinger, P. J., Maizel, J., and Sullivan, M. (1980) *Cold Spring Harbor Symp. on Quant. Biol.* **44**, 735–745.
10. Rechavi, G., Givol, D., and Canaani, E. (1982) *Nature (London)* **300**, 607–611.
11. Papkoff, J., Verma, I. M., and Hunter, T. (1982) *Cell* **29**, 417–426.

Enhanced Expression and Amplification of a Cellular Oncogene in Human Tumors

U. ROVIGATTI* AND S. M. ASTRIN

The Institute for Cancer Research
Fox Chase Cancer Center
Philadelphia, Pennsylvania

INTRODUCTION

HISTORICAL BACKGROUND

In 1911, Peyton Rous described Rous sarcoma virus, a filterable agent that produced sarcomas in inoculated chickens (1). It was not until many decades later that the gene responsible for tumor induction was identified in the viral genome. The fact that the product of a single gene appeared to be both necessary and sufficient for tumor formation was in itself a major revelation (2–6). In addition, it appeared that the transforming gene, subsequently called *src*, had a homologue in normal cells and that this cellular oncogene, named c-*src*, was present not only in the chicken genome but in the genome of all other vertebrates as well. Thus, we had the first evidence for the presence of potential transforming genes or "oncogenes" in normal cells (7–8).

The idea that such genes existed was not a new one. Prior to the description of the c-*src* gene, Huebner and Todaro put forth a very provocative theory, the virogene–oncogene hypothesis (9). According to this theory, normal cells, already known to harbor sequences homologous to the replicative genes of retroviruses (virogenes), were also postulated to contain genes homologous to viral transforming genes, i.e., oncogenes. Although Huebner and Todaro postulated that virogenes and oncogenes were genetically linked, a prediction we

---

*Present address: St. Jude Children's Research Hospital, Memphis, Tennessee

now know to be incorrect, the idea that normal cells contain sequences homologous to the retroviral oncogenes was amazingly accurate.

Can tumors arise as a consequence of activation of a cellular "oncogene?" An answer to this question did not come until more than a decade after the oncogene theory had been postulated and a few years after the discovery of cellular oncogenes. Not surprisingly, information was again obtained from the study of oncogenic retroviruses. The clues came this time, not from study of the rapidly transforming viruses such as Rous sarcoma virus (RSV), but rather from work on a tumor virus with a long period of latency: the avian leukosis virus (ALV) (10) of chickens. These viruses were shown to cause tumors by activation of a cellular "oncogene."

THE PROMOTER INSERTION MECHANISM OF ONCOGENESIS

Avian leukosis virus, when inoculated into 1-day-old chicks, causes bursal lymphomas in 4 to 12 months. The mechanism of tumor formation in this system is markedly different from the mechanism employed by the acute transforming viruses, typified by RSV. The acute viruses cause tumors in 2 to 3 weeks as compared with the period of months needed by ALV. Although the acute viruses transform virtually every infected cell producing a nonclonal tumor, transformation by ALV is a very rare event; the tumors produced are clonal, because they originate from a single transformed cell (11).

The properties of tumor induction by RSV and the other acute transforming viruses are explained by the presence of an oncogene in the viral genome. Genetic studies, employing mutants of RSV defective or temperature-sensitive for transformation, have rigorously demonstrated the existence of viral functions required for establishment and maintenance of transformation (2–4,11). To date, over 15 retroviral oncogenes have been identified; each has a homologue in normal cells. Experiments employing cells transformed by retroviruses and their revertant clones (11,12) have indicated that viral oncogenes, when expressed at high levels, are responsible for the transformed phenotype.

In the case of RSV, both the genes involved in replicative functions and the viral transforming *src* gene are under control of the viral promoter contained in a region called "long terminal repeat" (LTR) (11,13). There are two such LTRs in each integrated provirus, one at the 5' end and one at the 3' end of the protein coding sequences (see Fig. 1). LTRs are the general and unifying characteristic of all ret-

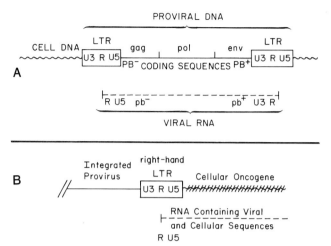

Fig. 1A. Schematic diagram of an ALV provirus. LTR, long terminal repeats. They are divided into three regions: U3 and U5 (two regions unique to the 3' end and to the 5' end respectively of the viral RNA) and a terminal redundancy R. PB, primer binding sites gag, pol, env, viral genes required for viral replicative functions.

Fig. 1B. Schematic representation of the promoter insertion model.

roviruses studied so far; therefore, they are also present in the ALV. Since ALVs are devoid of transforming sequences, the only way they could activate an oncogene would be if an ALV provirus were to integrate next to a cellular oncogene in the chicken genome. As indicated schematically in Fig. 1, transcription started from the right-hand promoter could continue and read into the flanking cellular sequences — sequences containing, as hypothesized above, a cellular oncogene (14,15).

As it turns out, this is exactly what happens. In the ALV-induced tumors, the viral promoter sequence is found integrated adjacent to the cellular *myc* gene, the homologue of the oncogene carried by the acute transforming retrovirus MC29. The level of expression of the c-*myc* gene in tumors is 30- to 100-fold greater than that found in normal tissue. In addition, the c-*myc* transcripts contain viral sequences, thus indicating that transcription is initiated within the viral promoter (16).

Integration adjacent to the c-*myc* gene is a rare event; hence, out of a large population of infected bursal cells, only one or very few contain a provirus integrated next to c-*myc* and therefore become transformed. The resulting tumors are clonal, consisting of progeny of those rare transformants. The mechanism of ALV tumorigenesis has been termed "promoter insertion" and serves as the first example of tumorigenesis by activation of a cellular oncogene.

ONCOGENE EXPRESSION IN HUMAN LEUKEMIC CELLS

Our experiments with human leukemic cells have been aimed at determining whether, in some cases, the oncogenic event leading to human neoplasia is enhanced expression of the human *myc* gene. Can leukemia and lymphoma result from an overabundance of *myc* gene product in cells at a particular stage of development in the lymphoid or myeloid series? If a genetic change occurs that results in the elevated expression of the *myc* gene, are cells then "frozen" at a particular stage of development and locked into a proliferative state? In order to answer these questions, it was necessary first to obtain relatively pure subpopulations of tumor cells. The following questions were then considered experimentally: (1) Do leukemic cells contain elevated levels of *myc* gene transcripts? (2) Is there a genetic change in the tumor cells that is associated with enhanced *myc* oncogene expression? (3) Is the genetic change sufficient to induce neoplastic transformation in the target cells?

We collected samples from 25 individual patients with lymphoma or leukemia for the purpose of assaying the level of *myc* gene transcripts in the tumor cells. We focused our attention on patients with chronic lymphocytic leukemia (CLL), acute lymphoblastic leukemia (ALL), and non-Hodgkin's lymphoma (NHL). Our samples consisted mainly of peripheral blood leukocytes (PBL) purified from the blood of leukemic patients with elevated white cell counts (from 27,000 to

TABLE I
Summary of the Data Obtained by Dot Blot Analysis of RNA Samples from Leukemia–Lymphoma Patients and from Normal Blood Donors

	Expression of c-*myc* sequences in RNA samples obtained from normal and tumor PBL	
	Number of positives	Number analysed
Chronic lymphocytic leukemia	3	11
Acute lymphoblastic leukemia	3	3
Non-Hodgkin's lymphoma	4	4
Total tumor samples	10	18
Total normal blood donors	0	9

480,000/μl of blood). In addition, we have collected a series of leukocyte samples from normal (nonleukemic) individuals; in these samples the white cell counts ranged from 5,000 to 10,000 per μl of blood. Total nucleic acids were extracted from the leukocytes and separated by CsCl gradient centrifugation (17). In order to estimate the amount of c-*myc* transcripts present in leukemic versus normal cells, we performed "dot blot" experiments (18,19). As illustrated in Fig. 2, differ-

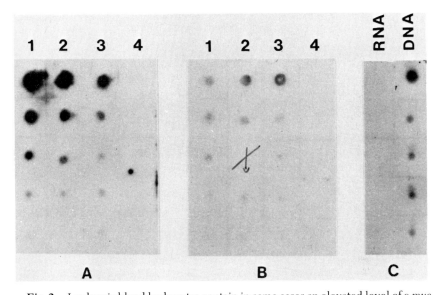

Fig. 2. Leukemic blood leukocytes contain in some cases an elevated level of c-*myc* transcripts in comparison to normal peripheral blood leukocytes. Total nucleic acids were extracted as previously described (14) and total cellular RNA was purified by CsCl centrifugation. The presence of c-*myc* transcripts was then detected by "dot-blot" hybridization, as it is illustrated here and according to Thomas (18). Total RNA was spotted in two-fold dilutions, starting from the highest amounts listed in the top dots, into nitrocellulose filter papers soaked in a 20× SSC salt solution. The filters were then baked for 2 hr at 80°C under vacuum. Hybridization conditions were as previously described [(20) Fig. 4—higher stringency] and the X-ray films were exposed for 24 hr. Two different probes were used in these experiments: a viral *myc* probe kindly provided by T. Papas (panels A and C) and a human c-*myc* probe, kindly provided by B. Neel and W. Hayward (panel B). The samples in panels A and B are PBL RNA extracted from two different ALL patients (lanes 1 and 3), a CLL patient (lane 2), and a normal blood donor (lane 4). Panel C shows a control experiment, where the two fractions, RNA and DNA, obtained by CsCl centrifugation of total nucleic acids from normal blood peripheral blood leukocytes were independently hybridized to our v-*myc* probe. As it appears, the DNA fraction, but not the RNA one, hybridizes to this probe.

ent dilutions of total cellular RNA are spotted on nitrocellulose filter paper and then hybridized with probes containing *myc* specific sequences (in this case v-*myc* or human c-*myc*). It is clear from the samples shown in Fig. 2 and from others we have analyzed that total cellular RNA extracted from normal blood donor PBL did not hybridize to our *myc* specific probes, whereas RNA extracted from lymphocytes of a large fraction of leukemic patients hybridized strongly to these probes (20). In order to rule out the possibility of DNA contamination in our RNA preparations, we hybridized two different fractions from our CsCl gradient, corresponding to RNA and DNA, to *myc* probes. As shown in Fig. 2C, for nucleic acid fractions from normal blood PBL, our v-*myc* probe hybridized to the DNA fraction only. Other control experiments have been performed in which RNA samples from leukemic blood were treated with RNAse prior to hybridization. In all the positive samples tested, hybridization to *myc* probes disappeared after RNAse treatment (data not shown). These experiments confirmed the interpretation that the probes did detect the presence of RNA transcripts.

AMPLIFICATION OF C-*myc* DNA SEQUENCES IN THE HUMAN PROMYELOCYTIC LEUKEMIC CELL LINE HL-60

In order to determine if there was a genetic change in the tumor cells and if this change was associated with enhanced *myc* oncogene expression, we used Southern Blots (21) to analyze the *myc* gene locus in several tumor samples with high levels of c-*myc* transcripts. Most of the tumor samples we analyzed so far did not show a pattern of hybridation different from that obtained with DNA from normal blood donor PBL or from different nonneoplastic tissues. However, an exceptionally intense signal was observed for one of the bands hybridizing to our v-*myc* probe in the DNA extracted from HL-60, a cell line established a few years ago from a human promyelocytic leukemia patient (22). As shown in Fig. 3, a band of relatively high intensity was present in both the *Eco* RI (lane 1) and *Hind* III (lane 2) digests of HL-60 DNA. These bands were clearly distinguishable in intensity from the other bands of lower molecular weight in the same digestion patterns as well as from the bands produced by digestion of DNA from normal leukocytes (compare the HL-60 sample in lane 3 with the normal sample in lane 4). The simplest explanation of these results is that one set of human sequences homologous to v-*myc* is amplified in this cell line. In addition to this amplification at the DNA level, we observed an elevated level of c-*myc* transcripts in HL-60. The level of transcription as assayed by dot blots was more than 20 times higher than that in normal

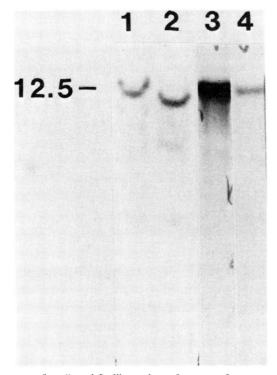

Fig. 3. Presence of an "amplified" number of copies of c-*myc* sequences in the DNA of HL-60. DNA has been extracted from the cell line HL-60 (lanes 1–3) and from normal blood PBL (lane 4) as previously described and digested with *Eco*Rl (lanes 1,3 and 4) or *Hin*dIII (lane 2) restriction enzymes. The digested DNA was fractionated by electrophoresis on a 0.8% agarose gel, then transferred to nitrocellulose filters according to Southern, and hybridized with different probes in the presence of dextran sulphate. The probes had a specific activities of $1-5 \times 10^8$ cpm/μg of DNA and were, respectively: Lanes 1 and 2: the v-*myc* probe already described in Fig. 2. Lanes 3 and 4: human c-*myc*, the Cla1–*Eco*Rl fragment of a clone containing the 12.5-kb *Eco*Rl fragment (M. Erisman and S. M. Astrin, unpublished results). Hybridization conditions were as previously described. Notice that the gels hybridized to the v-*myc* (1,2) have been exposed to an X-ray film for 7 days, instead of only 24 hours as in the case of the human c-*myc* probe.

leukocytes (data not shown). Similar results have been obtained independently by two other groups (23,24). In one case, a more detailed analysis of the human c-*myc* loci was obtained (25). Our results with HL-60 DNA indicated that the amplified sequence corresponds to the human c-*myc* locus with the most homology to viral *myc*. Other loci, probably containing incomplete c-*myc* genes, produced bands of lower intensity and therefore do not seem to be amplified [Fig. 3; lanes 1 and 2 and ref. (24)]. We confirmed these results by employing another probe: A *Cla*I-*Eco*RI fragment containing the 3′ exon of the human c-*myc* gene (M. Erisman and S. Astrin, unpublished results). This probe recognized human c-*myc* sequences present in the 12.5-kb *Eco*RI fragment detected by v-*myc* probes, but it did not hybridize to the lower molecular weight *Eco*RI fragments that were detected by v-*myc* (Fig. 3, lanes 3 & 4). As shown in Fig. 3, lane 3, the *Cla*I-*Eco*RI probe hybridized strongly to the amplified sequences in HL-60 DNA.

DISCUSSION

We presented data indicating that the cellular homologue of a retroviral oncogene is expressed in human leukemia or lymphoma peripheral blood leukocytes (PBL) at a level that is higher than that in PBL from normal blood. Furthermore, we have shown that, in at least one case, this enhancement of c-*myc* expression is correlated with a genetic change, i.e., the amplification of the sequences that probably code for the c-*myc* transcripts in HL-60 cells (25).

What is the function of the cellular homologues of the retroviral oncogenes? These genes are highly conserved in evolution. It appears that all vertebrate species as well as some invertebrates contain homologues for all or most of the known oncogenes (8,11,12,26). Substantial homology of the nucleic acid sequence of the human and viral genes exists for all the oncogenes studied so far. In some cases, this homology extends not only to the exons of the gene, but to some regions of the introns as well (27,28). When amino acid sequences for the same oncogene in different species are compared, the homologies are even more striking. Furthermore, different oncogenes that showed no homology at the nucleic acid level have been shown to have up to 50% homology in the amino acid sequence (29).

The ubiquity of the oncogenes and the great extent to which their protein sequence has been conserved argue that these are essential genes serving important functions. Current experimental evidence indicates that they may function as regulators of cell proliferation and/or differentiation. For example, the oncogenes seem to be expressed early in development of selected tissues and organs (30–32). The idea that oncogenes also play a role in the control of cell division is supported by at least two general observations. First of all, these genes are expressed early in development, when rates of cell proliferation are at their peak. Furthermore, elevated expression of viral oncogenes in infected cells leads to transformation, and loss of the transformed phenotype is usually accompanied by significantly lower oncogene expression (33). However, oncogene expression is not a mere concomitant of cell proliferation, because stimulation of cell proliferation itself does not lead to their expression. Resting and proliferating mouse liver cells or fibroblasts as well as resting and mitogen stimulated human lymphoblasts have the same level of oncogene expression, a very low one (8,19,33).

There is some evidence to indicate a role for the oncogenes in differentiation. Not only does there appear to be a tissue specificity for oncogene expression, but these genes seem to be expressed only in immature cells and not in the terminally differentiated cells of a given organ (19–21). In addition, in systems where cells expressing a given oncogene can be induced to differentiate *in vitro*, oncogene expression appears to be turned-off as the cells become terminally differentiated (32). Thus, certain oncogenes may be essential for the development of given tissues, not only to stimulate proliferation of immature stem cells, but also as regulators of differentiation.

However, several gaps are still present in the overall picture linking viral and cellular oncogenes to cellular differentiation and neoplastic transformation. One often touted theory is that neoplasia might involve a block of cellular differentiation, possibly caused by uncontrolled oncogene expression. Thus an immature progenitor, prevented from differentiating further, would continue to proliferate. In the normal situation, on the contrary, the very same cell is capable of undergoing differentiation either into a lineage without self-renewal (terminal differentiation) or into a stem cell still capable of self-renewal, but controlled in its proliferation. In other words, in this sort of model, neoplastic transformation would just freeze some stem cells at certain stages of differentiation without otherwise altering the phenotype of the cells.

There is now experimental evidence that contradicts this sort of generalization. When several parameters of differentiation are studied, a much more complex picture of the effects caused by neoplastic transformation emerges. First of all, new cell phenotypes appear, which do not seem to be normally present under physiological conditions. Second, the alteration in the cell phenotype is, in most instances, not appropriately defined as a "block of cell differentiation." Rather, several different phenotypes appear to accompany transformation of a certain stem cell. Usually, they correspond to the various steps of differentiation, from the more immature to the more mature and even to the most differentiated phenotypes. Thus far, attempts to correlate the cell type expressing a given oncogene with the target cell specificity or pathogenicity of the retrovirus that carries the homologous oncogene have failed. Also, the expression of certain viral oncogenes does not always lead to transformation. Erythroblasts, for example, can be infected and transformed by avian erythroblastosis virus (AEV). However, the same cells can also be infected by other acute leukemia viruses, such as MC-29, which replicate and even induce the synthesis of the specific viral oncogene product without transforming the cells (34).

Our data on the expression of the human oncogene c-*myc* suggest that the failure to properly control oncogene expressions might be associated with the pathogenesis of certain leukemias and lymphomas. What is the significance of elevated oncogene expression in terms of neoplastic transformation of human cells? Animal model systems have shown that (1) viral oncogene products are present in high amounts in virally transformed cells in comparison with levels of the endogenous cellular oncogene (proto-oncogene) products. Furthermore, when the synthesis of oncogene product falls by a factor of 10–100, the virally transformed cells revert to a normal phenotype (11); (2) the promoter insertion model of viral carcinogenesis described above is another example of increased oncogene expression that leads to neoplastic transformation (14,16,19); (3) enhanced expression of a cellular oncogene achieved by artificially providing a strong viral promoter (contained in the LTR) for the poorly expressed proto-oncogene also leads to neoplastic transformation (29,11).

Considering the above, our results indicate that activation of the c-*myc* oncogene is a causal element in some cases of human leukemias and lymphomas. As previously discussed, this event could lead to misreading of the signals controlling cell proliferation and/or differentiation. Further studies should indicate whether this activation is the initial event leading to neoplastic transformation or whether it is directly or indirectly associated with the neoplastic process.

REFERENCES

1. Rous, P. (1911) *J. Exp. Med.* **13**, 397–411.
2. Martin, G. S. (1970) *Nature (London)* **227**, 1021–1023.
3. Wyke, J. A., and Linial, M. (1973) *Virology* **53**, 152–161.
4. Bernstein, A., MacCormick, R., and Martin, G. S. (1976) *Virology* **70**, 206–209.
5. Kawai, S., Duesberg, P. H., and Hanafusa, H. (1972) *Virology* **49**, 302–304.
6. Duesberg, P. H., and Vogt, P. K. (1970) *Proc. Nat. Acad. Sci. U.S.A.* **67**, 1673–1680.
7. Stehelin, D., Varmus, H. E., Bishop, J. M., and Vogt, P. K. (1976) *Nature (London)* **260**, 170–173.

8. Spector, D. H., Varmus, H. E., and Bishop, J. M. (1978) *Proc. Nat. Acad. Sci. U.S.A.* **75**, 4102–4106.
9. Huebner, R. J., and Todaro, G. J. (1969) *Proc. Nat. Acad. Sci. U.S.A.* **64**, 1087–1094.
10. Ellermann, V., and Bang, O. (1908) *Zentralbl. Bakt.* **46**, 595–609.
11. Weiss, R., Teich, N., Varmus, H., and Coffin, J. (ed.) (1982) "RNA Tumor Viruses." Cold Spring Harbor Laboratory Press, Cold Spring Harbor, New York.
12. Coffin, J. M., Varmus, H. E., Bishop, M. J., Essex, M., Hardy, W. D., Jr., Martin, G. S., Rosenberg, M. E., Scolnick, E. M., Weinberg, R. A., and Vogt, P. K. (1981) *J. Virol.* **40**, 953–957.
13. Temin, H. (1981) *Cell* **27**, 1–3.
14. Neel, B. G., Hayward, W. S., Robinson, H. L., Fang, J., and Astrin, S. M. (1981) *Cell* **23**, 323–334.
15. Payne, G. S., Courtneidge, S. A., Crittenden, L. B., Fadly, A. M., Bishop, J. M., and Varmus, H. E. (1981) *Cell* **23**, 311–322.
16. Hayward, W. S., Neel, B. G., and Astrin, S. M. (1981) *Nature (London)* **290**, 475–480.
17. Maniatis, T., Fritsch, E. F., Sambrook, J. (1982) "Molecular Cloning." Cold Spring Harbor Laboratory Press, Cold Spring Harbor, New York.
18. Thomas, P. S. (1980) *Proc. Nat. Acad. Sci. U.S.A.* **77**, 5201–5205.
19. Rovigatti, U. G., Rogler, C. E., Neel, B. G., Hayward, W. S., and Astrin, S. M. (1982) *In* "Tumor Cell Heterogeneity", pp. 319–330. Academic Press, New York.
20. Rovigatti, U. G., and Astrin, S. M. *In* "Gene Transfer and Cancer" (N. L. Sternberg and M. L. Pearson, eds.). Raven Press, New York. In press.
21. Southern, E. (1975) *J. Mol. Biol.* **98**, 503–517.
22. Collins, S. J., Gallo, R. C., and Gallagher, R. E. (1977) *Nature (London)* **270**, 347–349.
23. Collins, S. J., and Groudine, M. (1982) *Nature (London)* **298**, 679–681.
24. Dalla Favera, R., Wong-Staal, F., and Gallo, R. (1982) *Nature (London)* **299**, 61–63.
25. Dalla Favera, R., Gelmann, E. P., Martinotti, S., Franchini, G., Papas, T. S., Gallo, R., and Wong-Staal, F. (1982) *Proc. Nat. Acad. Sci. U.S.A.* **79**, 6497–6501.
26. Shilo, B. Z., and Weinberg, R. A. (1981) *Proc. Nat. Acad. Sci. U.S.A.* **78**, 6789–6792.
27. Tabin, C. J., Bradley, S. M., Bargmann, C. I., Weinberg, R. A., Papageorge, A. G., Scolnick, E. M., Dhar, R., Lowy, D. R., and Chang, E. H. (1982) *Nature (London)* **300**, 143–149.
28. Reddy, E. P., Reynolds, R. K., Santos, E., and Barbacid, M. (1982) *Nature (London)* **300**, 149–152.
29. VanBeveren, C., VanStraten, F., Gallenshaw, J. A., and Verma, I. M. (1981) *Cell* **27**, 97–108.
30. Chen, J. H. (1980) *J. Virol.* **36**, 162–170.
31. Gonda, T. J., Sheiness, D. K., and Bishop, M. J. (1982) *Mol. Cell Biol.* **2**, 617–624.
32. Westin, E. H., Wong-Staal, F., Gelmann, E. P., Dalla Favera, R., Lautenberger, J. A., Eva, A., Reddy, P., Tronick, S. R., Aaronson, S. A., and Gallo, R. C. (1982) *Proc. Nat. Acad. Sci. U.S.A.* **79**, 2490–2494.
33. Bishop, J. M., Courtneidge, S. A., Levinson, A. D., Oppermann, H., Quintrell, N., Sheiness, D., Weiss, R., and Varmus, H. E. (1980) *Cold Spring Harbor Symposium on Quantitative Biology* **44**, 919–930.
34. Graf, T., Beug, H., and Hayman, M. J. (1980) *Proc. Nat. Acad. Sci. U.S.A.* **77**: 389–393.

Cellular DNA Sequences Involved in Chemical Carcinogenesis

PAUL T. KIRSCHMEIER, SEBASTIANO GATTONI-CELLI,
ARCHIBALD S. PERKINS, AND I. BERNARD WEINSTEIN

Department of Human Genetics and Development
Division of Environmental Sciences
and
The Cancer Center/Institute of Cancer Research
Columbia University
New York, New York

INTRODUCTION: GENES INVOLVED IN CARCINOGENESIS

In contrast to oncogenic viruses, carcinogens and tumor promoters can not introduce new genetic information into cells. During the transformation process, they must, therefore, call upon genes already present in the target cells to bring about and to maintain the transformed state. A major challenge in carcinogenesis research is to identify the cellular gene(s) involved in radiation and chemical carcinogenesis and to compare their sequences, state of integration and/or expression in normal and carcinogen-transformed cells. The RNA tumor viruses provide both a useful model and an important biological tool to help solve these problems.

Studies of the RNA acute leukemia and sarcoma viruses have led to the concept that these viruses arose by the recombination of RNA leukemia viruses with specific onc genes (also called proto-oncogenes) endogenous to normal vertebrate species (for review, see ref. 1). Infection of cells by sarcoma viruses leads to the integration of the viral onc genes into aberrant sites in the host genome, where they are expressed at high levels and thus lead to the transformed state. The proviral DNAs of retroviruses are flanked by long terminal repeat sequences (LTR). There is evidence that these sequences contain strong promoter signals for controlling transcription and that they might also play a role in gene transposition (2–4).

It is possible that the cellular homologs of onc or LTR sequences are involved in the transformation of cells by nonviral agents. Indeed, one of the genes that may be involved in human bladder carcinoma is homologous to the *ras* gene, the transforming gene of the Harvey sarcoma virus (5). We have postulated that in higher organisms DNA damage induced by chemicals or radiation might trigger alterations in the state of integration and/or cause switch-on of the constitutive expression of these DNA sequences in the absence of a virus vector. The results of our recent studies are described below.

STUDIES OF CELLULAR onc GENES IN NORMAL AND TRANSFORMED MURINE CELLS

We utilized a cloned DNA fragment pmos-1 representing a portion of the onc gene (v-*mos*) of Moloney murine sarcoma virus (Mo-MSV) as a probe to determine whether or not transformation of rodent cells by chemical carcinogens or radiation is associated with alterations in the state of integration or transcription of the normal cellular sequence (c-*mos*) that is homologous to the v-*mos* gene (6,7). However, after examining a number of normal rodent cell types and rodent cells trans-

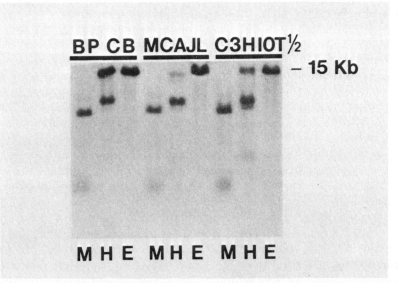

Fig. 1. DNAs from three mouse cell lines C3H 10T1/2, C3H 10T1/2 J.L. #3 (MCA-transformed), and C3H 10T1/2 C.B. #2 (BP-transformed) were digested with *Eco*R1, *Eco*R1 + *Hpa*II, and *Eco*R1 + *Msp*1, electrophoresed, and blot hybridized to the ^{32}P-labeled p-mos1 DNA. (For additional details, see ref. 8).

formed by either chemical carcinogens or radiation, we found no evidence that transformation by these agents was associated with rearrangement of the c-*mos* sequence within the cell genome (Fig. 1). In addition, in all of the normal and transformed cells, the c-*mos* sequence was hypermethylated and transcriptionally silent (Fig. 1). These results are in contrast to the situation in cells transformed by Mo-MSV in which the exogenous v-*mos* sequence is integrated at new sites within the host genome, is undermethylated, and is extensively transcribed (8). In related studies, we have found that another onc gene, c-*ras*, also fails to show any evidence for rearrangement in carcinogen or radiation transformed C3H 10T1/2 cells (unpublished studies). Because there is evidence that eukaryotic cells can contain at least 15 onc genes (1), our results do not rule out the possibility that 17 onc genes other than c-*mos* and c-*ras* may play a role in the transformation of murine cells by chemicals or radiation. Additional studies utilizing probes to other onc genes are required to evaluate this possibility.

In considering the possible involvement of cellular oncogenes (sometimes designated proto-oncogenes) in chemical carcinogenesis, it is important to stress that they differ in several respects from their viral homologs. For example, they usually contain introns, they are not necessarily flanked by LTR sequences, and they may differ in certain other respects from the oncogenes present in retroviruses. Therefore, they may be expressed at lower levels and under a greater degree of host control in carcinogen-induced tumors than the corresponding oncogenes introduced into cells by the acute transforming viruses. For this reason, cell transformation induced by chemical carcinogens might often depend on changes in the function of multiple cellular oncogenes, as well as other types of host genes, rather than a single dominant gene. The acute transforming viruses have had time to evolve so that their oncogenes can act as single dominant genes. In contrast, if the oncogenes present in normal cellular DNA do play a role in chemical carcinogenesis, it is unlikely that they do so by undergoing the full evolution in structure that occurred in the origin of viral oncogenes.

STUDIES ON THE EXPRESSION OF POLY(A)$^+$RNA-CONTAINING LTR-LIKE SEQUENCES

In another set of studies we have pursued an alternative approach to identifying retrovirus-related genes that might be involved in radiation and chemical carcinogenesis. Several types of evidence indicate that the LTR regions of retrovirus genomes play a crucial role in controlling transcription and that the specific sequences involved are present in the U3 portion of the LTR sequence (3). The R region of the LTR defines the site of initiation (at the 5' end) of viral RNA synthesis, and the U5 region represents the portion of the LTR sequence that is present at the 5' terminus of the mature viral RNA (3). The LTR sequence also has structural features similar to transposable elements of bacteria, suggesting that LTR sequences might also be involved in gene transposition (3,8). They could act, therefore, as mobile promoters capable of initiating the transcription of sequences adjacent to sites into which they might become integrated (4). Consistent with this possibility are recent studies indicating that the insertion of the LTR sequence of an avian leukosis virus into the host cell DNA activates transcription of a flanking host oncogene sequence designated c-*myc* and that this is responsible for the induction of lymphomas (4). Normal murine cells contain several copies of DNA sequences homologous to the LTR sequence of murine retrovirus proviral DNA (9). These findings suggest that, if damage to cellular DNA caused the rearrangement and/or activation of endogenous LTR sequences, this could lead to the constitutive expression of host sequences (oncogenes or other genes) whose products might be capable of inducing cell transformation. In theory, these events could occur in the absence of a replicating leukemia virus.

To test this hypothesis, we have examined whether there are differences between normal C3H 10T1/2 cells transformed by radiation or chemical carcinogens in terms of the expression of RNA species containing sequences homologous to a probe prepared to the LTR sequence of Mo-MSV (9). The poly(A)$^+$RNA fraction was purified from these cells, separated by gel electrophoresis, and then hybridized by the "Northern" blotting technique to a ^{32}P-labeled DNA probe for LTR sequences.

Figure 2 indicates that with the poly(A)$^+$RNA from normal C3H 10T1/2 cells there was negligible hybridization to the LTR probe (lane f). On the other hand, the poly(A)$^+$RNA from five different transformed C3H 10T1/2 cell lines (Fig. 3, lanes a–e) showed appreciable hybridization to this probe. At least five distinct poly(A)$^+$RNA species ranging from about 38 S to 18 S were detected in the transformed lines. We have analyzed a total of eight transformed cell lines that were originally derived from normal C3H 10T1/2 following exposure to chemical carcinogens, UV, or X rays, and all of these displayed RNAs homologous to the LTR probe, yielding profiles similar to those shown in Fig. 1. With both an early and late passage clone of normal C3H 10T1/2 cells, there was always undetectable or only slight hy-

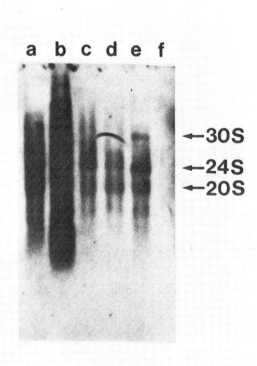

Fig. 2. Northern blot analysis showing hybridization of the LTR probe to the poly(A)+RNA from normal, transformed, and drug-induced murine cells. The gels contained poly(A)+RNAs from the following cell lines: lane a, C3H 10T1/2 JL #2 (UV-transformed); lane b, C3H 10T1/2 CB #2 (benzo[a]pyrene-transformed); lane c, C3H 10T1/2 JL#1 (X-ray transformed); lane d, C3H 10T1/2 CB #1 (X-ray transformed); lane e, C3H 10T1/2 JL #3 (methylcholanthrene transformed); and lane f, normal C3H 10T1/2. Poly(A)+RNA was isolated from these cells, denatured with 50% formamide–6% formaldehyde at 65°C, and separated on a 0.8% agarose gel containing 6% formaldehyde and 20 mM morpholinopropanesulfonic acid at pH 7.0. The RNA was then transferred to nitrocellulose sheets (BA 85, Schleichler and Schuell) and hybridized to a ^{32}P-labeled LTR probe. (For additional details, see ref. 9).

bridization to this probe. On the other hand, we have found that the exposure of normal C3H 10T1/2 cells to either bromodeoxyuridine (BrdUrd) or 5-azacytidine for 48 hr induced the expression of a series of poly(A)+RNAs that were homologous to the LTR probe and were similar in size to those found in the transformed C3H 10T1/2 cell lines (9). Thus, it appears that what is unusual in the transformed cells is the constitutive expression of these transcripts rather than the presence of LTR-containing DNA sequences that are unique to the transformed cells.

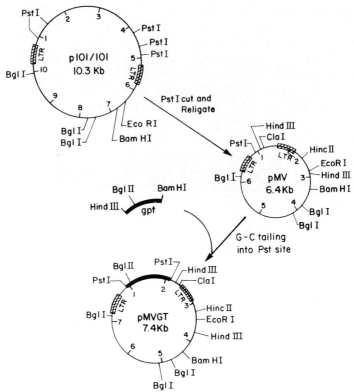

Fig. 3. Construction of vector. This figure illustrates the construction of the retrovirus-derived vector and insertion of the bacterial xanthine guanine phosphoribosyl transferase gene (*gpt*). Plasmid p101/101 (gift of D. Dina), containing the entire Mo-MSV genome poly(dA·dT) homopolymer tailed into the *Pst*I site of pBR322, was cleaved with *Pst*I, phenol extracted, and ethanol precipitated. 100 ng of the DNA fragments were diluted in 100 µl of ligase buffer (20 mM Tris, pH. 7.6; 20 mM MgCl$_2$; 5 mM DTT; 30 mM NaCl) containing 1 mM ATP. T4 DNA ligase (Miles, 2 units) was added and the mixture incubated at 4°C for 15 hr. Then 20 µl of the reaction mixture was used to transform *E. coli* HB101. Colonies containing the religated plasmid were selected on LB plates containing 20 µg/ml tetracycline. Tet^r colonies were picked, and plasmids were isolated. The plasmids were digested with *Pst*I, *Eco*R1, *Hin*dIII, or *Eco*R1 plus *Pst*I, and the products of the digestion were analyzed on 1% agarose in 89 mM Tris, 89 mM Borate, pH 8.1; 1 mM EDTA (TBE), or 7% acrylamide in TBE. A colony containing the plasmid that yielded appropriate restriction fragments (6.4 kb after *Pst*I or *Eco*R1 digestion; 1.9 and 4.1 kb after double digestion with *Eco*R1 plus *Pst*I were picked, and a large scale preparation of this plasmid, designated pMV-1, was made.

Plasmid pMO was digested with *Bam*H1 and *Hin*dIII, which removed a 1-kb fragment that contained the *E. coli gpt* gene. This fragment was separated by electrophoresis through low geling temperature agarose (BRL) and isolated from the agarose by phenol extraction as suggested by the manufacturer.

The poly(A)⁺RNA transcripts detected with the LTR probe in the transformed C3H 10T1/2 cell lines could originate from endogenous murine retrovirus genomes and/or from the expression of host genes unrelated to the retroviruses but flanked by LTR sequences. Therefore, we performed a set of experiments, utilizing the appropriate ³²P-labeled DNA probes, to determine whether these transcripts contained, in addition to LTR-like sequences, sequences homologous to the known retrovirus genes *gag, pol, env* and to the *U3* region of the LTR region. We found that when a probe for the *gag-pol* region of murine retroviruses was hybridized to the poly(A)⁺RNA of carcinogen transformed C3H 10T1/2 cell lines there was hybridization to the 30–38 S RNAs, but there was no detectable hybridization to the lower molecular weight RNAs (24-18S) detected with the LTR probe (9). The probe to the *env* region also hybridized to RNAs of about 30–38 S and, in addition, to an RNA of about 24 S, but it did not hybridize to the 20-18 S RNAs detected with the LTR probe in transformed C3H 10T1/2. We also utilized a probe specific to the U3 region of the LTR sequence. This region is usually contained in virally related messages because it is just proximal to the viral polyadenylation site (3). We found that the U3 probe was homologous to the 30–38 S and 24 S RNAs present in the carcinogen-transformed derivatives of C3H 10T1/2. However, the 20 S and 18 S transcripts recognized by the total LTR probe did not hybridize to the probe specific for the U3 region of the

---

pMV-1 was linearized by digestion with *Pst*I, phenol extracted, and ethanol precipitated. Poly(dG) tails were added to this fragment, and poly(dC) tails were added to the *Bam*H1–*Hin*dIII fragment from pMO as follows. The fragments were incubated in 140 mM potassium cacodylate, pH. 6.9; 30 mM Tris-OH; 1 mM $CoCl_2$; 0.1 mM DTT, with an excess of terminal transferase (Boehringer Mannheim) and a 100-fold molar excess of either dGTP or dCTP. The failed fragments were phenol extracted, ethanol precipitated, and dissolved in 5 mM Tris, pH 7.9; 0.2 mM EDTA. The fragments were mixed at an equimolar ratio at 5 ng/μl in 150 nM NaCl; 10 mM Tris, pH 7.6; 1 mM EDTA, and this solution was heated to 70°C for 5 min, then gradually cooled and maintained for several hours at room temperature. The reaction mixture was then chilled on ice and ethanol precipitated. The annealed DNA was dissolved in 10 mM Tris, pH 7.6; 10 mM NaCl; 1 mM EDTA. Competent HB101 cells were transformed with 5–10 ng of annealed vector-insert DNA and Tet^r clones were selected. These clones were picked and screened for those that carried the *gpt* insert by colony hybridization.

Plasmids from clones that gave positive hybridization results were isolated as described above and digested with *Pst*I and *Bgl*II plus *Eco*R1. The *Bgl*II plus *Eco*R1 digestion permitted orientation of the insert because the *Bgl*II site is located assymetrically (150 bp from the 5′ end) in the *gpt* gene. Colonies having the *gpt* gene in the correct transcriptionaly orientation relative to the LTR's of pMV-1 were picked, purified, and designated pMVGT-1, -2 and -3. Two clones with the *gpt* gene in the opposite orientation were also selected and designated pMVTG-1 and pMVTG-2.

LTR. Significant hybridization was not detected with either the *gag*, *pol*, *env*, or *U3* probes or with the poly(A)$^+$RNA from normal C3H 10T1/2 cells (9).

Although normal C3H 10T1/2 cells do not contain appreciable levels of poly(A)$^+$RNA homologous to the LTR probe, all of the transformed C3H 10T1/2 cell lines contain a series of poly(A)$^+$RNAs ranging in size from about 38–18 S. The results obtained with the additional probes suggest that the 30–38 S poly(A)$^+$RNAs present in the transformed cells are transcripts of endogenous retroviral genome(s). The 24 S RNA may represent a specific mRNA for the viral envelope glycoprotein. Althugh the 20 S and 18 S transcripts that we have detected in transformed C3H 10T1/2 cells contain sequences homologous to LTR, they do not appear to be due to transcription of known endogenous MuLV genomes. We suspect that these transcripts reflect the expression of nonvirally related host sequences that utilize virally related LTR sequences as promoters. Southern blot analyses indicate that the genome of normal C3H 10T1/2 cells contains over 30 copies of LTR sequences (9). It is possible that some of these flanking host sequences are unrelated to endogenous leukemia virus genomes. We are currently analyzing whether the transformation of C3H 10T1/2 cells is associated with changes in the state of integration and/or methylation in one or more of these endogenous sequences.

DEVELOPMENT OF AN EXPRESSION AND TRANSDUCTION VECTOR

The fact that acute transforming RNA tumor viruses arose by recombination between a parental RNA leukemia virus and host nucleic acid sequences suggested to us that these viruses would be useful eukaryotic cloning vectors that would permit expression and transduction of any given gene.

We tested this hypothesis by using a cloned Moloney Murine sarcoma virus genome and the cloned *E. coli* xanthine guanine phosphoribosyltransferase *(gpt)* gene (10). We removed 3950 basepairs (bp) between the 5' and 3' LTRs of the cloned provirus and then introduced a *Hin*dIII to *Bam*H1 fragment containing the entire *gpt* gene, by GC-tailing (11) (Fig. 3). We obtained clones containing the *gpt* gene in the same and opposite transcriptional polarity as the LTRs and tested the biological activity of these constructs.

DNA was introduced by the CaPO$_4$ precipitation technique to either mouse (NIH3T3) or rat (Rat 1) cells and clones selected by their ability to grow in the presence of mycophenolic acid (12). Clones were obtained from plasmids that contained the *gpt* gene with the same transcriptional polarity as the LTRs but not from plasmids in which the *gpt* gene was in the opposite polarity. Rat 1 cells yielded 4×10^2 clones/μg/10^5 and the NIH 3T3 cells yielded $5-10 \times 10^1$ clones/μg/10^5 cells. We have consistently found that the Rat 1 cells are more competent for transfection with this recombinant plasmid than NIH 3T3.

Several of the Rat 1 Gpt$^+$ clones were picked, and their DNAs and RNAs were analyzed by nucleic acid blotting techniques. All clones resistant to mycophenolic acid contained DNA sequences homologous to probes prepared to the LTR region of M-MSV, and the *E. coli gpt* gene and at least some of the clones contained 2.3 kb polyadenylated RNA that hybridized to both probes, a size consistent with induction of transcription within the 5' LTR and polyadenylation within the 3' LTR of our construct (Fig. 3).

To determine which regions of the plasmid were required for biologic activity, we cut our construct pMVGT-1 with different restriction endonucleases and then transfected the DNA onto Rat 1 cells and selected Gpt$^+$ clones. Three enzymes that cut within the LTRs virtually eliminated biologic activity. Thus *Pvu*II (which cuts within the 72 bp repeat of the LTR) lowered the efficiency of Gpt transfer to 1%; *Sac*I (which cuts just 5' to the AATTAA "promoter" sequence) cleavage lowered the efficiency to 1.5%, and cleavage with *Sma*I (which cut 3' to the promoter sequence and 5' to the polyadenylation signal) lowered the transfection efficiency to 2.5%, when compared to that obtained with the intact plasmid. These data indicate that the LTRs are required for expression of the *gpt* gene. We next examined the specific roles of the 5' and 3' LTRs contained in our plasmid with respect to transfection of the Gpt$^+$ marker. *Hin*dIII (which cuts between the *gpt* gene and 3' LTR) cleavage reduced the transfection efficiency to 23%, suggesting that the downstream LTR enhances but is not absolutely essential for biologic activity. Poly(A)$^+$RNA isolated from several of the latter Gpt$^+$ clones was analyzed by the "Northern" blotting technique. In several cases, discreet transcripts were found that hybridized to the Gpt probe, but not to a U3 probe, suggesting that termination and polyadenylation were achieved by signals from host cell sequences that flanked the integrated *Hin*dIII cut fragment. Thus it would appear that the 3' LTR present in our construct is required for maximal, but not complete, biologic activity of this vector.

TRANSDUCTION WITH MULV HELPER VIRUS

Figure 4 depicts three procedures we used to determine if we could obtain transduction of our LTR-*gpt* construct by utilizing Mo-MLV as a helper or rescue virus. In the first procedure, plasmid pMVGT-3 was transfected onto a clone of NIH 3T3 cells that had been previously infected with Mo-MuLV and had continuously produced this virus (Mo-MLV clone 1). Virus particles were then harvested from the medium at 18, 42, and 66 hr in the absence of XAT selection. Ordinary NIH 3T3 cells were then infected with this virus and were selected in XAT medium. Using 10^6 pfu/ml (determined by XC plaque assay), we obtained about 100 XAT resistant colonies with the 18-hr virus preparation, about 50 XAT resistant colonies with the 42 hr virus preparation, and no colonies with the 66 hr preparation. Thus, without prior XAT selection, the transiently expressed *gpt* gene was rescued by Mo-MuLV, but only with a relatively low efficiency that declined to zero by 66 hr. We believe that this reflects the relatively low competence of this strain of NIH 3T3 for transfection.

In the second procedure, the pMVGT-3 transfected MuLV-producer NIH 3T3 cells were transferred to the XAT selection medium 72 hr after adding the plasmid, and virus particles were then collected from a mass culture of XAT resistant cells. These were used to infect NIH 3T3 cells, and the number of Gpt^+ clones was determined by selection in XAT medium. With this procedure, we obtained 10^3 Gpt^+ colonies per 10^6 pfu (again pfu was determined by the XC plaque assay). Thus, stable integration and expression of the plasmid under conditions of XAT selection enhanced the process of transduction with Mo-MLV.

A third procedure we used to obtain Gpt^+ virions is also shown in Fig. 4. The plasmid pMVGT-3 was transfected onto ordinary NIH 3T3 cells, and the cells were then transferred to XAT medium. A XAT resistant colony was isolated, grown in mass culture, and then infected *de novo* with Mo-MLV (10^6 XC pfu). Two days later, the medium from these cells was harvested and assayed. With this procedure, the efficiency of transfer of Gpt^+ was extremely high because the 0.001 ml of virus-containing medium gave rise to 100 pfu (XC test) and more than 50 Gpt^+ colonies following infection of NIH 3T3 cells (Fig. 4).

Thus, by excising the major portion of the central region of a cloned proviral genome of Mo-MSV, we developed an expression vector, pMV, that functions in rodent cells to promote transcription of a newly inserted DNA sequence. The fact that RNA transcripts of this sequence can be encapsidated into viral particles makes this system an

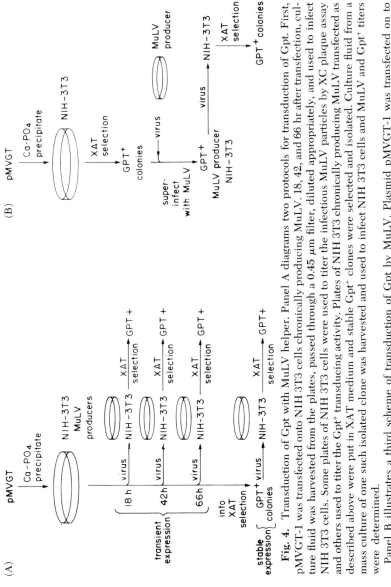

Fig. 4. Transduction of Gpt with MuLV helper. Panel A diagrams two protocols for transduction of Gpt. First, pMVGT-1 was transfected onto NIH 3T3 cells chronically producing MuLV 18, 42, and 66 hr after transfection, culture fluid was harvested from the plates, passed through a 0.45 μm filter, diluted appropriately, and used to infect NIH 3T3 cells. Some plates of NIH 3T3 cells were used to titer the infectious MuLV particles by XC plaque assay and others used to titer the Gpt+ transducing activity. Plates of NIH 3T3 chronically producing MuLV transfected as described above were put in XAT medium and stable Gpt+ clones were selected and isolated. Culture fluid from a mass culture of one such isolated clone was harvested and used to infect NIH 3T3 cells and MuLV and Gpt+ titers were determined.

Panel B illustrates a third scheme of transduction of Gpt by MuLV. Plasmid pMVGT-1 was transfected on to NIH 3T3 cells and stable Gpt+ clones were selected and isolated. Culture fluid from NIH 3T3 cells chronically producing MuLV was used to infect a mass culture of 1 Gpt+ NIH 3T3 clone. Three days after infection culture fluid was isolated and used to infect NIH 3T3 cells. Viral titer and Gpt+ titer were determined as above.

extremely efficient one for transfering specific genes into mammalian cells. We intend to insert specific genes for growth factors into this vector, and also to insert genes for other putative proteins involved in carcinogenesis, and then transfect these factors into normal cells with this vector. This approach should greatly facilitate the identification of genes and physiologic factors responsible for establishment and maintenance of the transformed state in cells exposed to chemical carcinogens, tumor promoters, and various co-factors.

CONCLUSION

Although considerable progress has been made in identifying the cellular targets that are involved in the primary encounter between cells and initiating carcinogens the subsequent genetic changes that lead to malignant transformation are not known. With respect to the hypothesis that carcinogens act by inducing specific gene rearrangements, we found no evidence that the proto-oncogene c-*mos* undergoes rearrangement within the genome during the transformation of rodent cells induced by chemical carcinogens or radiation. We believe, however, that this mechanism should be explored in detail utilizing probes for additional protooncogenes and other types of host genes. Our evidence that the transformation of rodent cells by chemical carcinogens or radiation is associated with the constitutive expression of a series of poly$(A)^+$RNAs that contain LTR-like sequences raises the possibility that carcinogenesis might involve disturbances in the structure and function of specific DNA sequences that regulate gene transcription. LTR-like sequences have been identified in the genomes of normal and other type of vertebrate cells. The precise mechanism by which these diverse types of DNA sequences enhance gene transcription is not known, but it is conceivable that carcinogenesis involves disturbances in the function of these regulatory sequences.

Finally, eukaryotic expression vectors provide a means to introduce specific genetic information into mammalian cells and to ascertain the effects of these genes on the cellular phenotypes.

ACKNOWLEDGMENTS

This research was supported by DHS and NCI Grants CA 021111 and CA 02656. The authors appreciate the valuable role of Dr. Dino Dina and his colleagues at the Albert Einstein College of Medicine, Bronx, New York in the studies on c-*mos* and LTR sequences and of Dr. R. Mulligan for providing plasmid pMO. We thank Patricia Kelly for assistance in preparing this manuscript. P. K. was a recipient of a postdoctoral fellowship from the Daymon-Runyon Walter Winchell Cancer Fund.

REFERENCES

1. Bishop, J. M. (1980) *N. Engl. J. Med.* **303**, 675.
2. Blair, D. G., Oskarsson, M., Wood, T. G., McClements, W. L., Fischinger, P. J., and Vande Woude, G. G. (1981) *Science* **212**, 941.
3. Dhar, R., McClements, W. C., Enquist, L. W., and Vande Woude, G. G. (1980) *Proc. Natl. Acad. Sci. U.S.A.* **77**, 3937.
4. Hayward, W. S., Neel, B. G., and Astrin, S. M. (1981) *Nature (London)* **290**, 475.
5. Parada, L. F., Tabin, C. J., Shih, C., and Weinberg, R. A. (1982) *Nature (London)* **297**, 474.
6. Oskarsson, M., McClements, W. L., Blair, D. G., Maizel, J. V., and Vande Woude, G. G. (1980) *Science* **207**, 1222.
7. Jones, M., Bosselman, R. A., Hoorn, V. D., Berns, F. A., Fan, H., and Verma, I. M. (1980) *Proc. Natl. Acad. Sci. U.S.A.* **72**, 2651.
8. Gattoni-Celli, S., Kirschmeier, P., Weinstein, I. B., Escobedo, J., and Dina, D. (1982) *Mol. Cell. Biol.* **2**, 42.
9. Kirschmeier, P., Gattoni-Celli, S., Dina, D., and Weinstein, I. B. (1982) *Proc. Natl. Acad. Sci. U.S.A.* **79**, 2773.
10. Mulligan, R. C., and Berg, P. (1980) *Science* **209**, 1422.
11. Perkins, A. S., Kirschmeier, P. T., Gattoni-Celli, S., and Weinstein, I. B. (1982) In preparation.
12. Mulligan, R., and Berg, P. (1981) *Proc. Natl. Acad. Sci. U.S.A.* **78**, 2072.

PART V

GENE AMPLIFICATION AND TRANSFECTION

Gene Amplification and Methotrexate Resistance in Cultured Animal Cells

ROBERT T. SCHIMKE, PETER C. BROWN, RANDAL N. JOHNSTON, BRIAN MARIANI, AND THEA TLSTY

Department of Biological Sciences
Stanford University, Stanford, CA

INTRODUCTION

Our laboratory has been studying the acquisition of resistance of cultured mouse and hamster cells to the 4-amino analog of folic acid, methotrexate (MTX). In particular, we have concentrated on resistance resulting from the overproduction of the target enzyme of MTX inhibition, dihydrofolate reductase (DHFR). Our studies (1,34) as well as others (8,26) have shown that the overproduction of DHFR results from a proportional increase in the number of DHFR genes, i.e., gene amplification. The MTX resistance can be either stable or unstable when cells are grown in the absence of MTX. In cells with a stable resistance phenotype, the DHFR genes are present on one or more chromosomes associated with expanded regions called "homogeneously staining regions" [(HSRs), see Nunberg et al. (29)]. In the unstable phenotype, the amplified DHFR genes are associated with extrachromosomal, self-replicating elements which do not contain centromeric staining regions, the so-called "double-minute" chromosomes (DMs) (11,18). By virtue of their lack of participation in cytokinesis, these small extrachromosomal elements can be distributed unequally into daughter cells and/or can undergo micronucleation (14,19). As a consequence of these phenomena, individual cells and cell populations can lose the amplified DHFR genes upon subsequent growth in the absence of MTX (34).

Although gene amplification in mammalian somatic cells was first reported with the dihydrofolate reductase gene (33), an increasing number of examples of such amplification phenomena have been reported in the past several years (see *Gene Amplification,* 1982, for many such examples). These include amplification of insect chorion genes in *Drosophila* ovaries at the time of egg laying (35), as well as amplification of the CAD gene (PALA resistance) (40), the metallothionine gene (cadmium resistance) (6), and reversion of leaky auxotrophs of Chinese hamster ovary DHFR (12), and mouse HPRT$^-$ (27). In addition, three different groups have shown that resistance of cultured mouse and hamster cells to vincristine (5,21,28) results from the overproduction of a specific protein, the consequence of which is cross-resistance to a number of toxic drugs. The mechanism involves exclusion of these drugs, i.e., an alteration in transport. Although the nature of the protein and the effect of its overproduction on membrane properties is not known, the karyological properties of such cells are consistent with specific gene amplification, i.e., the presence of HSRs or DMs. There are also a number of other examples of selection for resistance phenomena with overproduction of a protein in which the definitive evidence for gene amplification has not been provided as of yet. The above cited examples of gene amplification in cultured animal cells have a theme in common with gene duplications as studied in microorganisms (2). Thus, when cells are inhibited from growth and where the overproduction of a protein can impart growth advantages, the cells that survive and regain the capacity for growth characteristically are those with specific gene duplications–amplifications. This is particularly the case where evolution has not provided for regulatory mechanisms to insure the ability to modulate the content of specific proteins.

The amplification of the DHFR gene occurs not only in experimental laboratory systems. In our laboratory we have also studied a patient who received MTX and whose tumor has become resistant to MTX as a result of amplification of the DHFR gene (16). Thus, gene amplification is a mechanism whereby actual tumors become resistant to agents employed in cancer therapy.

THE MECHANISM OF GENE AMPLIFICATION

For the past several years our laboratory has been attempting to study the molecular and biochemical mechanism(s) underlying gene amplification in cultured somatic cells. The problem in undertaking such studies are twofold. The first problem is the fact that cells that emerge as clones following selection for MTX resistance constitute a heterogeneous population with varying degrees of gene amplification and varying properties with respect to stable and/or unstable amplification (20). Thus, the population of cells is not uniform. The second problem is the relatively low frequency with which gene amplification occurs within a population of cells (as determined by the number of colonies emerging as resistant—a frequency of the order of 10^{-4} to 10^{-6}. Thus, in order to study the process of amplification, it is neces-

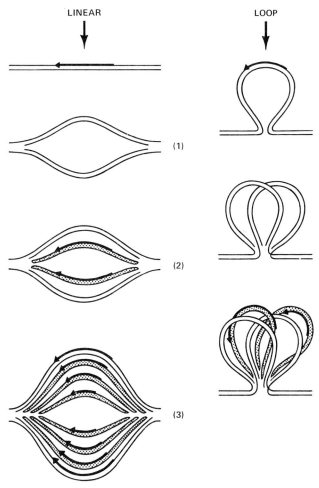

Fig. 1. Saltatory replication model.

sary to explore variables that can enhance the frequency of the amplification so that they can be studied as part of the initial event as opposed to studying cells several months after the event when sufficient cells can be grown for analysis.

The approaches we have made to enhancing the frequency of the amplification events are based on the model for gene amplification that we have favored for several years (34), the so-called "disproportionate (or saltatory) replication" model (Fig. 1). This model is based on the concept that, within the region of DNA that will be duplicated, one or more rounds of replication can be initiated within a single S phase of the cell cycle. The additional component of the model is the concept that replication elongation can slow or cease, resulting in one or multiple free DNA strands within the replication bubble (the so-called "onion skin" replication bubble of Botchan et al. (9). As depicted in Fig. 1, the replicating DNA can be envisaged as forming looped structures, a concept based on the work of Vogelstein et al. (39), which proposes that replication occurs within a matrix in which the replicating DNA is threaded through fixed replication complexes. The free DNA strands can be released from the chromosome, and if they contain replication origins and are retained in the nucleus during cytokinesis and mitosis, they constitute double minute chromosomes. If the free ends undergo 3' to 5' ligation or undergo rolling circle replication and are subsequently recombined into the chromosome structure, they would constitute sites of multicopy gene amplification at the site of the resident gene, i.e., a homogeneously staining region.

The disproportionate replication model suggests that any treatment of cells that will facilitate additional (aberrant) rounds of DNA replication within a single S phase of the cell cycle might increase the frequency of gene amplification. It also questions the basic assumption that a segment of a chromosome is replicated only once during a single cell cycle. We have found that a number of different treatments of cells can markedly enhance the frequency of MTX resistance. These agents include drugs that inhibit DNA replication, i.e., hydroxyurea (HU), MTX, as well as several agents that damage DNA, including ultraviolet light and the chemical carcinogen, acetyl aminofluorene (AAF).

SINGLE-STEP SELECTION FOR METHOTREXATE RESISTANCE

Figure 2 shows the basic experimental design in which a subclone of 3T6 cells, denoted S5 (selected for uniform morphology and high

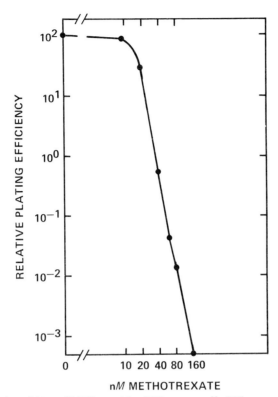

Fig. 2. A cloned line of MTX sensitive 3T6 mouse cells (S5) was grown in DMEM containing 10% newborn calf serum, sodium pyruvate, additional glutamine, and penicillin and streptomycin in 10% CO_2 at 37°C. Cells were plated in 10-cm plates at varying densities (from 1×10^2 to 1×10^6 cells per plate) and placed in media containing varying concentrations of MTX. Media were changed every other day. The cells were grown for 2–3 weeks until discrete colonies could be detected. Relative plating efficiency was calculated by using the plating efficiency of cells grown in the absence of MTX (70–90% plating efficiency) as the 100% value.

plating efficiency), is grown in varying concentrations of MTX, i.e., single-step selection, and the number of surviving colonies is recorded. When the surviving colonies are subcloned and examined for amplification of the DHFR gene (Fig. 3) we find that only a fraction of the MTX-resistant colonies have amplified DHFR genes. At 80 nM MTX 20% of colonies have increased DHFR genes (we can detect only a fourfold increase in DHFR genes over sensitive cells). At 120 and 160 nM of MTX, approximately 60% of colonies have amplified DHFR genes, whereas only 25% of colonies resistant to 200 nM of MTX have DHFR gene amplification. Other mechanisms for resist-

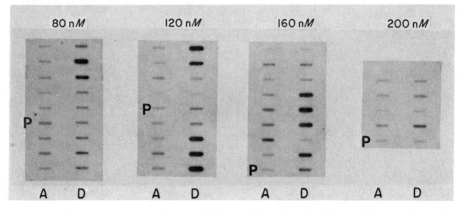

Fig. 3. Determination of gene amplification frequency. DHFR amplification in colonies selected in different concentrations of MTX. MTX-resistant colonies were subcloned and 2×10^5 cells were prepared by the dot hybridization method of Kafatos et al. (1979) with minor modifications (11a). Equal portions of DNA were placed in a template backed with nitrocellulose paper. Hybridization to a ^{32}P-labeled cDNA clone of DHFR (D) or of α-fetoprotein (A) (a gift of Dr. A. Dugaiczyk) was as described by Kaufman and Schimke (20). The intensities of the paired bands were compared to determine if the DHFR gene was amplified relative to sensitive cell DNA. S, DNA isolated from sensitive cells; A, α-fetoprotein probe; D, DHFR probe; P, parental MTX sensitive cells.

ance to MTX include altered affinity of DHFR for MTX (13), as well as alterations in inward MTX transport. We are currently studying the mechanisms of resistance that are not DHFR gene amplification. In particular, we are interested in the possibility that some of the resistant colonies may have alterations in plasma membrane properties resulting from gene amplification, the result of which is a putative "transport" alteration (see above for drug resistance and gene amplification resulting from alterations in surface properties of cells).

One of the major parameters involved in the frequency of DHFR gene amplification is the cell type employed in studies. For instance, MTX resistance and gene amplification is readily obtained with HeLa cells, whereas in mouse and hamster cells the frequency is of the order of 10^{-5} to 10^{-6}. Within species, there are also marked differences inasmuch as we have obtained MTX resistance and DHFR gene amplification in mouse 3T3 (NIH Swiss Webster) cells readily, whereas in the same protocol we have been unable to obtain MTX resistant 3T3 (BALB/c) cells. In addition, we have found markedly different frequencies within subcloned populations of both 3T6 and CHO cells. Thus, it is important in studying the frequency of amplification to uti-

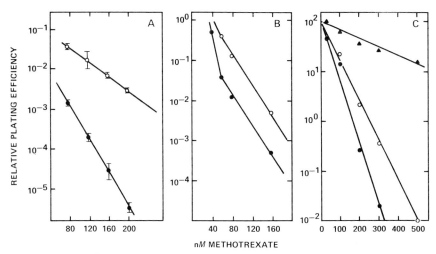

Fig. 4. (A) Effect of hydroxyurea (Hu) pretreatment on the generation of methotrexate resistant colonies. Subcloned 3T6 mouse cells were exposed to medium containing 0.3 mM hydroxyurea for 16 hr, grown in preconditioned control medium for 6 hr, trypsinized, and plated at 5×10^4 to 5×10^5 cells per 10-cm plate in medium containing concentrations of MTX from 8×10^{-8} to $2 \times 10^{-7}M$ (—○—). Control cells were treated identically with the omission of hydroxyurea (—●—). The cells were selected in MTX as described in the legend in Fig. 1. The results are expressed in terms of relative plating efficiency where cells pretreated with Hu and grown in the absence of MTX showed a 2% survival which was taken as the 100% value. (B) Effect of UV pretreatment on the generation of MTX resistant colonies. Subcloned 3T6 mouse cells were plated at 2×10^5 cells per 10-cm plate and allowed to attach for 12 hr. Monolayers of cells were exposed to 1 J/m² UV light (—○—) by removing the media, irradiating the cells, and replacing the conditioned media. Control samples (—●—) were manipulated in the same manner with the omission of the irradiation. All samples were incubated for an additional 6 hr and then fresh media containing MTX were added. The results are expressed in terms of relative plating efficiency where cells irradiated with UV and grown in the absence of MTX showed a 10% survival which was taken as the 100% value. (C) Effect of Methotrexate pretreatment on generation of MTX-resistance. Uncloned CHO K_1 cells were plated (1-6 × 10³ cells/cm²) in control medium for 24 hr and then exposed to media containing MTX in one of three ways: (1) cells in some (—▲—) were treated with the specified concentration of MTX for 48 hr and then allowed to recover in medium without MTX. (2) In other plates (—○—), MTX was restored, at the original concentration, after the 48-hr recovery period. (3) In a third group of plates (—●—), cells were treated continuously with MTX. Colonies were counted after 3–5 weeks, and their frequency expressed as percentage of controls in 0 MTX, but without otherwise correcting for plating efficiency.

lize subcloned populations of cells, inasmuch as a population of cells has inherent differences in the frequencies of growth at specified MTX concentrations.

PRETREATMENT REGIMENS AND ENHANCED FREQUENCIES OF MTX RESISTANCE

Figure 4 shows representative results in which sensitive mouse 3T6 cells (subclone S5) have been subjected to pretreatments with various agents prior to single-step selection for MTX resistance. In such experiments, there are a large number of variables, including whether cells are treated prior to or after plating, the dose of agent used, the time after treatment when cells are subjected to MTX selection, the concentration of MTX employed for selection, the plating density of cells, as well as the nature of the subclone employed. Each of these variables has optimal conditions, and they vary for the cell types as well as the treatments employed. Only a few comments concerning various of the variables will be mentioned below.

Figure 4a shows the effect of pretreatment of 3T6 cells with hydroxyurea, an inhibitor of DNA synthesis by virtue of its inhibition of ribonucleotide reductase. In such experiments, the cells are placed in media containing HU for 16 hr. The media is then replaced with HU-free media, and subsequently the number of surviving colonies at various MTX concentrations is determined (see Figure 4a legend for details). Such treatment enhances the frequency with which MTX-resistant colonies are obtained. The extent of enhancement relative to untreated cells is a function of the MTX concentration used for selection. Treatment with HU for times less than 16 hr results in less enhancement of resistance. In the experiment reported here, the concentration of HU employed (0.3 mM) inhibited labeled thymidine incorporation by 99%. In other experiments where the concentration of HU was 0.1 mM and thymidine incorporation was inhibited only 70%, the number of MTX-resistant colonies was 3–5 times greater. Thus, partial inhibition of DNA synthesis enhances the frequency of MTX resistance more than does complete cessation of DNA synthesis during the period of HU treatment. The enhancement of MTX resistance declines when the recovery period following release of the HU inhibition is greater than 6 hr. Thus, if the cells are placed in MTX selection medium 96 hr after release from the HU inhibition, the frequency of MTX resistance is essentially the same as in the control, non-HU treated cells. Colonies subjected to HU treatment and with an enhanced frequency of MTX resistance have been analyzed for

DHFR gene amplification (see Fig. 3). The fraction of colonies with amplified DHFR genes is approximately 0.3 to 0.4. Thus, HU treatment does, indeed, markedly increase the frequency of DHFR gene amplification; it also increases proportionately resistance by other mechanisms.

Figure 4b shows experiments in which 3T6 cells are first exposed to UV (6 J/m^2), resulting in a 90% killing of cells. Treatment with UV increased the frequency of resistant colonies. We have found that optimal enhancement occurs at approximately 3 J/m^2, a dose that kills only some 30% of cells, and the enhancement is approximately 5 times greater than that seen in Fig. 3b. We have also carried out studies in 3T6 cells with the carcinogen, acetylaminofluorene, and have obtained enhanced MTX resistance similar to that described for UV light. As with UV light, a concentration of AAF that kills approximately 20–30% of cells is optimal for the enhancement phenomenon. In addition, with both UV and AAF, the time of placing cells in MTX is critical and appears to be optimal some 16–48 hr after treatment. Thus, we conclude that optimal doses for observing the effect are those that result in minimal killing of cells. In addition, if the cells are not allowed to recover from the treatment for a certain time period prior to addition of MTX (presumably to allow DNA replication to commence) no enhancement of MTX resistance is observed. Figure 4c shows experiments with Chinese hamster ovary cells (CHO) subjected to brief inhibition of DNA replication with MTX. Inhibition of DNA replication, followed by subsequent relief of replication inhibition followed by selection in MTX also enhances the frequency of MTX resistance.

More recently we have been studying the regulation of DHFR enzyme levels and DHFR gene replication in the cell cycle in CHO cells. DHFR enzyme levels decline during G_1, and in the first and second hour of S, they increase rapidly to G_2 levels, which are twice those of early G_1 (25). We have now found (24a) that the DHFR gene is replicated in the first hour of S (but not in the G-S boundary time). Most interesting are preliminary results (T. Tlsty) which indicate that it is only in the first part of the S phase of the cell cycle (i.e., just when the DHFR gene is replicated) that treatment of cells with UV light enhances MTX resistance and gene amplification.

DHFR GENE AMPLIFICATION UNDER NONSELECTING CONDITIONS

We have shown that various treatments that block and relieve DNA replication and treatments that introduce damage into DNA can increase the frequency of DHFR gene amplification in mouse and hamster cells. Another question that arises is the frequency of the spontaneous process of amplification of the DHFR gene in a population of cells. To answer this question, we have employed a population of CHO K_1 cells sensitive to MTX (half-killing of cells is 20 nM MTX). Two types of experiments have been performed. The first is to obtain random clones from the population and determine the half-killing. In 10 such subclones obtained, the half-killing MTX concentration varied by 20-fold. When the two subclones showing the 20-fold difference in half-killing were subcloned (10 subclones of each), these subclones also varied markedly, and the range of half-killing increased to 100-fold. These cells had been grown and cloned in the absence of any MTX; this indicates the rapid rate at which cells diverge in relation to MTX resistance. The second type of experiment performed employed the Fluorescence Activated Cell Sorter and the fluorescein conjugate of MTX which has allowed us to determine the amount of DHFR enzyme in individual cells (17). In this study the MTX-sensitive CHO K_1 cell population was sorted for DHFR enzyme content, and the cells of the upper end of the fluorescence distribution were obtained. These cells were then grown in nonselective media until such time as they could again be analyzed for fluorescence distribution. Again, the most highly fluorescent 10% of the cell population was sorted and again grown until such time as the population was sufficiently large to analyze. Figure 5 summarizes these results. After 10 such sorts (S-10 of Fig. 5), the cells show a wide distribution of fluorescence, and the mean fluorescence per cell is approximately 50 times that of the original (MTX-sensitive) population. This population is resistant to 50 times the MTX concentration of the sensitive population and has 50 times the DHFR gene copy number.

We interpret these results to indicate that DHFR gene amplification occurs spontaneously in cells at a relatively high frequency. In the study of Fig. 5, the only time that cells are in contact with MTX (as the fluorescein derivative) is during the 24 hours required for the cells to equilibrate with the fluorescein MTX conjugate. At all other times, the cells are grown in MTX free media. In addition, we have undertaken the same experiments in which glycine, hypoxanthine, and thymidine have been present in the medium during the time that the cells are exposed to MTX–fluorescein. Under these conditions MTX does not inhibit cell growth (since these are the products whose synthesis is inhibited by MTX), and the same progression of increase in fluorescence is observed. In these experiments—although cells are selected for a specific fluorescence per cell—when the cells are

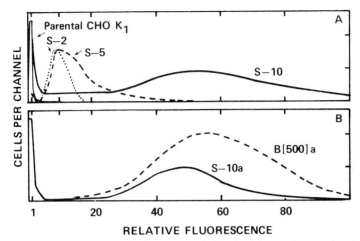

Fig. 5. Sorting of CHO cells for increased dihydrofolate reductase levels under nonselective conditions. Cell variants that spontaneously produce excess DHFR, even without exposure to mutagens or to selection by MTX, can be detected and isolated using FACS (17). An uncloned population of CHO K_1 cells (ID_{25} = 50 nM MTX) was stained for DHFR with F-MTX (17), and a 5% fraction of the cells, displaying maximal fluorescence, retained for growth as a new population (designated S-1). At 10–14 day intervals, the cells were restained, and new populations (S-2.....S-10) were sequentially derived. Metabolic inhibition of the cells by F-MTX was relieved by exposure during staining to glycine, hypoxanthine, and thymidine. Thus, the cells were at no time under physiological stress. (A) The mean fluorescence of each derived population is greater than that of its predecessors. By the tenth round of sorting, fluorescence has increased 50-fold with respect to parental cells. The S-10 cells are also highly resistant to MTX (ID_{25} = 10 μM), and are 20–30-fold amplified for DHFR gene. (B) The newly derived fluorescent S-10 cells are unstable. When subcloned populations (designated S-10$_{a\cdots f}$) are analyzed after 14-day growth, many cells in each population have reverted to low, parental intensity of stain. In contrast, subcloned cells of a population (CHOK$_1$ B$_1$,500) maintained at 500 μM MTX for 3 years (25) do not rapidly lose fluorescence.

grown subsequently, there is extensive spread in the fluorescence distribution, and some of the cells rapidly revert to the fluorescence per cell of the sensitive (parental) population (see Fig. 5). This finding is in keeping with the concept of redistribution of extra chromosomal elements (DMs) containing amplified DHFR genes (19,20).

DISCUSSION

Although the phenomenon of gene amplification and loss has been known for many years, in particular in relation to ribosomal genes in

Drosophila (36) and in amphibian oocytes (10), an appreciation of the frequency of gene amplification in somatic cells is of recent occurrence (32). This discussion will be directed at the possible role of gene amplification in only one aspect of biology, i.e., cancer biology. The questions to be addressed are related to the possible role of gene amplification in the emergence of resistance to various agents employed to kill cancer cells, and the possible role of gene amplification in the malignant process itself.

In the laboratory, it is clear that cultured cells can become resistant to various agents employed in cancer chemotherapy, and in certain instances, e.g., methotrexate resistance and vincristine resistance, this resistance is associated with overproduction of specific proteins. In the case of MTX, the protein is the specific target of MTX inhibition. In the case of vincristine resistance, the specific protein(s) are not identified, but resistance is the result of exclusion of the drug from entry into cells. One wonders, then, whether certain of the so-called "transport" defects associated with multiple cross-resistances as observed clinically might also be the result of amplification events the consequence of which is altered membrane permeability. Clearly this is an area of future investigation. What is clear, however, is the fact that clinical MTX-resistance can be associated with amplification of the DHFR gene (16,37). Thus, gene amplification is potentially a significant mechanism for clinical resistance.

The studies described briefly and in preliminary fashion herein concerning the enhancement of MTX resistance may have clinical importance in terms of the types of treatment regimens employed. Here we are particularly concerned with the problem of resistance which ultimately is the reason tumors kill. Our studies suggest that partial inhibition of DNA replication, or inhibition of DNA replication followed by relief of that inhibition, may well facilitate gene amplification and the emergence of MTX resistance. This may imply that episodic and repeated use of a single drug treatment regimen, in particular where DNA replication is not completely inhibited, may actually facilitate the emergence of cells with specific or multiple cross resistances.

In addition to agents (drugs) which inhibit DNA replication, we find that agents that induce DNA damage (AAF and UV light) also facilitate gene amplification and MTX resistance. Our preliminary results indicate that treatment with UV light enhances MTX resistance and gene amplification only when the UV damage is introduced within the S phase of the cell cycle, and most likely within the time when the DHFR gene is normally replicated. Thus MTX resistance is increased only in cells that are actively cycling, i.e., synthesizing DNA. Such results might be the basis for considering how DNA damaging agents should be employed, and in particular the combination of use of X-rays and cancer chemotherapeutic agents. If we assume that X-ray and UV damage are comparable (clearly they are not in the specific type of lesion, but they may be comparable in terms of facilitation of gene amplification), then our results would suggest that X-ray treatment of tumors should be employed only in cells which are not actively undergoing DNA synthesis.

Perhaps a more basic question is whether gene amplification is related to the progression of cells to an increasingly malignant state. One of the current concepts concerning malignancy is that such cells overproduce the product of a normal cellular gene, so-called "oncogenes" (7). Such overproduction of a protein can be the result of increased transcriptional activity, either by the "capture" of an oncogene by a retrovirus or by integration of a retrovirus next to the normal cellular gene so that the gene is now under the more active promotion function of the retrovirus (15). Such a general mechanism is comparable to so-called "up-promoter" mutations in bacteria, mutations that are extremely rare. The other general mechanism for overproduction of a protein is gene amplification, and this is a frequent process in bacteria (2). What evidence, then, can be marshalled to support the concept that gene amplification may be involved in malignancy? First, is the finding in a number of solid tumors of the presence of the karyological consequences of gene amplification, i.e., HSRs or DMs (3,4, 30,31). It is not presently known whether DNA sequences contained in such structures are causally related to the uncontrolled growth of the tumors or whether they are the consequences of malignant growth without any relevance to the growth characteristics of the cells. On the other hand, as studied by Levan *et al.* (24) and Levan and Levan (23), DMs would appear to play a role in the growth of a mouse ascites tumor (SEWA). These workers have shown that when the tumor cells are grown under tissue culture conditions the DMs are lost. When such cells are grown in an ascitic form in the mouse, the cells that emerge contain multiple DMs. The phenomenon is analogous to growth of unstably MTX-resistant cells in the presence or absence of growth selection, and it suggests that the DMs in the SEWA cells contain genes the consequence of which is the capacity of the cells to overcome normal growth constraints as imparted by growth in the mouse, a type of restraint(s) not encountered when cells are grown under tissue culture conditions.

In addition, it is interesting, although clearly not proof, that agents that facilitate amplification are the types of agents that are involved in tumorigenesis. Varshavsky (38) has shown that the phorbal ester, TPA, markedly enhances DHFR gene amplification, and our laboratory has both confirmed his finding as well as shown that TPA enhances the UV-induced increase in DHFR gene amplification (36a). In addition, Lavi (22) has shown that a number of carcinogens facilitate extensive amplification of integrated SV40 sequences in CHO cells. Thus, it appears that the physical agents that are carcinogenic can enhance gene amplification, as well as introduce other types of alterations in DNA, i.e., point mutations and deletions. It remains to be determined if gene amplification and overproduction of proteins will turn out to be important in the onset and/or progression of the malignant state.

Clearly the possible implications and ramifications of gene amplification in a number of areas of biology are basically unexplored, and the next few years will be interesting in determining whether some of the speculations expressed in this discussion might have some validity.

REFERENCES

1. Alt F. W., Kellems, R. E., Bertino, J. R., and Schimke, R. T. (1978) *J. Biol. Chem.* **253**, 1357–1370.
2. Anderson, R. P., and Roth, J. R. (1977) *Annu. Rev. Microbiol.* **31**, 473–504.
3. Balaban-Malenbaum, G., and Gilbert, F. (1977) *Science* **198**, 739–742.
4. Barker, P. E., Lau, Y.-F., and Hsu, T. C. (1980) *Cancer Genet. Cytogenet.* **1**, 311–319.
5. Baskin, F., Rosenberg, R. N., and Dev, V. (1981) *Proc. Natl. Acad. Sci. U.S.A.* **78**, 3654–3658.
6. Beach, L. R., and Palmiter, R. D. (1981) *Proc. Natl. Acad. Sci. U.S.A.* **78**, 2110–2114.
7. Bishop, J. M. (1981) *Cell* **23**, 5–6.
8. Bostock, C. J., and Tyler-Smith, C. (1981) *J. Mol. Biol.* **153**, 219–236.
9. Botchan, M., Topp, W., and Sambrook, J. (1979) *Cold Spring Harbor Symp. Quant. Biol.* **43**, 709–719.
10. Brown, D. D., and Dawid, I. B. (1968) *Science* **160**, 272–280.
11. Brown, P. C., Beverley, S. M., and Schimke, R. T. (1981) *Mol. Cell. Biol.* **1**, 1077–1083.
11a. Brown, P. C., Tlsty, T., and Schimke, R. T. (1983) *Mol. Cell Biol.* (in press).
12. Chasin, L. A., Graf, L., Ellis, N., Landzberg, M., and Urlaub, G. (1982) *In* "Gene Amplification" (R. T. Schimke, ed.), pp. 161–165. Cold Spring Harbor Lab., Cold Spring Harbor, New York.
13. Haber, D. A., Beverley, S. M., Kiely, M. L., and Schimke, R. T. (1981) *J. Biol. Chem.* **256**, 9501–9510.
14. Haber, D. A., and Schimke, R. T. (1981) *Cell* **26**, 355–362.
15. Hayward, W. S., Neel, B. G., and Astrin, S. M. (1981) *Nature (London)* **290**, 475–480.

16. Horns, R. C., Jr., Dower, W. J., and Schimke, R. T. (1982) In preparation.
17. Kaufman, R. J., Bertino, J. R., and Schimke, R. T. (1978) *J. Biol. Chem.* **253**, 5852–5860.
18. Kaufman, R. J., Brown, P. C., and Schimke, R. T. (1979) *Proc. Natl. Acad. Sci. U.S.A.* **76**, 5669–5673.
19. Kaufman, R. J., Brown, P. C., and Schimke, R. T. (1981) *Mol. Cell. Biol.* **1**, 1084–1093.
20. Kaufman, R. J., and Schimke, R. T. (1981) *Mol. Cell. Biol.* **1**, 1069–1076.
21. Kuo, T., Pathak, S., Ramagli, L., Rodriquez, L., and Hsu, T. C. (1982) In "Gene Amplification" (R. T. Schimke, ed.), pp. 53–57. Cold Spring Harbor Lab., Cold Spring Harbor, New York.
22. Lavi, S. (1981) *Proc. Natl. Acad. Sci. U.S.A.* **78**, 6144–6148.
23. Levan, G., and Levan, A. (1982) In "Gene Amplification" (R. T. Schimke, ed.), pp. 91–97. Cold Spring Harbor Lab., Cold Spring Harbor, New York.
24. Levan, G., Mandahl, N., Bengtsson, B. O., and Levan, A. (1977) *Hereditas* **86**, 75–90.
24a. Mariani, B. D., and Schimke, R. T. (1983) To be published.
25. Mariani, B. D., Slate, D. L., and Schimke, R. T. (1981) *Proc. Natl. Acad. Sci. U.S.A.* **78**, 4985–4989.
26. Melera, P. W., Lewis, J. A., Biedler, J. L., and Hession, C. (1980) *J. Biol. Chem.* **255**, 7024–7028.
27. Melton, D. W. (1981) *Somatic Cell Genet.* **7**, 331–334.
28. Meyers, M. B., and Biedler, J. L. (1981) *Biochem. Biophys. Res. Commun.* **99**, 228–235.
29. Nunberg, J. N., Kaufman, R. J., Schimke, R. T., Urlaub, G., and Chasin, L. A. (1978) *Proc. Natl. Acad. Sci. U.S.A.* **75**, 5553–5556.
30. Guinn, L. A., Moore, G. E., Morgan, R. T., and Woods, L. K. (1979) *Cancer Res.* **39**, 4914–4924.
31. Reichman, A., Riddel, R. H., Martin, P., and Levin, B. (1980) *Gastroenterology* **79**, 334–339.
32. Schimke, R. T. (1982) In "Gene Amplification" (R. T. Schimke, ed.), pp. 317–333. Cold Spring Harbor Lab., Cold Spring Harbor, New York.
33. Schimke, R. T., Alt, F. W., Kellems, R. E., Kaufman, R., and Bertino, J. R. (1978) *Cold Spring Harbor Symp. Quant. Biol.* **42**, 649–657.
34. Schimke, R. T., Brown, P. C., Kaufman, R. J., McGrogan, M., and Slate, D. L. (1981) *Cold Spring Harbor Symp. Quant. Biol.* **45**, 785–797.
35. Spradling, A. C., and Mahowald, A. P. (1980). *Proc. Natl. Acad. Sci. U.S.A.* **77**, 1096–2002.
36. Sturtevant, A. H. (1925) *Genetics* **10**, 117–147.
36a. Tlsty, T. (1983) To be published.
37. Trent, J. M., Buick, R. N., and Olson, S. (1982) In preparation.
38. Varshavsky, A. (1981) *Cell* **25**, 561–572.
39. Vogelstein, B., Pardoll, D. M., and Coffey, D. S. (1980) *Cell* **22**, 79–85.
40. Wahl, G. M., Padgett, R. A., and Stark, G. R. (1979) *J. Biol. Chem.* **254**, 8679–8689.

Identification, Isolation, and Characterization of Three Distinct Human Transforming Genes

M. WIGLER, M. GOLDFARB, K. SHIMIZU, M. PERUCHO,
Y. SUARD, O. FASANO, E. TAPAROWSKY, AND J. FOGH*

Cold Spring Harbor Laboratory
Cold Spring Harbor, New York
and
*Sloan Kettering Institute,
Rye, New York

DETECTING HUMAN TRANSFORMING GENES

The progression of a cell from normalcy to malignancy may result from the presence of transforming genes of cellular origin. Such transforming genes have been detected in genomic DNAs from certain tumors and cell lines by their ability to induce morphologically altered foci in NIH 3T3 cells, a growth controlled mouse fibroblast line (9). Various investigators report transforming genes in rodent cell lines (24,25) and in human cell lines derived from bladder carcinoma, colon carcinoma, lung carcinoma, breast carcinoma, osteosarcoma, promyelocytic leukemia, lymphoma, and neuroblastoma (10–14, 18,24).

We have surveyed DNA from 21 human tumor cell lines. DNA from five cell lines were capable of reproducibly transforming NIH 3T3 (see Table I and Fig. 1). To distinguish the transforming genes contained in these cells, we employed both restriction endonuclease analysis combined with gene transfer and a technique for visualizing human sequences in mouse cells (14,18). We were thus able to demonstrate that the five human tumor cell lines contained three different transforming genes: one present only in T24, derived from a bladder carcinoma; one present only in SK-N-SH, derived from a neuroblastoma; and one present in each of three lines: SK-CO-1, derived from

TABLE I
Transformation Efficiencies of DNA from
Human Tumor Cell Lines[a]

Cell line	Origin	Foci
T24	Bladder carcinoma	2.0 ± 1.0
Calu-1	Lung carcinoma	0.15 ± 0.1
SK-LU-1	Lung carcinoma	0.08 ± 0.02
SK-CO-1	Colon carcinoma	0.05 ± 0.03
SK-N-SH	Neuroblastoma	0.05 ± 0.03
SK-HEP-1	Liver adenocarcinoma	<0.005
SW-1088	Astrocytoma	<0.005
J82	Bladder carcinoma	<0.005
RT4	Bladder carcinoma	<0.005
C-4-II	Cervix carcinoma	<0.005
RMPI-2650	Nasal septum carcinoma	<0.005
IMR-32	Neuroblastoma	<0.005
575A	Bladder carcinoma	<0.005
SK-BR-3	Breast carcinoma	<0.002
734 B	Breast carcinoma	<0.003
Cama-1	Breast carcinoma	<0.003
HT-29	Colon carcinoma	<0.005
SW-594	Fibrosarcoma	<0.005
T-98	Glioblastoma	<0.005
SK-MES-1	Lung carcinoma	<0.005
Caki-2	Renal carcinoma	<0.005
Controls	Human placenta DNA	<0.008
	NIH 3T3 DNA	<0.0005

[a] Transformation assays were performed as previously described using DNA prepared from the indicated cell line. Induction of foci is the average number of foci per microgram DNA calculated from several experiments. For more details, see Perucho et al. (18).

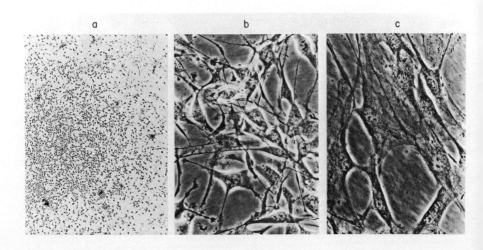

colon carcinoma, SK-LU-1, and Calu-1, each derived from lung carcinomas. In this article, we concern ourselves only with the three genes found in our survey.

ISOLATING TRANSFORMING GENES FROM HUMAN SOURCES

We have isolated portions or the entirety of three human transforming genes. We have employed two strategies—one useful for isolating "small" genes [i.e., less than 10–15 kilobase pairs (kbp)] and one useful for cloning fragments of large genes.

The strategy for cloning small genes depends on linking the desired gene to a marker sequence. We have used as marker the bacterial tRNA suppressor gene *Sup*F isolated by Drs. Rajbhandary and Kudo at Massachusetts Institute of Technology. Linkage to the tRNA suppressor allows the rapid cloning of the transforming sequences into bacteriophage λ cloning vectors that contain amber mutations in genes vital for lytic growth. Details of the methods can be found in Goldfarb *et al.* (7) and Shimizu *et al.* (28). We cloned biologically active transforming genes of T24 and SK-N-SH using this method. A partial characterization of the structures of the T24 gene can be found in Goldfarb *et al.* (7), Santos *et al.* (22), Pulciani *et al.* (19), and Taparowsky *et al.* (32). A characterization of the SK-N-SH gene can be found in Shimizu *et al.* (28,29).

The rescue strategy cannot conveniently be employed on large-sized genes such as the transforming gene found commonly in lung and colon carcinoma-derived cell lines. We suspected this gene to be large because the total length of all the *Eco*R1 fragments containing Alu human repeated sequences reproducibly present in NIH 3T3 cells transformed with this gene were in excess of 30 kbp. However, the presence of these human repeat sequences can be exploited to clone fragments of this gene from mouse cells. This strategy is a modification of the strategy described by Gusella *et al.* (8) and employed by others (19,26). We made a λ phage library from secondary transformants and screened this library with human repeat sequence DNA. One candidate λ phage was isolated this way. Unique sequence DNA from

Fig. 1. Morphology of NIH 3T3 cells transformed by DNA from human tumor cell lines. (A) Low magnification (12.5×) of focus induced by T24 DNA. Transformed focus is in center left, and normal untransformed cells occupy upper and lower right. (B) Higher magnification (200 ×) of transformed cells. (C) Higher magnification (200 ×) of untransformed cells.

the phage insert was then used in turn as a probe for further library screening, and several phages were isolated containing overlapping DNA inserts. A composite structure of part of this gene, spanning 26 kb, is described in Shimizu et al. (29). Although we have cloned most of this gene, parts of the gene remain uncloned.

THE RELATIONSHIP OF HUMAN TRANSFORMING GENES TO KNOWN RETROVIRAL ONCOGENES

Another class of genes capable of transforming cells is found in animal tumor viruses. The oncogenes of the retroviruses (v-*onc*) appear to have arisen by the transduction of cellular genes (c-*onc*) (4,15,21,23, 30,33). The c-*onc* genes, which number in excess of 10, are highly conserved throughout mammals and birds and presumably have normal physiological function when residing unperturbed in their usual chromosomal setting. Several groups have shown that genes detectable by DNA transfer in tumor cell DNAs are cellular homologs of the oncogenes found in Harvey and Kirsten sarcoma viruses, the Harvey-*ras* and Kirsten-*ras* genes, respectively (3,17,22). The viral genes arose by transduction of distinct rat genes and code for immunologically cross-reactive and structurally related 21,000-dalton proteins (5).

Were the human transforming genes we detected related to any of the v-*onc* genes? Because most of the v-*onc* genes have been cloned into prokaryotic vectors, we used the cloned human transforming genes as probes against a panel of filter blotted v-*onc* genes (29). The result indicated that all three human genes were homologous to the H-*ras* and the K-*ras* genes. H-*ras* had strongest homology to the bladder carcinoma T24 transforming gene, whereas K-*ras* had the greatest homology to the common transforming gene of the lung/colon carcinomas. The transforming gene of the neuroblastoma cell line SK-N-SH was only distantly related by nucleotide homology to both *ras* genes. It represents a previously unknown member of the *ras* gene family.

The relationship between these human transforming genes and the H-*ras* and K-*ras* genes can be shown another way. Monoclonal antisera, which are broadly reactive with the 21,000 dalton (p21) product of both Harvey and Kirsten viral *ras*, have been developed by Furth *et al*.(6). Using one such broadly reactive monoclonal antiserum Y13-259, we precipitated [^{35}S]methionine-labeled extracts either from untransformed mouse cells or mouse cells transformed with various human transforming genes. The results clearly indicate that a protein in the weight range of the *ras* p21 is precipitated in NIH 3T3 transformed with DNA from either T24, SK-N-SH, or the lung/colon carcinoma cell

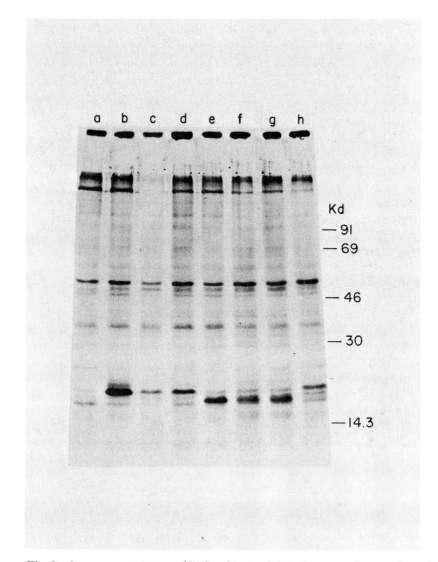

Fig. 2. Immunoprecipitation of [^{35}S]methionine-labeled extracts of untransformed and transformed NIH 3T3 cells. NIH 3T3 cells were incubated for 2 hr in [^{35}S]methionine, and total cellular protein extracts were prepared, incubated with broadly reactive anti-*ras* p21 monoclonal antibody, precipitated with staph A, washed, and run on an SDS polyacrylamide slab gel. (Lane a) NIH 3T3; (b) NIH 3T3 cells containing and expressing the Harvey viral *ras* gene; (c,d) two NIH 3T3 cells independently transformed with Calu-1 DNA; (e,f,g) three NIH 3T3 lines independently transformed with SK-N-SH DNA; (h) NIH 3T3 transformed with SK-CO-1 DNA. A ^{14}C-labeled protein mixture provided molecular weight standards: from top to bottom, 91, 69, 46, 30, and 14.3 kilodaltons.

lines (see, for example, Fig. 2). Comparable amounts of protein of that size range are not seen in immune precipitates of untransformed NIH 3T3 cells.

ACTIVATION OF NORMAL CELLULAR GENES INTO TRANSFORMING GENES

The transforming genes found in tumor cell DNA have counterparts in normal human DNA. In what way do they differ? This question has been resolved for the transforming gene of T24 by comparing the cloned transforming gene (and that of EJ as well) to its counterpart cloned from normal DNA (20,31,32). A combined approach of sequence and genetic analysis indicates that the H-*ras* gene in T24 has a single critical sequence change that results in its coding a *ras* p21 with valine in position 12 instead of glycine. Studies of the expression of the normal and transforming alleles in NIH 3T3 cells confirm this conclusion: The difference in expression of the normal and transforming gene is not the amount of protein made but the potency of the protein made (31,32). The transforming H-*ras* p21 is perhaps 100 to 1000-fold more efficient at inducing the transformed phenotype than an equivalent amount of normal H-*ras* p21. Elevated levels of the normal p21 can result in the induction of the transformed phenotype (1,2) but this type of mechanism does not appear to be involved in the present instance.

CONCLUSION

One striking theme emerging from current work is the presence of common human transforming genes in several tumor cell lines. Weinberg's lab has reported that one transforming gene was common to independently derived mouse fibroblasts transformed by chemical carcinogens (27) and that another gene may be commonly involved in rat neuroblastomas (16). Similarly, Cooper's lab has suggested that a common gene may be involved in breast tumors of mice and humans (11) and yet another in lymphomas (12). Moreover, the three human genes we are studying appear to be related members of the *ras* gene family.

A wide variety of tumor cells contain activated *ras* genes, detectable by gene transfer into NIH 3T3 cells. Several factors probably contribute to this observation: Activated *ras* genes may be readily detected by the NIH 3T3 focus assay; *ras* genes may be easily activated by mutation; and *ras* genes may have critical functions in a wide variety of cell types. Based on the study of the mechanism of activation of the T24 transforming gene, we speculate that alteration of *ras* gene products may be a common step in many forms of cancer.

REFERENCES

1. Chang, E. H., Furth, M. E., Scolnick, E. M., and Lowy, D. R. (1982) *Nature (London)* **297**, 479–483.
2. DeFeo, D., Gonda, M. A., Young, H. A., Chang, E. H., Lowy, D. R., Scolnick, E. M., and Ellis, R. W. (1981) *Proc. Natl. Acad. Sci. U.S.A.* **78**, 3328–3332.
3. Der, C. J., Krontiris, T. G., and Cooper, G. M. (1982) *Proc. Natl. Acad. Sci. U.S.A.* **79**, 3637–2640.
4. Ellis, R. W., DeFeo, D., Margak, J. M., Young, H. A., Shih, T. Y., Ghang, E. H., Lowry, D. R., and Scolnick, E. M. (1980) *J. Virol.* **36**, 408–420.
5. Ellis, R. W., DeFeo, D., Shih, T. Y., Gonda, M. A., Young, H. A., Tsuchida, N., Lowy, D. R., and Scolnick, E. M. (1981) *Nature (London)* **292**, 506–511.
6. Furth, M. E., Davis, L. J., Fleurdelys, B., and Scolnick, E. (1982) *J. Virol.* **43**, 294–304.
7. Goldfarb, M. P., Shimizu, K., Perucho, M., and Wigler, M.H. (1982) *Nature (London)* **296**, 404–409.
8. Gusella, J. F., Keyus, C., Barsayi-Breiner, A., Kao, F. T., Jones, C., Puck, T., and Housman, D. (1980) *Proc. Natl. Acad. Sci. U.S.A.* **77**, 2829–2834.
9. Jainchill, J. S., Aaronson, S. A., and Todaro, G. J. (1969) *J. Virol.* **4**, 549–553.
10. Krontiris, T. G., and Cooper, G. M. (1981) *Proc. Natl. Acad. Sci. U.S.A.* **78**, 1181–1184.
11. Lane, M. A., Sainten, A., and Cooper, G. M. (1981) *Proc. Natl. Acad. Sci. U.S.A.* **78**, 5185–5189.
12. Lane, M. A., Sainten, A., and Cooper, G. M. (1982) *Cell* **28**, 873–880.
13. Marshall, C. J., Hall, A., and Weiss, R. A. (1982) *Nature (London)* **299**, 171–173.
14. Murray, M., Shilo, B., Shih, C., Cowing, D., Hsu, H. W., and Weinberg, R. A. (1981) *Cell* **25**, 355–361.
15. Oskarsson, M., McClements, W. L., Blair, D. G., Maizel, J. V., and Vande Woude, G. F. (1980) *Science* **207**, 1222–1224.
16. Padhy, L. C., Shih, C., Cowing, D., Finkelstein, R., and Weinberg, R. A. (1982) *Cell* **28**, 865–871.
17. Parada, L. F., Tabin, C. J., Shih, C., and Weinberg, R. A. (1982) *Nature (London)* **297**, 474–478.
18. Perucho, M., Goldfarb, M. P., Shimizu, K., Lama, C., Fogh, J., and Wigler, M. H. (1981) *Cell* **27**, 467–476.
19. Pulciani, S., Santos, E., Lauver, A. B., Long, L. K., Robbins, K. C., and Barbacid, M. (1982) *Proc. Natl. Acad. Sci. U.S.A.* **79**, 2845–2849.
20. Reddy, E. P., Reynolds, R. K., Santos, E., and Baracid, M. (1982) *Nature (London)* **300**, 149–152.
21. Rousell, M., Saule, S., Lagrov, C., Rommens, C., Bevy, H., Graf, D., and Stehelin, D. (1979) *Nature (London)* **281**, 452–455.
22. Santos, E., Tronick, S. R., Aaronson, S. A., Pulciani, S., and Barbacid, M. (1982) *Nature (London)* **298**, 343–347.

23. Scherr, C. J., Fedele, L. A., Donner, L., and Turek, L. P. (1979) *J. Virol.* **32**, 860–875.
24. Shih, C., Padhy, L. C., Murray, M., and Weinberg, R. A. (1981) *Nature (London)* **290**, 261–264.
25. Shih, C., Shio, B.-Z., Goldfarb, M. P., Dannenberg, A., and Weinberg, R. A. (1979) *Proc. Natl. Acad. Sci. U.S.A.* **76**, 5714–5718.
26. Shih, C., and Weinberg, R. A. (1982) *Cell* **29**, 161–169.
27. Shilo, B.-Z., and Weinberg, R. A. (1981) *Nature (London)* **289**, 607–609.
28. Shimizu, K., Goldfarb, M., Perucho, M., and Wigler, M. (1983) *Proc. Natl. Acad. Sci. U.S.A.* **80**, 383–387.
29. Shimizu, K., Goldfarb, M., Suard, Y., Perucho, M., Li, Y., Kamata, T., Feramisco, J., Stavnezer, E., Fogh, J., and Wigler, M. (1983) *Proc. Natl. Acad. Sci. U.S.A.* (in press).
30. Spector, D. H., Varmus, H. E., and Bishop, J. M. (1978). *Proc. Natl. Acad. Sci. U.S.A.* **75**, 4102–4106.
31. Tabin, C. J., Bradley, S. M., Bargmann, C. I., Weinberg, R. A., Papageorge, A. G., Scolnick, E. M., Dhar, R., Lowy, D. R., and Chang, E. H. (1982) *Nature (London)* **300**, 143–148.
32. Taparowsky, E., Suard, Y., Fasano, O., Shimizu, K., Goldfarb, M., and Wigler, M. (1982) *Nature (London)* **300**, 762–765.
33. Witte, O. N., Rosenberg, N., and Baltimore, D. (1979) *Nature (London)* **281**, 396–398.

PART VI

TRANSFORMATION-RELATED PROTEINS AND TRANSCRIPTS

Do Variant SV40 Sequences Have a Role in the Maintenance of the Oncogenic Transformed Phenotype?

ROBERT POLLACK
CAROL PRIVES
JAMES MANLEY
Department of Biological Sciences
Columbia University
New York, New York

INTRODUCTION

SV40 is a small virus, but its approximately 5200 base pairs (bp) contains enough information either to productively grow in a permissive cell or to convert a normal cell into a transformed one (1). The conversion is usually genetically stable enough to generate clonable lines of transformed cells. Some of these lines lose so much of their normal growth control that they acquire the capacity to grow as tumors (2). Since the total DNA sequence of SV40 is known (1), it is somewhat surprising that the mechanism of action of SV40 in maintaining the various transformed states *in vitro* is not at all well understood. Also, and probably not coincidentally, precisely which SV40-encoded molecules are necessary for maintenance of any of these states also is an unsettled issue (1).

SV40 has been utilized as a model system both to study eukaryotic gene expression and regulation and to further understand viral-mediated neoplastic transformation. In this respect, the virus is ideal because it is small, coding for only six known polypeptides (VP-1, VP-2, VP-3, t and T antigens, and the agno protein); it has been entirely sequenced (1) and expresses its genes in a manner similar to eukaryotic chromosomal gene expression (Fig. 1). There are viral and cellular parallels for DNA replication, chromatin structure, RNA transcription,

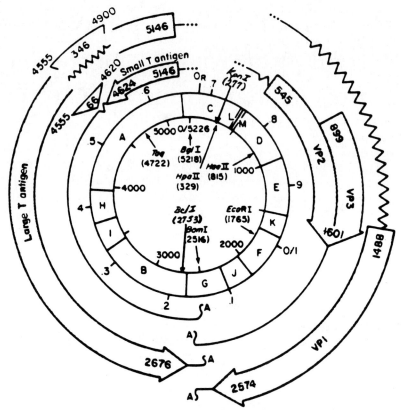

Fig. 1. Schematic representation of SV40 genome indicating locations of some principal features. The genome is represented as a circle with the origin of replication at the top. Outer boxed areas are *Hin*dII and III fragments. Numbers outside the circle refer to locations in fractional genome length relative to the single *Eco*RI change site within SV40 DNA. Residue numbers and labels refer to SV40 sequence numbers (see refs, Appendix A). Sequence numbers for codons include termination codons. Shaded area with arrowheads indicate coding portions of mRNAs, with arrowheads pointing in 5′–3′ direction. Introns are indicated by zig-zag lines. (This figure was taken, with minor modification, from ref. 1.)

modification, splicing, and translation (1). Depending on the host cell, SV40 induces either a lytic or nonlytic response. The lytic response makes more complete use of the viral genetic information. The kidney cells of the African green monkey (and derived cell lines) are the permissive host for the lytic response to SV40. In these cells, the infectious cycle can be divided into early and late phases. In the early phase E strand-specific RNA is expressed, presumably on a single unspliced transcript which is then subjected to two alternate splicing pathways. This generates mRNAs for the large T and small t antigen. After T antigen is synthesized, there is a burst of viral and cellular DNA synthesis, the viral DNA being replicated on a circular minichromosome. The role of the large T antigen in these processes has been demonstrated by genetic and biochemical analyses. The T antigen binds with high affinity to sequences at the viral origin of replication *ori*. The promoter for early strand RNA synthesis is within the *ori* region, and T antigen, in binding to this region, brings about considerable down regulation of the early mRNAs. After viral DNA replication begins, L strand specific transcription initiates, and after the later 19 S and 16 S viral mRNAs are processed, their translation producing the capsid proteins. VP-1 (abundant) and VP-2 + VP-3 (minor) are synthesized. Considerably more L strand than E strand mRNA, and mRNA translation products are synthesized during the late phase. The simplest explanation for this is that E strand mRNAs are autogeneously regulated by T antigen, and L strand mRNAs levels are increased due to an increase in DNA template number brought about by the combination of T antigen function, but this has not yet been proved. Electron microscopic and restriction enzyme analysis have demonstrated that although the SV40 minichromosome resembles cellular chromatin in the classical histone–DNA "bead" structure, there is a nucleosome-free gap in the noncoding regulatory sequences which include the 300 bp *ori* region and sequences extending in the late direction upward until approximately 0.72 μm.

In nonpermissive cells, there is little or no viral DNA replication, presumably due to the lack of the viral DNA replication-specific factors present in AGMK cells. The remainder of the genome, the early region, is expressed in transformed cells.

SV40 VIRAL GENES AND TRANSFORMATION

The SV40 early region encodes two well-characterized proteins, the large (94K) and small (17K) T antigens. Although the large T is found in SV40 transformed cell lines and although DNA encoding the early region of SV40 is necessary for transformation, the actual mechanism(s) by which SV40 T antigens maintained the transformed state remain poorly understood (1). The problem of mechanism is especially thorny because normal fibroblasts display distinct serum- and anchorage-dependent growth controls and because both of these controls can be independently disrupted by SV40 transformation (3).

TABLE I
Assays for Selecting and Characterizing SV40 Transformed Cells

Assay	Change in transformed cell
Physiological	
Serum	Grows in (<1%) serum
Anchorage	Grows in agar as a physical colony
Focus	Grows to high cell density
Nonselective	All possible stable changes
Tumorigenicity	Grows in animal directly as a tumor
Chemical	
Insulin	Grows in defined medium lacking insulin
Calcium	Grows in chelex-serum plus calcium-free medium

In molecular terms, transformation must be the consequence of an interaction between SV40-encoded molecules and molecules of the transformed cell. Further complexity is therefore introduced into the role of SV40 in transformation by the choices of which cell to transform and which selective assay to use in generating transformed clones (Table 1).

Given all these possible variations on the theme of the capability of SV40 to transform and to maintain transformation, it is perhaps inevitable that the virus, although small, makes a great deal of use of each base pair. SV40 encodes a number of mRNAs and proteins (Fig. 1) even when integrated into the genome of the nonpermissive cell. One of these proteins, the large T antigen, is itself a molecule with many functions and domains (Table II). Thus, a major unanswered question at this time is whether the phenotypic events in transformation and lytic expression are linked to different domains of T antigen.

Simple genetics has not been able to complete this task, because only mutations in SV40 sequences needed for virus growth can be detected by classic selection techniques. A set of recent technical developments has made it possible now to map SV40 domains in a new and complete way. The construction of SV40 mutants using restriction enzymes and recombinant technology has allowed the analysis of SV40s which are totally defective for growth (e.g., origin minus) but which can transform cells quite well at least in certain assays. At the same time, the major transforming gene product of WT SV40, the large T antigen, has been purified and has been found to display many different activities in cell-free preparations. Many monoclonal antibodies to

TABLE II
Properties of Partially and Highly Purified
SV40 T Antigens

Activities
   Binds to double- and single-stranded DNA
   Binds to chromatin
   Binds specifically to sites at the viral origin of replication
   ATPase Activity [stimulated by poly(dT)]
   Specific SV40 DNA Unwinding Activity
   Binds to and stabilizes the 54K host protein
   Exists as 5–7 S, 18 S and 22 S oligomers
   Stimulates cell DNA synthesis (after microinjection)
   Induces host tumor rejection response
Posttranslational modifications
   N-terminal acetylated
   Phosphorylated (*ser* and *thr*; not *tyr*)
   ADP ribosylated

WT T antigen have been prepared; these recognize different specific regions of the T antigen protein (4).

SV40 sequences integrated in transformed cells have been mapped, cloned, and sequenced (5). These rescued sequences have also been used to transform cells, thus constructing a novel genetic cycle from eukaryotic host to plasmid vector and back. If specific regions of SV40 T carry out specific functions, then this cycle is a powerful way to apply the knowledge acquired from analysis of a new functional domain.

SV40 T antigen regulates both SV40 DNA replication and SV40 transcription. As SV40 DNA has become so malleable, its replication and transcription also have therefore become subject to experimental manipulation, for example, by directed methylation of specific restriction-enzyme sites or by deletion of sequences required for transcription (6).

INITIAL HYPOTHESIS

The SV40 DNA sequence has a second open coding frame in the distal portion of the early transcript (7). If permitted, a splice into this frame would generate a new sort of protein, with a very hydrophobic C-terminus. This postulated hydrophobic T might be related to the 100K T we have observed in SV101 and its anchorage-independent

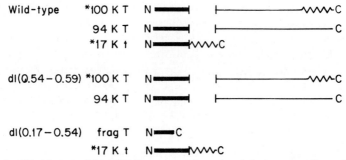

Fig. 2. Working model of SV40 gene-products in anchorage-transformation. Hypothesis: (1) To maintain anchorage independence, a cell must contain at least one molecule of the class N━━━┫—hydrophobic—C, in order to bring the N-terminus common domain to its functional site. (2) To first efficiently initiate and then to maintain anchorage independence, a cell needs both a molecule of the class N━━━┫—hydrophobic——C and a wild-type 94K T.

* A molecule of the class: N━━━┫—hydrophobic—C.

derivatives (8). How might such a T antigen be generated? So far, the search in wt-infected lytic cells for an mRNA with this splice has not been productive. However, although mutant viruses and virus sequences in transformed cells have not been examined, there is no need to necessarily impose a second splice on the transformed cells in order to generate a 100K T. Alternative mutational methods exist. These form the basis of our hypothesis. We hypothesize first that the 100K T antigen in SV101 is the product of a mutated SV40 within SV101 and that this mutation is a deletion (as small as a base pair) that brings the mRNA into the second reading frame for the C-terminal, resulting in hydrophobic C-terminus for large T (Fig. 2).

At first sight, a difficulty with this hypothesis appears to arise from the fact that extensive C-terminal deletion mutants can transform in the anchorage assay. However, small t connects also the common N-terminus to a hydrophobic C-terminus. We propose that both 17K and 100K have the same function in transformation. Recently, the laboratories of K. Danna (personal communication) and T. Shenck (Cambridge T.V. meeting, Abstract 1981) have used transfection of PBR-cloned short SV40 sequences to show that the stretch of SV40 DNA from the origin to 0.54 is sufficient to transform in the stringent anchorage assay, albeit with a lower limitation frequency (~1%) compared to one full wild-type sequence. When proteins in such transformed cells are examined by immunoprecipitation, small t and a fragment of large T are both detected, but the C-terminal of large T has been deleted.

A second set of data that must be included in any hypothesis of anchorage transformation comes from deletions in 0.54–0.59 map units. These remove the coding sequence for the C-terminal of small t. In some deletions a splice function as well is removed, eliminating even N-terminal t fragments. Anchorage-transformed mouse lines generated by the 54/59 mutant dl 884 have 100K T and 94K T (8).

We may now ask, what parts of the SV40 genome, if any, do all three kinds of anchorage-transformed cells share? Providing our initial hypothesis about the 100K T is correct, then the answer is: *All anchorage transformed lines have the N-terminal common domain of the early region connected to a hydrophobic C-terminal domain* (Fig. 2). For WT transformants, two molecular forms, 100K T and 17K T, meet this requirement. For dl 884 transformants, 100K T meets it, and for transformants made from infection with extensive C-terminal-T deletion mutants, the 17K small t meets it. If this hypothesis is correct, we may then say:

1. The 94K T may not be the molecular species responsible for anchorage transformation, because it lacks the hydrophobic C-terminus and because we have seen no mouse anchorage transformants without such a link of N-terminus to hydrophobic C-terminus.

2. The 94K T is necessary for *efficient initiation* of anchorage transformation. We hypothesize that this is so because it enables SV40 to replicate during initiation. Mutation to second frame in one copy of a tandem array would not cause total loss of 94K T function.

3. In species other than mouse, the hydrophobic C-terminus might be unnecessary due to different and more efficient transforming interactions of these host molecules (such as 54K) with their 94K T. This would explain why WT transformed rat cells, for instance, both seem to lack 100K T and to be very efficiently transformed by dl 54/59 mutants (1).

If our frameshift theory is correct, we would expect the 100K protein to have a new, hydrophobic C-terminal domain in addition to the amino acid sequence present in the 94K T antigen. If this is what we find, we can ask: Does such a domain cause the protein to localize in the plasma membrane, where it might operate like transforming proteins from RNA oncoviruses? We can test this hypothesis in several ways. If there is a convenient enzymatic site, the domain of interest could be cleaved off the protein and isolated on a polyacrylamide-SDS gel. Antisera, monoclonal or conventional, could be raised against this domain alone. The 100K protein could then be traced in SV101 and its sublines by immunofluorescent second antibodies. Alternatively, the protein itself—or the new domain alone—could be microinjected into the nuclei of mouse embryo fibroblasts and followed by anti-T sera. Or the protein could be traced, with the same anti-sera and fluorescent second antibodies, in mouse embryo fibroblasts transformed with 100K T coding sequences.

Such a new domain could also be studied using recombinant DNA techniques. For instance, most of the sequences between the promoter and the new domain could be deleted from the pBR322 subclone, and the new construct used in an attempt to produce copies of the hydrophobic peptide. Such a construct, lacking an intron, might or might not be expressed in mouse cells. To test if the membrane localization is due to the new domain, another well-characterized, intron-containing coding sequence (such as the gene for β-globin) could be spliced in to replace the N-terminal T antigen sequence. It could then be observed if the new hybrid molecule accumulated in the plasma membrane.

Since the SV101 sublines which no longer grow in agar still produce 94K protein—though not 100K protein—it might be argued that a gene dosage—or domain dosage—effect is at work (8). It is possible that we will find, not a frameshift mutation producing a new domain, but rather a duplication of a domain already present in large T antigen. If so, we could use binding and protection studies to see if that domain binds either DNA or 54K nonviral T antigen with high affinity.

Cellular studies involving the function of the 100K protein as a whole would involve either microinjection or calcium-phosphate mediated transfer of cloned sequences into mouse embryo fibroblasts. It could then be determined if 100K protein without 94 or 17K protein can transform mouse cells in a quantitative manner. It would be especially intriguing to subculture lines transformed by injected 100K coding sequences to see if viral DNA rearrangements occur with high frequency.

REFERENCES

1. Tooze, J. (1981) "Molecular Biology of Tumor Viruses," 2nd ed., Chapters 1–4 and Appendix. Cold Spring Harbor Laboratory, Cold Spring Harbor, New York.
2. Freedman, V., and Shin, S. (1974) *Cell* **3**, 355.
3. Risser, R., and Pollack, R. (1974) *Virology* **59**, 477.
4. Scheller, A., Covi, L., Barnet, B., and Prives, C. (1982) *Cell* **29**, 375.

5. Botchan, M. et al. (1980) *Cell* **20,** 143; Gluzman, Y. (1981) *ibid.* **23,** 175.
6. Fradin, A., Manley, J., and Prives, C. (1982) *P.N.A.S.* **79,** 5142.
7. Mark, D., and Berg, P. (1979) *Cold Spring Harbor Symp. Quant. Biol.* **44,** 55.
8. Chen, S., Verderame, M., Lo, A., and Pollacle, R. (1981) *Mol. Cell Biol.* **1,** 994.

Protein Kinases Encoded by Avian Sarcoma Viruses and Some Related Normal Cellular Proteins

R. L. ERIKSON,* TONA M. GILMER,* ELEANOR ERIKSON,*
AND J. G. FOULKES†

*Department of Pathology
University of Colorado School of Medicine
Denver, Colorado
and
†Massachusetts Institute of Technology
Center for Cancer Research
Cambridge, Massachusetts

INTRODUCTION

In this paper we will review some of the progress that has been made over the past several years in the understanding of viral transformation of cells and particularly transformation caused by RNA tumor viruses. Individual RNA tumor viruses have been isolated that cause specific types of neoplastic disease, specific types of leukemia, and/or fibrosarcomas. The first RNA tumor virus to be isolated and identified as such was the Rous sarcoma virus (RSV), which causes transformation of cells in culture as well as fibrosarcomas when injected into suitable host animals (for reviews, see 1 and 2). The genome of RSV has been intensely studied for the past 10–12 years, and the organization of the genes along the RNA of the RSV genome is shown below. The transforming gene is denoted *src*, for sarcoma, and indicates that the virus itself carries a gene responsible for malignant transformation. This gene was identified genetically about 12 years ago through the isolation of temperature-sensitive mutants that were conditional for cell transformation in culture. These mutants were able to transform cells at a temperature of 35°C, the permissive temperature, but were unable to transform cells morphologically at 41°C, the nonpermissive temperature. The existence of these temperature-sensitive mutants implied that the product of the viral transforming gene, in the case of RSV, was a protein (3). In addition to temperature-sensitive mutants, nonconditional mutants were isolated that had deletions of the *src* gene. These mutants are unable to transform cells in culture or to cause fibrosarcomas under most conditions.

About 4 years ago, the product of the *src* gene was identified as a phosphoprotein of $M_r = 60,000$; this protein was denoted pp60^{src} (4). Very soon after the identification of the protein, we and others ascribed to this molecule the function of protein phosphorylation (5,6). However, there was considerable concern, on our part, that the enzymatic activity observed could potentially be due to contamination of preparations of the viral *src* gene product by one of the many protein kinases encoded by the cells used for infection and transformation by RSV, the starting material for our preparations of pp60^{src}. Thus, a great deal of our effort over the past several years has been devoted to attempting to determine unambiguously the source of the enzymatic activity observed, because in order to understand cell transformation by RNA tumor viruses, it is important to define the exact nature of the functions that are viral as opposed to cellular. The RSV genome and the expression of the *src* gene is illustrated below.

Rous sarcoma virus genome

$5'—gag—pol—env—src—3'$ 39 S RNA

$5'—src—3'$ 21 S mRNA
↓
$$\begin{array}{cc} P & P \\ | & | \\ N—ser—tyr—C & pp60^{src} \end{array}$$

RESULTS AND DISCUSSION

EXPRESSION OF THE RSV *src* GENE IN *E. coli*

One approach to determining whether or not pp60^{src} was indeed a protein kinase was to obtain expression of the RSV *src* gene in a prokaryote, such as *Escherichia coli*. These organisms are believed to produce few, if any, protein kinases that carry out reversible protein phosphorylation. Thus if the product of the *src* gene expressed in *E. coli* exhibited phosphotransferase activity, it would be strong evidence that the enzymatic activity observed was actually encoded by the *src* gene. The construction of plasmids for the generation of p60^{src} in *E. coli* is shown in Fig. 1 (7). In this case the RSV *src* gene was

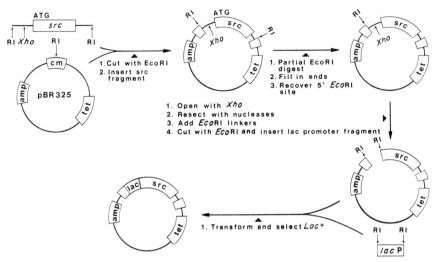

Fig. 1. Plasmid construction. The EcoRI fragment containing the src gene was isolated from a plasmid containing the entire Prague A-RSV genome and inserted into the EcoRI site of pBR325. The recombinant plasmid was subjected to partial EcoRI digestion, treatment with DNA polymerase I to fill in the EcoRI ends, and ligation with T4 DNA ligase to recircularize the molecule. After transformation of HB101, a plasmid that retained only the 5′ EcoRI site was recovered. This plasmid was digested with XhoI and resected with ExoIII and S1 nuclease. EcoRI linkers were added and the mixture was digested with EcoRI. The EcoRI fragment containing the UV5 lac promoter and the first seven amino acids of β-galactosidase was inserted by ligation. HB101 was transformed with the ligation mixture and plated on 5-chloro-4-bromo-3-indolyl-β-D-galactoside plates. Colonies containing the lac promoter fragment were identified by their blue color. From ref. (7).

placed under the control of the lac promoter-operator in the hope of obtaining efficient transcription and, presumably, translation. Bacteria that expressed $p60^{src}$ were detected by immunoprecipitation of ^{35}S-labeled proteins and those E. coli which were expressing $p60^{src}$ were used for purification of the src gene product. The purification of $p60^{src}$ from an expressor culture of E. coli and a similar preparation from a control culture carrying only the lac promoter-operator but not the src gene is shown in Fig. 2. When these preparations were tested for protein kinase activity as shown in Fig. 3, we found that, indeed, the protein produced in E. coli did display protein kinase activity. It had the capacity to phosphorylate casein, α- and β-tubulin, and anti-$pp60^{src}$ IgG. Thus, it had many characteristics identical to the protein, which is produced in eukaryotic host cells, that we had characterized previously. Furthermore, in each of the cases shown in this figure, the

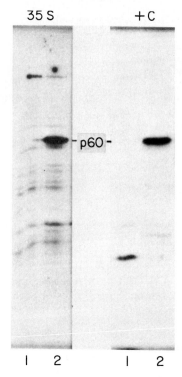

Fig. 2. Polyacrylamide gel electrophoretic analysis of bacterial extracts after immunoaffinity chromatography. *Escherichia coli* expressing p60 was detected by immunoprecipitation of ^{35}S-labeled extracts from cells containing the *lac* UV5 promoter-operator, and one of these cultures was used for the studies described here. *E. coli* carrying the *lac* UV5 promoter in pBR325 in the same orientation as in the *src*-containing clone served as a control for the expression of p60. Lysates were prepared and subjected to immunoaffinity chromatography. ^{35}S-labeled proteins were visualized by fluorography and ^{32}P-labeled proteins by autoradiography with the aid of Dupont Lightning Plus intensifying screens. Left panel: Proteins purified by immunoaffinity chromatography of lysates prepared from bacteria labeled in culture with $^{35}SO_4^{2-}$. Track 1, *lac* pBR325; 2, *lac-src* pBR325. Right panel: The preparations depicted in the left panel were incubated with CAT in the presence of 5 mM MgCl$_2$ and 1 μM [γ-^{32}P]ATP before analysis. Track 1, *lac* pBR325; 2, *lac-src* pBR325. The catalytic subunit of cyclic AMP-dependent protein kinase (CAT) was a gift of James L. Maller. From Gilmer and Erikson, see ref. (8), with permission from the publisher.

amino acid phosphorylated in the protein substrates was tyrosine (8–10). Thus, a number of characteristics suggest that the previously observed phosphotransferase activity closely associated with pp60^{src} isolated from eukaryotic host cells was, in fact, an intrinsic property of the molecule itself. These data, taken together with those previously published by our laboratory and others, lead to the near certain conclusion that pp60^{src} encodes a protein kinase.

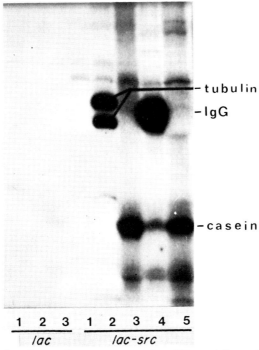

Fig. 3. Phosphorylation of various proteins by p60^{src} partially purified from E. coli. 10μl (10 ng) of the preparations described in Fig. 2 were incubated at room temperature for 10 min with the indicated additions in a total volume of 25 μl. MgCl$_2$ and [γ-^{32}P]ATP (1000–6000 Ci/mmol) were added to a final concentration of 5 mM and 1 μM, respectively, and incubation was continued for 30 min at room temperature. Proteins were then resolved by polyacrylamide gel electrophoresis and autoradiography. Left side (lac): Track 1, lac pBR325 alone; 2, plus α- and β-tubulin; 3, plus casein. Right side (lac-src): Track 1, lac-src pBR325 alone; 2, plus α- and β-tubulin; 3, plus casein; 4, plus casein and anti-pp60^{src} IgG; 5, plus casein and nonimmune IgG. Casein was present at 1 mg/ml, the IgGs at 350 μg/ml and tubulin at 100 μg/ml.

A number of other avian sarcoma viruses (ASVs) have been now identified and characterized in the laboratories of Professor Hanafusa at the Rockefeller Institute and Professor Peter Vogt at the University of Southern California. To date, there are at least three other classes of ASV which have transforming genes distinct from that of RSV and which encode distinct transforming gene products. Although they are antigenically distinct from pp60^{src}, these transforming gene products apparently have a functional similarity to pp60^{src} in that they are able to phosphorylate protein substrates at tyrosine residues. As illustrated in Table I, all four classes of ASVs have a protein kinase activity associated with them that when present in an immune complex with anti-

TABLE I
Transforming Gene Products of Avian Sarcoma Viruses

Class	Avian sarcoma virus	Transforming gene product	Immunoprecipitated by		Phosphorylates TBR IgG on tyrosine
I	Rous	pp60^{src}	TBR	+	+
			anti-gag	−	
II	Fujinami	P140$^{gag-fps}$	TBR	−	+
	PRCII	P105$^{gag-fps}$	anti-gag	+	
III	Y73	P90$^{gag-yes}$	TBR	−	+
	Esh	P80$^{gag-yes}$	anti-gag	+	
IV	UR2	P68$^{gag-ros}$	TBR	−	+
			anti-gag	+	

[a] See references 11–19.

src IgG from a RSV tumor-bearing rabbit results in the efficient phosphorylation of the heavy chain of IgG. Thus, anti-pp60^{src} IgG is a good substrate for the protein kinases encoded by other classes of ASV. There is no clear-cut explanation as to why this IgG is such a good substrate, but the result does demonstrate that it is likely that all classes of ASVs studied to date encode a protein kinase specific for tyrosine residues. In addition, the transforming proteins themselves become phosphorylated in the reaction and tyrosine appears to be the sole phosphorylated residue.

AUTOPHOSPHORYLATION OF pp60^{src}

As shown previously (10), pp60^{src} purified from eukaryotic sources is able to undergo self-phosphorylation or autophosphorylation when [γ-^{32}P]ATP is added to the purified protein. In addition, in the immune complex reaction, all the transformation-specific proteins indicated in Table I also undergo self-phosphorylation on tyrosine residues. However, there is no evidence to date that p60^{src} produced in *E. coli*, the prokaryotic host carrying the recombinant *src*-containing plasmid, is able to self-phosphorylate. Since we believe that the phosphorylation of pp60^{src} itself on a tyrosine residue is likely to be important in the regulation of the enzymatic activity of the protein, we have attempted to elucidate the source of the enzyme responsible for the phosphorylation of the tyrosine residue of pp60^{src}.

As shown in Fig. 4, the autophosphorylation reaction of enzyme purified from temperature-sensitive transformation mutant-infected cells is more thermolabile than that from wild-type infected cells.

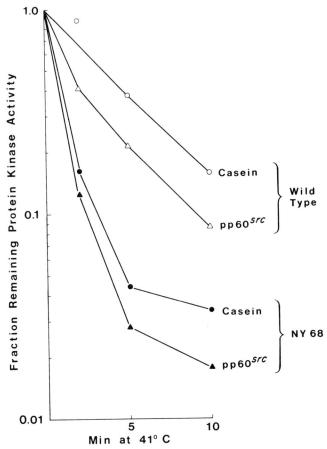

Fig. 4. Thermolability of pp60^{src} protein kinase activity. Chicken embryo fibroblasts were transformed with either wild-type Schmidt-Ruppin strain of RSV or the *ts* mutant of this strain, NY68. The cells were maintained at the permissive temperature, 35°C. Immunoaffinity chromatography was used to prepare pp60^{src} from these cells. These preparations were heat-inactivated at 41°C for the indicated time and then phosphotransferase reactions were carried out in the presence of 5 mM Mg^{-2} and 1 μM [γ-^{32}P]ATP (350 Ci/mmol) in the absence of any exogenously added substrates or in the presence of casein (1 mg/ml). After polyacrylamide gel electrophoresis, the phosphorylated proteins were localized by autoradiography, the bands were excised, radioactivity was determined, and percentage activity was calculated from the amount of radioactivity in the bands. From R. L. Erikson and A. F. Purchio (1982) *Adv. Viral Oncol.* 1, 43–57.

These results could be interpreted to mean that pp60^{src} is responsible for its own phosphorylation. Unfortunately, an alternative explanation is that the molecule, when denatured, is no longer a good substrate for a putative pp60^{src}-specific kinase. One way to resolve this issue would be to isolate a kinase that is specific for phosphorylating the correct tyrosine residue in pp60^{src}. However, to date, although we have identified other tyrosine-specific protein kinases in eukaryotic host cells infected by RSV, we have been unable to show with certainty that any of these protein kinases are able to phosphorylate pp60^{src} and are, in fact, the pp60^{src}-specific kinase. Thus, this issue is still unresolved, although it would appear, from the study of the eukaryotic enzyme, that, most likely, pp60^{src} is capable of self-phosphorylation.

PRODUCTION OF ANTIBODY AGAINST pp60^{src} USING ANTIGEN PRODUCED IN PROKARYOTES

Escherichia coli expressing p60^{src}, under the control of the *lac* promoter-operator, produce a substantial amount of protein, thus making

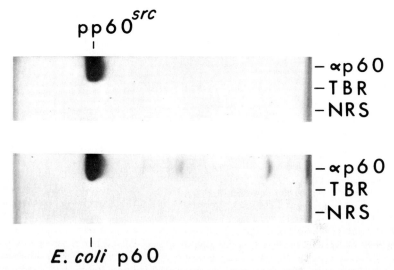

Fig. 5. Immunoprecipitation of denatured p60 from eukaryotic and prokaryotic cells. A preparation of pp60^{src} from eukaryotic cells was phosphorylated by the addition of [γ-^{32}P]ATP as described in the legend of Fig. 4, and p60^{src} from *E. coli* was phosphorylated as described in the legend to Fig. 2. After SDS-polyacrylamide gel electrophoresis the p60 bands were visualized by autoradiography and were eluted from the gel slices. The samples were then immunoprecipitated and analyzed by SDS-polyacrylamide gel electrophoresis as indicated.

it relatively easy to purify large amounts of antigen in a denatured form for subsequent injection and production of antibody. This exercise was carried out and adult rabbits, injected with denatured p60^{src}, produced antibody that was able to recognize not only native pp60^{src} but also denatured pp60^{src}, unlike tumor-bearing rabbit serum that had been available to us previously. This result is shown in Fig. 5. Normal rabbit serum and tumor-bearing rabbit serum are unable to precipitate denatured pp60^{src}, whereas the serum taken from rabbits injected with the denatured *E. coli* protein readily immunoprecipitates denatured pp60^{src}.

NORMAL CELLULAR HOMOLOGUES OF VIRAL TRANSFORMING GENE PRODUCTS

Some rabbits bearing RSV-induced tumors produce an antibody that recognizes, as shown in Fig. 6, a phosphoprotein of $M_r = 60,000$ from normal uninfected cells. This protein, which has been denoted pp60^{c-src}, is presumed to be the product of the nucleic acid sequences homologous to the RSV *src* gene that are present in normal uninfected cells (20), although no direct firm genetic evidence is yet available regarding this question. This molecule, the pp60 homologue of viral *src* in normal uninfected cells, has also been purified in our laboratory and has been shown to have, at least qualitatively, all of the properties of the viral transforming gene product. The only major difference is quantative. We find that virus-infected cells produce 40–50-fold more pp60^{src} than would normally be expressed in these cells, if only the cellular gene were functioning. Thus, one might imagine that transformation occurs because of a dosage phenomenon, because of the increased level of a protein which is very much like a normal cell protein. The *c-src* protein is very highly conserved, being very similar in both avian and mammalian cells, and, because of this high degree of conservation, it is presumed to play some crucial or essential role in normal cell metabolism (20). Recently, a normal cellular protein has been identified that is related to the Fujinami sarcoma virus-transforming protein (21).

A summary of our understanding of the structure of pp60^{c-src} and pp60^{v-src} is shown below. Both molecules have a serine residue that is phosphorylated by a cyclic AMP-dependent protein kinase present in all host cells of RSV. We presume, but do not know for sure, that the phosphorylation of this serine residue has important implications concerning the function of pp60^{src}. To date, however, no strong quantita-

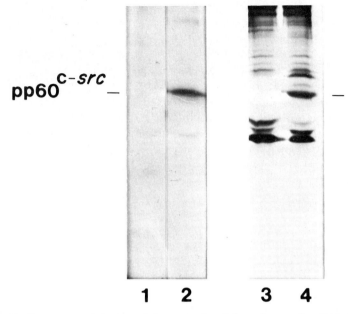

Fig. 6. Immunoprecipitation of the normal cell homologue of pp60^{src}. Chicken embryo fibroblasts were labeled with either [^{32}P] (Lanes 1 and 2) or [^{35}S]methionine (Lanes 3 and 4). After immunoprecipitation of extracts with either normal serum (Lanes 1 and 3) or cross-reacting TBR serum (Lanes 2 and 4), the samples were electrophoresed in SDS-polyacrylamide gels and autoradiographed.

tive data are available to suggest whether phosphorylation at the serine residue increases or decreases the enzymatic activity of the molecule. Furthermore, both molecules contain a phosphorylated tyrosine residue located somewhere in the carboxy terminus.

$$N \text{———} \overset{\overset{P}{|}}{\text{Ser}} \text{———} \overset{\overset{P}{|}}{\text{Tyr}} \text{———COOH}$$
$$\underset{\text{cAMP-dependent}}{\uparrow} \quad \underset{\text{cAMP-independent}}{\uparrow}$$

EPIDERMAL GROWTH FACTOR (EGF)-STIMULATED PHOSPHORYLATION AND COMPARISON TO ASV-INDUCED PHOSPHORYLATION

One of the proteins that we and others have studied that appears to be directly phosphorylated by the activity of pp60^{src} in the infected

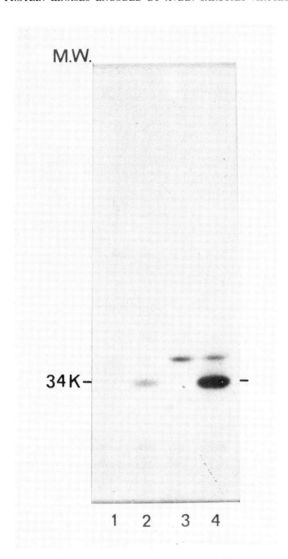

Fig. 7. Stimulation of phosphorylation of the $M_r = 34,000$ protein in EGF-treated cells. The medium was removed from cultures of A-431 cells that had just reached confluency. [^{32}P]Orthophosphate-containing medium was added, and after 30 min EGF was added to one culture to a final concentration of 400 ng/ml. Two hours later the cells were harvested, the lysates were passed through DEAE-Sephacel, and samples of the flow-through fractions were immunoprecipitated with preimmune or anti-34K serum. The immunoprecipitated proteins were resolved by SDS-polyacrylamide gel electrophoresis and visualized by autoradiography. Lanes 1 and 2, cells radiolabeled in the absence of EGF; lanes 3 and 4, cells radiolabeled in the presence of EGF; lanes 1 and 3, preimmune serum; lanes 2 and 4, anti-34K serum. From ref. (25).

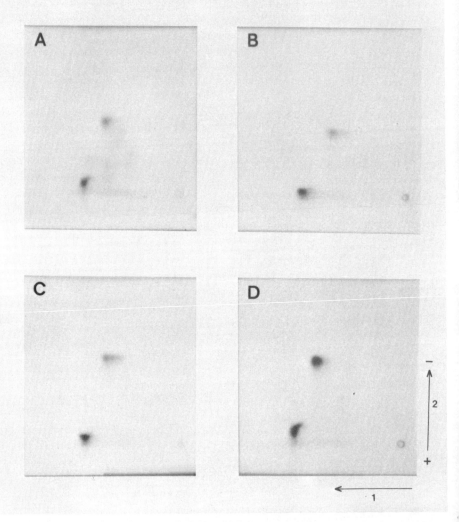

Fig. 8. Two-dimensional analysis of tryptic phosphopeptides from the $M_r = 34,000$ protein radiolabeled in culture in EGF-stimulated cells or phosphorylated *in vitro* by A-431 pp60^{c-src} or by RSV pp60^{src}. Radio-labeled $M_r = 34,000$ protein was isolated from EGF-treated A-431 cells by immunoprecipitation and SDS-polyacrylamide gel electrophoresis as shown in Fig. 7, lane 4. In addition, the $M_r = 34,000$ protein purified from A-431 cells was radiolabeled *in vitro* by A-431 pp60^{c-src} or by RSV pp60^{src} and then resolved by SDS-polyacrylamide gel electrophoresis. Tryptic digests were prepared and fractionated by ascending chromatography and electrophoresis at pH 3.5. (A) $M_r = 34,000$ protein radiolabeled in culture in EGF-stimulated cells; (B and C) $M_r = 34,000$ protein phosphorylated *in vitro* by A-431 pp60^{c-src} or by RSV pp60^{src}, respectively; (D) equal counts of (A) and (B) were mixed and fractionated together.

cell or *in vitro* is a molecule of $M_r = 34,000$ (22,23). It is a fairly abundant protein in cultured fibroblasts. Because of the tyrosine-specific nature of EGF-stimulated phosphorylation (24) and the tyrosine-specific phosphorylation of proteins by ASV transforming gene products, we attempted to determine the specificity of these two protein kinase activities (25). As shown in Fig. 7, when EGF is added to growing cells, the 34,000-dalton protein shows increased phosphorylation. Further, we wanted to know whether the sites of phosphorylation, in the case of EGF stimulation or ASV transformation were similar or different. The 34,000-dalton protein was purified in the unphosphorylated form from the same cells that were used for the studies shown in Fig. 7. Then purified $pp60^{v-src}$ or $pp60^{c-src}$ were used to phosphorylate the protein *in vitro* using [γ-^{32}P]ATP. Phosphopeptide maps were prepared and compared to the phosphopeptide maps of the protein that had been phosphorylated after the addition of EGF to growing cells. As shown in Fig. 8, we found two major phosphopeptides after tryptic digestion, and, indeed, in all three cases these peptides were the same. Thus, although this experiment does not identify the enzyme responsible for phosphorylation of the 34,000-dalton protein in EGF-stimulated cells, it does show that the kinase responsible for this phosphorylation has the same specificity as $pp60^{src}$. This illustrates the very similar nature of the two types of phosphorylation and suggests that the 34,000-dalton protein is perhaps a common substrate for the different types of kinases under study.

Clearly, in the identification of substrates for any of the growth-regulated phosphorylations that we are discussing here, one must show some functional change associated with the phosphorylation observed. In order to understand the functional significance of the phosphorylation, one must, of course, first understand the function of the particular substrate under study. To date, we have no clear understanding of the function of the 34,000-dalton substrate described in the previous two figures, and without an understanding of that function, it is difficult to assign any significance to its phosphorylation. Our best evidence is that in fibroblasts the 34,000-dalton protein is a strong RNA-binding protein, seemingly associated with messenger RNA in the cytoplasm of growing fibroblasts. Another way of assessing the significance of a substrate for the protein kinase activities observed is to determine the distribution of the particular protein in various tissues of the host that would normally be infected by a virus such as RSV. When we looked at the distribution of the 34,000-dalton protein in a variety of tissues of the adult bird, we found that there is a vast difference in the expression of the protein. For instance, in red blood cells, as shown in Fig. 9, it is undetectable, and there is only a low level of

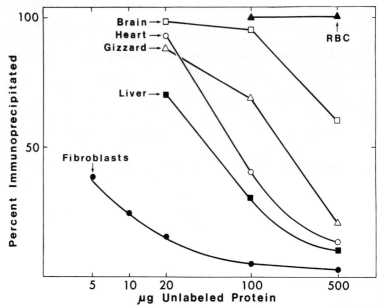

Fig. 9. Distribution of the M_r = 34,000 protein in various tissues. The indicated organs were removed from a 6-week-old chicken. Lysates were prepared from these tissues and from cultured chicken embryo fibroblasts, and samples containing from 5–500 μg of soluble protein were incubated for 1 hr with 1 μl of anti-34K serum. Then aliquots of a lysate prepared from [^{35}S]methionine-labeled chicken embryo fibroblasts were added and the immunoprecipitated proteins were analyzed by polyacrylamide gel electrophoresis. The 34K bands were excised, and the radioactivity determined by liquid scintillation spectrometry. The 34K content of the various lysates is reflected by the ability of that lysate to block the immunoprecipitation of the radiolabeled protein.

this protein in brain cells. On the other hand, there are intermediate levels of the protein in other tissues that were examined. This fact, of course, raises an obvious point that should be made when considering the significance of transforming proteins and their substrates, which is, that in order for viruses to be able to carry out cell transformation, one must not only have expression of the viral transforming gene product but must also have the presence of a suitable substrate for the activity of that product in the susceptible host cell.

The phosphorylation state of a particular protein will be a balance of the protein kinase activities and the protein phosphatase activities present in a particular cell (for review, see 26). The phosphorylation state, and presumably the functional behavior, of a particular protein will be greatly influenced by these two levels, and, in order to understand the circuits involved in phosphorylation–dephosphorylation of proteins on tyrosine residues, one must be concerned not only about the specific kinases responsible for the phosphorylation of these proteins but also about the phosphoprotein phosphatases that are involved in their dephosphorylation.

Figure 10 illustrates results recently obtained by Gordon Foulkes, working in Denver, who studied the distribution of tyrosine-specific phosphoprotein phosphatases taken from chicken brain and fractionated on DEAE-cellulose. These results indicate that the phosphotyrosine-specific phosphatases are probably unique and quite different from the phosphoserine-specific phosphatases previously described.

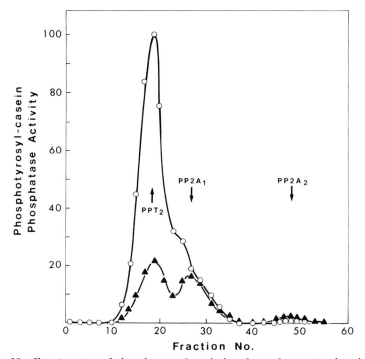

Fig. 10. Fractionation of phosphotyrosyl– and phosphoseryl–protein phosphatases on DEAE-cellulose. An extract prepared from chicken brain was loaded onto DEAE-cellulose and eluted with a linear gradient of 50–450 mM NaCl in 10 mM Tris, pH 7.0, 0.1 mM EDTA, 30 mM β-mercaptoethanol. Fractions were assayed for phosphatase activity using [^{32}P]phosphotyrosylcasein as substrate. Assays were carried out in the presence of either 1 mM EDTA (○) or 1 mM Mn^{2+} (▲). The positions of elution of the phosphoseryl–protein phosphatases $PP2A_1$ and $PP2A_2$ are indicated. From J. G. Foulkes, E. Erikson, and R. L. Erikson (1983) *J. Biol. Chem.* **258**, 431–438.

SEARCH FOR OTHER SUBSTRATES

The 34,000-dalton protein was identified by work in Denver, as well as by others, as a hyperphosphorylated normal cellular protein in RSV-transformed cells. However, when one examines phosphorylation patterns in transformed versus untransformed cells, many quantitative changes are observed, and most of these occur not at tyrosine residues but rather at serine residues. One dramatic example of that is the work of Decker (27), which shows that the ribosomal protein S 6 is heavily phosphorylated on serine residues in RSV-transformed cells. Even under conditions of serum starvation the phosphorylation of S 6 seems to be under the control of the *src* gene product. In order to further assess the significance of this phenomenon, in collaboration with Jim Maller in the Department of Pharmacology at this institution, we undertook the microinjection of partially purified pp60^{src} into *Xenopus* oocytes. As an assay for the activity of pp60^{src} in oocytes, we chose the phosphorylation of the ribosomal protein S 6. Our preliminary results show that 12 hr after the microinjection of either pp60^{src}, pp60^{src} that had been inactivated by treatment at 60°C for 2 min, or progesterone, a hormone known to stimulate S 6 phosphorylation, unheated pp60^{src} is also able to stimulate the phosphorylation of S 6. This result suggests other mechanisms for amplifying the activity of a viral transforming gene product by its capacity to phosphorylate other proteins. One might imagine in this case that pp60^{src} activates a protein kinase specific for serine and that the increased activity of this serine-specific protein kinases results in phosphorylation of S 6 which, in turn, leads to an increased efficiency of protein synthesis. Alternatively, pp60^{src} could phosphorylate a phosphoprotein phosphatase specific for S 6 dephosphorylation, inactivating it so that the protein kinases that are normally present now are able to cause an increased level of phosphorylation of S 6. Either of these two mechanisms is possible at the moment, but they suggest obvious approaches to further study the process of transformation by viral transforming gene products that encode protein kinases. Clearly, one of the most important areas of investigation in the near future will be directed at a biochemical description of the pathways that lead to neoplasia initiated by these and related viruses.

REFERENCES

1. Hanafusa, H. (1977) *In* "Comprehensive Virology" (H. Fraenkel-Conrat and R. P. Wagner, eds.), pp. 401–483. Plenum, New York.

2. Bishop, J. M. (1978) *Annu. Rev. Biochem.* **47**, 35-88.
3. Kawai, S., and Hanafusa, H. (1971) *Virology* **46**, 470-479.
4. Purchio, A. F., Erikson, E., Brugge, J. S., and Erikson, R. L. (1978) *Proc. Natl. Acad. Sci. U.S.A.* **75**, 1567-1571.
5. Collett, M. S., and Erikson, R. L. (1978) *Proc. Natl. Acad. Sci. U.S.A.* **75**, 2021-2024.
6. Levinson, A. D., Oppermann, H., Levintow, L., Varmus, H. E., and Bishop, J. M. (1978) *Cell* **15**, 561-572.
7. Gilmer, T. M., Parsons, J. T., and Erikson, R. L. (1982) *Proc. Natl. Acad. Sci. U.S.A.* **79**, 2152-2156.
8. Gilmer, T. M., and Erikson, R. L. (1981) *Nature (London)* **294**, 771-773.
9. Hunter, T., and Sefton, B. M. (1980) *Proc. Natl. Acad. Sci. U.S.A.* **77**, 1311-1315.
10. Collett, M. S., Purchio, A. F., and Erikson, R. L. (1980) *Nature (London)* **285**, 167-169.
11. Breitman, M. L., Neil, J. C., Moscovici, C., and Vogt, P. K. (1981) *Virology* **108**, 1-12.
12. Feldman, R. A., Hanafusa, T., and Hanafusa, H. (1980) *Cell* **22**, 757-765.
13. Feldman, R. A., Wang, L.-H., Hanafusa, H., and Balduzzi, P. C. (1982) *J. Virol.* **42**, 228-236.
14. Ghysdael, J., Neil, J. C., and Vogt, P. K. (1981) *Proc. Natl. Acad. Sci. U.S.A.* **78**, 2611-2615.
15. Ghysdael, J., Neil, J. C., Wallbank, A. M., and Vogt, P. K. (1981) *Virology* **111**, 386-400.
16. Kawai, S., Yoshida, M., Segawa, K., Sugiyama, H., Ishizaki, R., and Toyoshima, K. (1980) *Proc. Natl. Acad. Sci. U.S.A.* **77**, 6199-6203.
17. Lee, W.-H., Bister, K., Pawson, A., Robins, T., Moscovici, C., and Duesberg, P. H. (1980) *Proc. Natl. Acad. Sci. U.S.A.* **77**, 2018-2022.
18. Neil, J. C., Ghysdael, J., and Vogt, P. K. (1981) *Virology* **109**, 223-228.
19. Pawson, T., Guyden, J., Kung, T.-H., Radke, K., Gilmore, T., and Martin, G. S. (1980) *Cell* **22**, 767-775.
20. Collett, M. S., Brugge, J. S., and Erikson, R. L. (1978) *Cell* **15**, 1363-1369.
21. Mathey-Prevot, B., Hanafusa, H., and Kawai, S. (1982) *Cell* **28**, 897-906.
22. Radke, K., and Martin, G. S. (1979) *Proc. Natl. Acad. Sci. U.S.A.* **76**, 5212-5216.
23. Erikson, E., and Erikson, R. L. (1980) *Cell* **21**, 829-836.
24. Cohen, S., Carpenter, G., and King, L., Jr. (1980) *J. Biol. Chem.* **255**, 4834-4842.
25. Erikson, E., Shealy, D. J., and Erikson, R. L. (1981) *J. Biol. Chem.* **256**, 11381-11384.
26. Krebs, E. G., and Beavo, J. A. (1979) *Annu. Rev. Biochem.* **48**, 923-959.
27. Decker, S. (1981) *Proc. Natl. Acad. Sci. U.S.A.* **78**, 4112-4115.

The Control of Cellular Transcripts in Transformed Cells

ARNOLD J. LEVINE, TED SCHUTZBANK,
AND ROBIN ROBINSON[1]

Department of Microbiology
School of Medicine
State University of New York at Stony Brook
Stony Brook, New York

INTRODUCTION

Transformed cells exhibit a very large number of differences when compared with their nontransformed counterparts (1). Those differences encompass changes in cell morphology (2) and cytoskeletal structures (3) as well as the properties of altered growth control such as growth in agar (4) or low serum concentrations (5), growth on top of monolayer cell cultures (6), or the production of tumors in animals (7). On the biochemical level a variety of enzyme activities such as proteases (8), glycosyltransferases (9), and glycosidases (10) can be altered in transformed cells, often resulting in changes in the plasma membrane (11) or alterations in the transport of molecules through the plasma membrane (12,13). An additional source of some confusion in this type of analysis of the transformed phenotype has been the acknowledged heterogeneity of these properties in different transformed cell lines. The altered growth properties of transformed cells are not necessarily coordinately controlled (14) and can even exhibit quantitative differences (15). This often bewildering array of properties called the transformed phenotype must also be reconciled with the observation that, in some viral transformed cells, a single virus encoded protein, such as the SV40 large T antigen (16) or the Rous sarcoma virus (RSV) pp60^{v-src} (17), can regulate most, if not all, of these parameters in a transformed cell. Although the mechanisms by which such different viral proteins as the SV40 large T antigen and pp60^{v-src} regulate the transformed phenotype may well be different (18), it is

---

1. Present address: Department of Pathology, School of Medicine, State University of New York at Stony Brook, Stony Brook, New York

clear that both of these viral proteins result in an alteration of cellular gene expression. These viral proteins could ultimately act upon one or more processes such as the transcription patterns of cellular genes, RNA processing, mRNA stability and transport events, the translation of proteins, or even upon posttranslational events such as protein stability (19) or protein modification (20). The study presented here provides evidence that the steady-state levels of some cellular mRNAs are dramatically increased when transformed cells are compared to their nontransformed counterparts. In addition, the relative abundance of cellular mRNAs can differ between transformed cell lines, reflecting the heterogeneity of the transformed phenotype when different transformed cell lines are compared. Thus, many of the qualitative and quantitative differences both between transformed cell lines and between the transformed and nontransformed cell counterparts are reflected in the levels of several specific cellular mRNAs detected in these cells.

RESULTS

THE REGULATION OF CELLULAR TRANSCRIPTS IN SV40 TRANSFORMED CELLS

To investigate the transcriptional patterns of cellular genes in SV40 transformed cells a cDNA library of the mRNA species found in an SV3T3-T2 cell line was prepared (21). This cDNA library (430 clones) was employed to search for cellular transcripts that were present in higher levels in SV40-transformed 3T3 cells than in nontransformed 3T3 cells. Radioactively labeled cytoplasmic RNA from SV3T3 and 3T3 cells was hybridized to one of two duplicate filters containing the bacterial colonies with the cDNA plasmids (22). In this manner, a number of cDNA clones were obtained that detected higher abundance levels of cellular mRNA in SV40 transformed cells when compared to nontransformed cells (23).

Four of these cDNA clones were chosen for more intensive characterization. As a first step, the plasmids containing these cDNAs, termed 104, 403, 218 and 85, were purified and employed as probes for hybridization to cellular mRNA from SV40 transformed and nontransformed cell lines using Northern gel electrophoresis as an assay (24). Total cytoplasmic RNA was isolated (23) from SV3T3-T2 cells, 3T3 cells, SV40tsA58cb cells grown at 32° or 39°C and SV40tsA7 3T3 cells grown at 32° or 39°C. These two SV40tsA-transformed cell lines exhibit a transformed phenotype at 32°C and a nontransformed phenotype at 39°C (16,25). The cytoplasmic RNAs from these cell lines was fractionated on agarose gels (26) and was transferred to nitrocellulose paper (24). Nick-translated radioactive cDNA clones (27) were employed as probes for hybridization to this RNA. Figure 1 presents the autoradiograms of each filter probed with the cDNA clones 104 (Fig. 1A), 403(Fig. 1B), 218 (Fig. 1C) and 85 (Fig. 1D).

The clone 104 cDNA detected two mRNA species in SV40tsA58 cells at 32°C (transformed) and two-to-four species of mRNA in SV40tsA7 cells at 32°C. The levels of these mRNA species detected in these same two cell lines grown at 39°C (nontransformed) were much reduced or absent. Similarly, the levels of mRNA complementary to the clone 104 cDNA insert in SVT2 transformed cells (although somewhat degraded) was much higher than that detected in nontransformed 3T3 cells (Fig. 1A). The clone 403 cDNA (Fig. 1B) detected a single mRNA species in SVT2 cells and SV40tsA58 cells grown at 32°C, and no RNA species could be found in either the 3T3 cells or the SV40tsA58 cells grown at 39°C. Similarly, clone 218 cDNA hybridized with one or two mRNA species in SVT2 cells and SV40tsA58 cells grown at 32°C, but detected little or no RNA species in 3T3 cells or SV40tsA58 grown at 39°C (Fig. 1C). Clone 85 cDNA hybridized with a slightly greater level of mRNA (somewhat degraded sample) in SVT2 cells compared to 3T3 cells. In SV40tsA58 cells, much more RNA complementary to clone 85 was detected at 32° than at 39°C, although the magnitude of this differential varied from experiment to experiment (see Table I, for example). In SV40tsA7 cells grown at 32°C at least five mRNA species were detected by clone 85 DNA, one of which could be found in relatively high abundance when these same cells were grown at 39°C.

Collectively, these experiments demonstrate several points: (1) Some cellular transcripts are greater than 100-fold more abundant in SV40 transformed cells than in their nontransformed counterparts. By employing SV40tsA transformed cell lines, the same cell line can be compared when it expresses the transformed phenotype (32°C) or the nontransformed phenotype (39°C). In this case the abundance levels of some cellular mRNA are regulated in a temperature-sensitive fashion as is the SV40 large T antigen (29). (2) In some cases the same cellular transcript (cDNA clone 403 and 218) can vary 10–50-fold in abundance levels when two different SV40 transformed cell lines were compared (SVT2 and SVtsA58 at 32°C). (3) Some cDNA clones (clones 104 and 85) detect multiple RNA species in the cytoplasm of transformed cells. The abundance levels of these mRNAs can vary be-

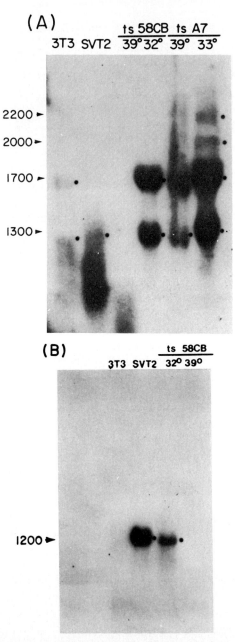

Fig. 1. Northern gel hybridization of the cytoplasmic RNA species complementary to the cDNA 104, 403, 218, and 85. Cytoplasmic RNA was extracted from transformed cells SV3T3-T2, SVtsA58 grown at 32°C and SVtsA7 grown at 32°C and nontransformed cells 3T3, SVtsA grown at 39°C. The RNA (50 μg/lane) was electrophoreses through agarose gels (26) and transferred to nitrocellulose paper (24). Nick-translated DNA (27) with a specific activity of 1×10^8 cpm/μg DNA was hybridized to the RNA. After washing the filters an autoradiogram was obtained (23) and a photograph of this X-ray film is presented. (A) clone 104. (B) clone 403. (C) clone 218. (D) clone 85.

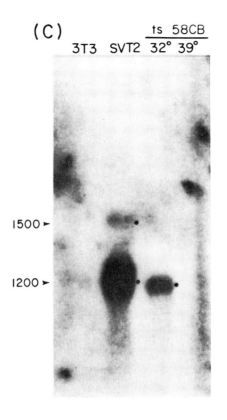

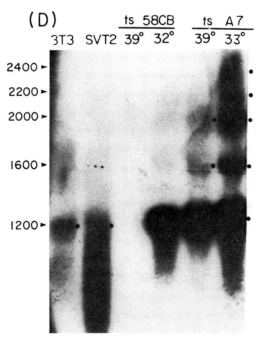

TABLE I
The Steady State Levels of RNA Species in Transformed and Nontransformed Cell Lines[a]

	Ratio of cpm hybridized							
Clone No.	SV3T3 / 3T3	SVtsA 32°C/39°C	Py3T3 / 3T3	RSV-3T3 / 3T3	RSVtsLA90 32°C/39°C	Meth A / 3T3	F9 / 3T3	SV40 tumor/ liver
104	>100	>100	>43	>96	>87	>100	>100	>100
403	>100	>100	—	—	—	—	—	>100
218	>100	>100	>100	>100	4	>100	>100	>100
85	5	6–19	2	2	9	1	14	7

[a] Cytoplasmic RNA was extracted from all of the transformed cell lines (for tumor or normal tissue total RNA was employed). The RNA was labeled with [γ-^{32}PO$_4$] ATP and polynucleotide kinase (28), and the RNA was hybridized to nitrocellulose filters containing the cDNA clones (23). After washing the filters, the RNA cpm hybridized were determined by scintillation counting. The ratio of RNA (cpm) hybridized from the transformed cells/nontransformed cell is presented. The sign > indicates that the divisor in each case contained no detectable cpm of background levels which were divided into the numerator. The sign — indicates that the numerator in each case contained no detectable cpm as did the divisor in these cases. The SV40tsA cell line employed was SV40tsA58cb. The SVtsA 32°C/39°C ratio with clone 85 provides the range (6–19) of ratio values detected in several experiments.

tween transformed cell lines and between transformed and nontransformed cell lines.

None of the four cDNA clones under study here hybridize with SV40 nucleotide sequences (23), and all four clones did not hybridize with each other indicating they detect distinct cellular transcripts.

THE REGULATION OF CELLULAR TRANSCRIPTS IN CELLS TRANSFORMED BY A WIDE VARIETY OF AGENTS

To investigate whether the mRNA species complementary to the cDNA clones under study here were also more abundant in cells transformed by agents other than SV40, the following experiments were carried out. Cytoplasmic RNA was extracted and purified from a wide variety of transformed cell lines: SV3T3T2, SVtsA58 at 32°C, Py3T3, RSV-3T3, RSVtspp60^{v-src} (LA90) at 32°C, a methylcholanthrene transformed cell line (Meth A), and an embryonal carcinoma cell line (F9). The RNA was also extracted from the appropriate nontransformed counterpart in each case, i.e., 3T3, SVts58 grown at

39°C; RSV-3T3ts-pp60^{v-src} grown at 39°C, etc. These RNAs were sheared with mild alkali treatment and labeled with [γ-^{32}PO$_4$]ATP using polynucleotide kinase (28). The labeled RNA preparations where hybridized to the four cDNA clones immobilized on nitrocellulose filters (23), and the filters were counted in a liquid scintillation counter. The ratio of the RNA cpm from the transformed cells hybridized to the cDNA clones over the RNA levels detected in the nontransformed cells is presented in Table I. RNA complementary to clone 104 DNA was detected in greater than 40–100-fold higher abundance levels in all the transformed cell lines when compared to their nontransformed counterparts. The high RNA levels complementary to clone 104 in SV40tsA and RSVtspp60^{v-src} transformed cell lines (at 32°C) were temperature regulated, declining 87–100-fold in the nontransformed cells (39°C). Clone 218 cDNA detected greater than 100-fold more RNA in all of the transformed cell lines except the RSVtspp60^{v-src} cell line (32°C) when compared to the nontransformed cell lines (39°C). The cDNA clone 403 detected high abundance levels (greater than 100-fold) of RNA in SV40-transformed cells (SVT2 and SVtsA58 at 32°C) but failed to detect any RNA species in nontransformed cells and in the cells transformed by agents other than SV40. Clone 85 cDNA showed only small differentials of RNA levels between transformed and nontransformed cell lines for all of the cell lines tested.

To extend these results to another system, tumors were induced in BALB/c mice by an injection of SVT2 cells (30). The RNA extracted from these SV40-induced tumors and mouse liver tissue from these same mice was purified and labeled as described previously (28). The cDNA clones 104, 403, and 218 hybridized greater than 100-fold more RNA from tumor tissue than liver tissue (Table I), whereas clone 85 cDNA detected only a small differential in RNA levels (Table I). Indeed, the cDNA clones 104, 403, and 218 failed to detect RNA species in several normal tissues of the adult mouse or in primary mouse kidney fibroblast cells (23). These results extend this characterization to *in vivo* tissue and eliminate the objection that the 3T3 cells employed in some of these studies are a poor choice or are an atypical example of a nontransformed counterpart cell line.

These studies provide several additional conclusions: (1) Some RNA species (clone 104) are more abundant (40–100-fold) in a wide variety of transformed cells than their nontransformed cell counterpart, independent of the original transforming agent. (2) Other RNA species (clone 403) appear to be preferentially found in SV40 transformed cell lines or tumors and have not been detected in either normal mouse tissue or cell lines transformed by other agents.

STIMULATION OF THE LEVELS OF CELLULAR mRNA AFTER SV40 INFECTION

To further examine the ability of SV40 to stimulate the levels of some cellular mRNAs, a different experimental system was exploited. Cell cultures of 3T3 cells were infected with SV40, and at various times after infection, the cells were labeled with [^{3}H]uridine for 1 hr. The RNA was then extracted from these cells (23) and hybridized to nitrocellulose filters containing one of the four cDNA clones under study. The same RNA preparations were also hybridized to filters containing the SV40 early region DNA to follow the synthesis and accumulation of SV40 mRNA encoding the viral T antigen. These filters were then counted in a liquid scintillation counter. The results of this experiment are presented in Fig. 2 where the cpm hybridized/µg of RNA is plotted as a function of time after infection. The zero labeling time point was obtained from a mock-infected control culture. The synthesis of SV40 T antigen mRNA can be detected by 4 hr postinfection with a maximum level found at 8 hr. There was a slight decline in the level of stable T antigen mRNA synthesized at 24 hr postinfection. The kinetics of production of the mRNAs complementary to the four cDNA clones under study varied in their time of appearance, and final level stimulated after virus infection. Small levels of RNA comple-

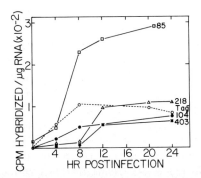

Fig. 2. The kinetics of production of SV40 T antigen mRNA and the cellular RNA complementary to the cDNA clones, after SV40 infection of 3T3 cells. Confluent monolayer culture of 3T3 cells were infected with SV40. At various times after infection the cells were labeled with [^{3}H]uridine for 1 hr. Cytoplasmic RNA was then extracted from these cell cultures and hybridized to nitrocellulose filters (23) containing the cDNAs or SV40 DNA to detect SV40 T antigen mRNA. The levels of RNA from uninfected 3T3 cells complementary to the cDNA clones are given by the zero time point. The results are plotted as the amount (cpm) of labeled RNA/µg of RNA in the reaction that hybridized to each cDNA on the filters verses the time in hours after SV40 infection. Clone numbers and SV40 T antigen mRNA are indicated in the figure.

mentary to clone 85 cDNA was detectable in mock-infected 3T3 cells and rapidly rose in amounts (the 1 hr pulse with [^{3}H]uridine measures some unspecified combination of rate of synthesis and stability of the RNA) by 4–8 hr after infection. RNAs complementary to clones 104, 403, and 218 increased with delayed kinetics and to different extents.

These experiments demonstrate that (1) SV40 infection of nontransformed cells results in an alteration of cellular gene expression at the mRNA level with the kinetics of appearance and levels of cellular RNA species differing depending upon the cellular gene under study. (2) The RNA complementary to clone 85 cDNA appears to be more rapidly (more directly) stimulated by SV40 infection. Some RNA species complementary to clone 85 appear to be regulated in transformed cells (Fig. 1D), but these appear to be the minor abundance species, and, when total cytoplasmic RNA is examined (Table I), the RNAs complementary to clone 85 DNA do not appear to be in higher abundance than in their nontransformed cell counterparts. These data may suggest that the early events in the SV40 transformation process may not necessarily be reflected in the final transformed cell line which is isolated, cloned, and examined.

THE REGULATION OF CELLULAR mRNA LEVELS AFTER SERUM STIMULATION OF NONTRANSFORMED CELLS

Some cellular functions may be regulated, not by the transformed phenotype, but by the physiological state of cell growth. Thymidine kinase activities are very high in actively growing cells in culture but decline in resting cells (31). Transformed cells often continue to grow under conditions where nontransformed cells enter a resting state, and the higher thymidine kinase activities detected in transformed cells have erroneously been reported to be a property of the transformed cell itself (32). To avoid these complications, the cDNA clones under study here were originally selected (23), based upon their abilities to hybridize with greater levels of RNA from actively growing SV40 transformed cells than from actively growing nontransformed 3T3 cells. Indeed, all the studies presented here were carried out with growing transformed and nontransformed cell cultures, although the growth rates (generation times) differ greatly between different cell lines.

To more directly examine the regulation of RNA species complementary to the four cDNA clones under study here as a function of cell growth, the following experiment was performed. Low serum arrested cultures of 3T3 cells were stimulated into a cell cycle and division by the addition of 10% fresh fetal calf serum. Control experiments demonstrated that about 2% of the serum arrested cells were in S phase, while 50% of these cells entered S phase by 18 hr after exposure to the serum. DNA synthesis was monitored by [^{3}H]thymidine incorporation, and the cells began to enter S phase by 10 hr; DNA synthesis reached a maximal rate by 18 hr after the addition of fresh serum. Cytoplasmic RNAs were extracted from these cells at various times after serum stimulation. These RNA preparations were labeled with [γ-^{32}PO$_4$]ATP and polynucleotide kinase (28) and hybridized to nitrocellulose filters containing the four cDNA clones studied here (23). The same quantities of RNA were hybridized to all filters. Figure 3 presents the results of this study. The cDNA clones 104, 403, and 218 failed to detect any RNA (by this filter hybridization technique) in resting (zero time) or growing (2–18 hr) 3T3 cells. Under these same conditions, all three of these cDNA clones readily hybridized with high RNA levels derived from several transformed cell lines (Table I). On the other hand, small levels of RNA complementary to clone 85 could be detected in resting 3T3 cells, and these levels were stimu-

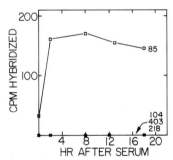

Fig. 3. The kinetics of production of cellular RNAs complementary to the cDNA clones, after serum stimulation of 3T3 cells. Nongrowing cultures of 3T3 cells that had been serum arrested for growth in 1% serum were stimulated into a growth phase by the addition of 10% fresh fetal calf serum. In resting cultures about 2% of the cells were in S phase, whereas in serum stimulated cells about 50% of the cells were in S phase. S phase, as measured by the incorporation of [^{3}H]thymidine into these cells, began about 10 hr after the addition of serum and was maximal at 18 hr after serum addition. At various times after the serum stimulation, cytoplasmic RNA was isolated and labeled with [γ-^{32}PO$_4$]ATP and polynucleotide kinase (28). Equal amounts of these RNA preparations were hybridized to nitrocellulose filters containing the four cDNA clones. The levels of RNA complementary to these cDNAs in nonserum stimulated resting cells are indicated at the zero time point. The results are plotted as the cpm hybridized to each filter versus the time after serum stimulation. The clone numbers are provided in the figure.

lated fivefold by 2 hr after the addition of fresh serum to the cells (Fig. 3). This rapid increase in the steady state levels of RNA complementary to clone 85 DNA after serum stimulation was similar to that observed after SV40 infection (Fig. 2).

These experiments indicate that the higher levels of RNAs complementary to clones 104, 403, and 218 found in transformed cells compared to their nontransformed counterparts do not appear to be accounted for by the physiological state of growth of these cells. The assay employed here (filter hybridization) is less sensitive than the Northern gel analysis (Fig. 1), and it remains possible that cellular growth could effect these RNA levels to some degree, but it is unlikely that the abundance levels of these RNAs are as great as those found in transformed cells. RNA complementary to clone 85 on the other hand may well be regulated by or with the growth rate of a cell.

DISCUSSION

Four cDNA clones (104, 403, 218, and 85), prepared from the mRNA of SV40 transformed cells, were employed to probe the levels of complementary cellular mRNA in transformed and nontransformed cells. Each of these clones demonstrates a number of properties of cellular mRNA in transformed and nontransformed cells. With clones 104 and 218 (1) a wide variety of transformed cells, independent of the transforming agent, contained 40–100-fold higher levels of mRNA complementary to these cDNA, than their nontransformed counterparts. (2) SV40 induced tumors of mice contained greater than 100-fold higher levels of RNA complementary to these clones than did normal mouse tissue. (3) SV40tsA (104 and 218) and RSVtspp60^{v-src} (104 only) transformed cells (grown at 32°C) contained greater than 80–100-fold higher levels of mRNA complementary to these clones than these same nontransformed cells grown at 39°C. (4) Although SV40 infection of nontransformed 3T3 cells stimulated the levels of 104 and 218 mRNA above background, serum stimulation and cell growth of 3T3 cells failed to detect any mRNA complementary to these cDNA clones. With clone 403 (1) four different SV40-transformed cell lines (23) all contained greater than 100-fold high levels of mRNA complementary to this cDNA clone than did their nontransformed cell counterparts. (2) SV40 tumor tissue also contained greater than 100-fold higher levels of RNA complementary to clone 403 DNA than did normal mouse tissue. (3) SV40tsA transformed cell lines at 39°C contained higher levels of mRNA complementary to 403 cDNA than these same nontransformed cells grown at 39°C. (4) mRNA complementary to cDNA 403 was only detected in SV40 transformed cells or tumor tissue. No RNA complementary to this cDNA clone was detected in Py, RSV, Meth A, or embryonal carcinoma transformed cell lines, or was RNA complementary to this cDNA clone detected in normal mouse tissues and nontransformed cell lines. These high levels of this mRNA appear to be specific to SV40 transformed cells and tumors. (5) The relative abundance levels of mRNA complementary to cDNA 403 may vary 10–50-fold between different SV40 transformed cell lines. (6) mRNA complementary to cDNA 403 was not detected in serum stimulated or actively growing nontransformed (3T3) cell lines. With clone 85 cDNA: (1) Total cellular RNA complementary to clone 85 cDNA showed little or no evidence of differential regulation of levels in transformed and nontransformed cell lines. (2) Clone 85 cDNA hybridizes with one major abundant mRNA species and four minor abundant mRNA species in the cytoplasm of some transformed cell lines. The four minor abundance mRNA species are differentially regulated and are found in higher amounts in transformed cells (SV40tsA regulated) than in nontransformed cells. (3) There is a good deal of heterogeneity between the levels and regulation of these five mRNA species between different transformed cell lines. (4) In SV40 infected 3T3 nontransformed cells there is a dramatic increase in the levels of mRNA complementary to clone 85 cDNA by 4–8 hr after infection. (5) In serum stimulated nontransformed 3T3 cells (entering a growth cycle), the levels of mRNAs complementary to clone 85 cDNA increase (fivefold) by 2 hr after exposure to serum. Thus, clone 85 detects a series of mRNAs whose levels are dramatically and rapidly stimulated by entry into a growth phase, induced by either SV40 infection or exposure to fresh serum.

It is clear from these results that the levels of some cellular mRNAs can be dramatically altered by cell growth or cellular transformation. Furthermore, different cellular genes (cDNA clones) are affected differentially by promotors of cell growth (such as serum) or different transforming agents. The studies presented here suggest the possibility that all transformed cells, independent of the transforming agent, may express higher levels of some mRNAs not detectable in nontransformed cells or normal adult tissues (clone 104). The high levels of mRNA complementary to clone 104 cDNA detected in all transformed cells were not expressed in actively growing or serum stimulated nontransformed cells, which tends to eliminate cell growth as the common variable regulating this gene. Further studies employing more sensitive techniques will be required to substantiate these ideas.

Most of the techniques employed in this study measure the steady state levels of cellular mRNAs in transformed and nontransformed cells. Higher levels of mRNA could be achieved either by an enhanced rate of transcription or an increase in the stability of these cellular mRNAs. Experiments will have to be designed to choose between these alternative mechanisms. It is interesting that two diverse transforming agents such as the SV40 large T antigen (tsA mutations) and the RSVpp60^{v-src} (tspp60^{v-src}) each can regulate, in a temperature dependent fashion, the levels of mRNA derived from the same gene (clone 104). Although the mechanisms of action of these two viral proteins may differ and the manner in which they affect the levels of mRNA from gene 104 could even differ, the result is the same, i.e., an increased level of mRNA in the transformed cells. Perhaps different transforming agents, in spite of the diversity of the transformed phenotype, may show some common events which will lead to a basic understanding of cellular transformation and tumorigenesis.

ACKNOWLEDGMENTS

The authors thank R. Pashley and A. K. Teresky for able technical assistance and G. Urban for aid with the manuscript. The research reported here was supported by a grant from the NCI CA28146-02 and 03. Some of the experiments reviewed in this paper have first been reported elsewhere (23).

REFERENCES

1. Tooze, J., ed. (1981) "Molecular Biology of Tumor Viruses, DNA Tumor Viruses," 2nd rev. ed. Part 2, pp. 205-296. Cold Spring Harbor Lab., Cold Spring Harbor, New York.
2. Todaro, G. J., Green, H., and Goldberg, B. D. (1964) *Proc. Natl. Acad. Sci. U.S.A.* **51**, 66-69.
3. Pollack, R. E., Osborn, M., and Weber, K. (1975) *Proc. Natl. Acad. Sci. U.S.A.* **72**, 994-998.
4. MacPherson, I., and Montagnier, L. M. (1964) *Virology* **21**, 291-292.
5. Holly, R. W., and Kiernan, J. A. (1968) *Proc. Natl. Acad. Sci. U.S.A.* **60**, 300-304.
6. Todaro, G. H., and Green, H. (1964) *Virology* **23**, 117-119.
7. Shin, S. I., Freedman, V. H., Risser, R., and Pollack, R. (1975) *Proc. Natl. Acad. Sci. U.S.A.* **72**, 4435-4439.
8. Qugley, J. P., Ossowski, L., and Reich, E. (1974) *J. Biol. Chem.* **249**, 4306-4314.
9. Grimes, W. J. (1970) *Biochemistry* **9**, 5083-5091.
10. Bosmann, H. B. (1969) *Exp. Cell Res.* **54**, 217-224.
11. Hakamori, S., and Murakami, W. T. (1968) *Proc. Natl. Acad. Sci. U.S.A.* **59**, 254-258.

12. Forster, D. O., and Pardee, A. B. (1969) *J. Biol. Chem.* **244**, 2675-2684.
13. Cunningham, D. D., and Pardee, A. B. (1969) *Proc. Natl. Acad. Sci. U.S.A.* **64**, 1049-1053.
14. Risser, R., and Pollack, R. (1974) *Virology* **59**, 477-484.
15. Pollack, R., Risser, R., Conlon, S., Freedman, V., Shin, S.-I., and Rifkin, D. B. (1975) *Cold Spring Harbor Conf. Cell Proliferation* **2**, 885-894.
16. Brockman, W. W. (1978) *J. Virol.* **25**, 800-874.
17. Martin, G. S., Venuta, S., Weber, M., and Rubin, H. (1971) *Proc. Natl. Acad. Sci. U.S.A.* **68**, 2739-2743.
18. Levine, A. J. (1982) *Adv. Cancer Res.* **37**, 75-109.
19. Oren, M., Maltzman, W., and Levine, A. J. (1981) *Mol. Cell. Biol.* **1**, 101-110.
20. Collett, M. S., and Erickson, R. L. (1978) *Proc. Natl. Acad. Sci. U.S.A.* **75**, 2021-2024.
21. Villa-kormaroff, L., Efstrodiadis, A., Broome, S., Lomedico, P., Tizard, R., Naber, S. P., Chick, W. L., and Gilbert, W. (1978) *Proc. Natl. Acad. Sci. U.S.A.* **75**, 3527-3731.
22. Thayer, R. E. (1979) *Anal. Biochem.* **98**, 60-63.
23. Schutzbank, T., Robinson, R., Oren, M., and Levine, A. J. (1982) *Cell* **30**, 481-490.
24. Thomas, P. S. (1980) *Proc. Natl. Acad. Sci. U.S.A.* **77**, 5201-5205.
25. Maltzman, W., Linzer, D. I. H., Brown, F., Teresky, A. K., Rosenstraus, M., and Levine, A. J. (1979) *J. Int. Cytol., Suppl.* **10**, 179-189.
26. Rave, N., Crkvenjaker, R., and Boedtker, H. (1979) *Nucleic Acids Res.* **6**, 3559-3567.
27. Rigby, P. W. J., Dieckmann, M., Rhodes, C., and Berg, P. (1977) *J. Mol. Biol.* **113**, 237-251.
28. Spradling, A. L., Fijan, M. E., Mahowald, A. P., Scott, M., and Craig, E. K. (1980) *Cell* **19**, 905-914.
29. Tegtmeyer, P., Schwartz, M., Collins, J. R., and Rundell, K. (1975) *J. Virol.* **16**, 168-178.
30. Aaronson, S. A., and Todaro, G. J. (1968) *J. Cell. Physiol.* **72**, 141-148.
31. Postel, E. H., and Levine, A. J. (1975) *Virology* **63**, 404-420.
32. Bull, D. L., Taylor, A. T., Austin, D. M., and Jones, W. O. (1974) *Virology* **57**, 279-284.

PART VII

NUCLEAR MATRIX, CYTOSKELETON, AND MITOCHONDRIA IN CARCINOGENESIS

The Concept of DNA Rearrangement in Carcinogenesis: The Nuclear Matrix

DONALD S. COFFEY
Departments of Urology and Pharmacology
Oncology Center
The Johns Hopkins University
School of Medicine
Baltimore, Maryland

INTRODUCTION

The purpose of this paper is to discuss the concept of DNA rearrangement as a molecular mechanism in the process of carcinogenesis. It is proposed that DNA rearrangement is an integral part of normal differentiation and that this process is reactivated during the events of carcinogenesis. This reinitiation of DNA rearrangement in carcinogenesis may be a continuing process and could be responsible for the subsequent development of a wide variety of tumor cells with different properties (tumor cell heterogeneity). As this tumor cell heterogeneity occurs, selection pressures under therapeutic or environmental conditions would select for continuing growth the most resistant or appropriate tumor cell clones (1).

The processes that initiate or control DNA rearrangement are unknown at present, but it appears that the nuclear matrix (skeleton) plays a central role in DNA organization and replication within the mammalian nucleus, and this nuclear structure could be a prime factor in the cellular events associated with DNA rearrangement.

DNA REARRANGEMENT IN CANCER

DNA rearrangements certainly occur in many types of human cancers, and these karyotypic changes can be observed by visual alterations in the tumor chromosomes (2). More detailed cytogenetic studies, utilizing high resolution chromosome banding or hybridization techniques indicate that these tumor-associated changes in DNA rearrangement may be far more common than first suspected (3). At present, it is still difficult to assess fully the extent of tumor cell DNA rearrangement that may be occurring at the submicroscopic level.

DNA rearrangement may be defined as a change in the linear order of the DNA sequence and may be accomplished through deletion or movable genetic elements occurring within the genome. The insertion of external DNA such as viral insertion would also accomplish DNA rearrangement. In the broadest sense, DNA rearrangement is a form of mutation, and this mechanism of DNA control has been proposed to explain the many paradoxes of genetic versus epigenetic (nongenetic) phenomena in carcinogenesis (1).

The possible role of DNA rearrangement in cancer was first noted by Temin, who suggested that tumor proviruses might act similarly to transposable elements and that their location at specific sites in the host genome might modify cellular expression and lead to cancer (4). Tumor viruses can change the linear order of the DNA by insertion into the host genome, and recent experimental evidence supports the hypothesis that a specific DNA arrangement or proviral DNA within the host genome leads to cancer in two animal tumor virus systems (5–7). Many cytogenetic studies show that consistent chromosomal rearrangements are often found in cancer cells, and from these studies, it has been suggested that DNA rearrangement may lead to cancer (8–11). Any DNA rearrangement could be preserved during subsequent cellular replication and would appear as a genetic or mutational event. In addition to point mutations, gross chromosomal aberrations or sister chromatid exchanges can also be induced by most carcinogenic agents (12–14). DNA repair disorders that predispose to cancer also manifest chromosomal rearrangements (2,15,16). Thus, if specific DNA rearrangements are shown to cause cancer, these carcinogenic rearrangements may be the genetic events attributed to mutation.

Recently, Feinberg and Coffey (1) reported an analysis of the published case histories of genetic disorders that predispose to cancer to determine the cytogenetic types and chromosomal locations of chromosomal rearrangements reported in the *nontumor cells* of patients with these disorders (1). Nontumor cells were chosen in order to determine whether these rearrangements might precede clinical cancer rather than result from the cancer *per se*. It appeared from their analysis that people with genetic diseases predisposed to cancer had chromosomal instabilities occurring in their *nontumor* cells in chromosomal areas that have been reported to be frequent sites of karyotypic changes associated with cancer.

Another indirect argument for DNA rearrangement rather than point mutation as a cause of human cancer is derived from Cairns' recent genetic epidemiological study (17). Cairns examined published data on mortality from fatal internal cancers among patients with two inherited disorders that predispose to cancer. Cairns' analysis convinced him that the incidence and pattern of internal cancers in these epidemiological diseases were more consistent with a mechanism of DNA rearrangement than with point mutations.

Reversal of the DNA arrangement in a tumor cell, in order to restore the original linear order of DNA existing in normal cells, may underlie the apparent reversibility of cancer in several systems (teratocarcinomas, neuroblastomas, Lucké tumors, and plant galls). In these systems DNA rearrangement could account for the high frequency of reversal from the malignant state to the normal state. For example, a transposable element that controls the mating type of yeast (18) can cause interconversion between two phenotypic states after a single cell division (19). The high frequency of transformation of normal cells on exposure to carcinogens also might be explained by DNA rearrangement rather than by point mutation (17,20,21). Thus, DNA rearrangement provides a potential unifying mechanism to encompass both genetic and apparently epigenetic factors in carcinogenesis (1). The linear order of DNA segments could be rearranged by transpositions, insertions, deletions, or inversions, although in the case of deletions, reversibility would not seem possible unless the deleted segment were episomal.

In summary, a number of investigators have proposed DNA rearrangement as an important concept to be considered in both the initiation of carcinogenesis (1,4,8,17,20,23–24a) and in the subsequent development of tumor cell heterogeneity during tumor progression (1).

As intriguing DNA rearrangement appears, there are obviously many critical questions that require resolution through rigorous investigations: (1.) Is the process of DNA rearrangement an initiation factor or is it essential to the subsequent multistage carcinogeneic events or could it only be an epiphenomenon? (2) Is DNA rearrangement a common factor in all forms of carcinogenesis? (3) Is the tumor rearrangement initiated as the result of stress on the cell, which has reactivated a normal cellular developmental process that has remained quiescent since embryogenesis and development? (4) What are the cellular mechanism that control DNA rearrangement?

DNA REARRANGEMENT IN NORMAL DEVELOPMENT AND EXPRESSION OF THE ONCOGENE

The concept of chromosomal rearrangement as a normal aspect of phenotypic expression was introduced in 1911 by Morgan, who proposed that an equal exchange between homologous chromosomes during meiosis could explain genetic linkage (25). McClintock, in 1950, discovered that genetic rearrangement affects phenotypic expression in the seeds of maize during development by movement of transposable genetic controlling elements (26). Such transposable elements have subsequently been observed in both prokaryotes and eukaryotes (27). The extent of DNA rearrangement in the normal differentiation of mammalian cells has not been determined. However, the development of antibody diversity associated with the formation of specific immunoglobulins, which occurs during the differentiation of B lymphocytes, results from DNA deletions (28–31) which in effect rearrange the DNA. Similar types of rearrangements have been noted in yeast, fruit flies, and many other nonprokaryotes, and in each case they control important developmental functions.

If the role of movable genetic elements in the control of normal development and differentiation in mammals can be extended from the specific case of immunoglobulin diversity to a more general role in or-

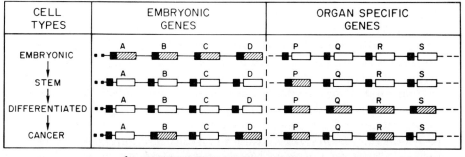

Fig. 1. A schematic of DNA rearrangement in normal development, differentiation, and carcinogenesis. Each horizontal line represents segments of a cellular genome. Genes denoted A–D are related to the embryonic state and genes P–S to a specific organ development. The black square represents a movable control element that activates a gene through position and could involve different types of mechanisms (promoter, repressor, or enhancer).

ganogenesis, it would be easier to visualize how DNA rearrangement could explain many diverse properties of the cancer cell. For example, in the schemata shown in Fig. 1, a linear array of DNA in a cell is represented in the horizontal line, and a series of genes are noted as A to S blocks. Genes A-B-C-D represent functions related to embryonic functions and are active when represented by the shaded boxes. This activation of the genes is represented in this schemata as the placement of a transposable control unit (promoter or enhancer sequence denoted by the dark square) in proximity to the (A-B-C-D) gene. The inactivation of the embryonic genes and the subsequent activation of organ specific genes (P-Q-R-S) is accomplished by moving the transposable genetic units away from the embryonic genes and toward the organ specific genes. If such rearrangements were activated by carcinogenesis, it could reform cells with new arrangements back toward the embryonic state that could produce clones with increased growth advantages. Such cells may on occasion express embryonic functions or express inappropriate differentiated functions as observed in the expression of certain ectopic hormones by some tumor cells. If this reactivation of the genomic arrangement continued, it would produce, in effect, a genetic instability that would result in the observed tumor cell heterogeneity (1,21). The central feature of the schematic is the movable genetic control elements. These movable elements are represented as being located as flanking regions to the gene in the example, but they do not necessarily require the proximity noted here. The nearest analogy is the promotor effects of the large terminal repeat (LTR) sequences associated with some of the retroviruses (32). The model of promotional insertion is intriguing, and it does appear that such events can activate oncogenes. Whether carcinogenesis is the result of increased transcription of a preexisting oncogene following promotion or amplification has not been resolved.

At present, it appears that oncogenes or proto-oncogenes exist in all mammalian cells, are active during periods of development, and are expressed at very low levels during normal differentiation or in mature cells. If so, what are termed oncogenes may only be quiescent normal fetal or developmental genes reactivated at a later state during oncogenesis, similar to the mechanisms depicted in the schemata in Fig. 1 or as a result of promotional or retroviral insertion. In contrast, the recent reports of a point mutation in the oncogene causing activation for transformation are obviously of great interest (33,34). These studies appear to suggest that the gene product may be qualitatively different in cancer. Other mechanisms of gene control, such as the extent of DNA methylation, have also been implicated in carcinogenesis

(24,35). At present, the first step in initiation of carcinogenesis is not known; however, any model of carcinogenesis must include the multistep and temporal events that are required before carcinogenesis is completed. It seems feasible that DNA rearrangement may be one of the important events. Indeed, Ruth Sager (24) visualizes genomic rearrangement in tumor cells as having activated a micro or cellular evolutionary event that provides the tumor cell with great diversity and adaptability (24).

In summary, DNA rearrangements do occur in cancer cells, and if analogous to other types of rearrangements observed in nature, they could have very important biological consequences that could reasonably explain many properties of the cancer cell. The exact importance and extent of these processes remain to be proven, but those concepts form the background for an exciting new area of oncological research.

What controls DNA rearrangements? It appears that structural elements (matrix) within the nucleus organize DNA and participate in DNA function. These nuclear structures will be the focus of the following discussion.

NUCLEAR STRUCTURE AND FUNCTION

There is increasing evidence that chromatin structure may be involved directly in DNA function. Our understanding of chromatin structure has progressed from the nucleosomal bead concept of DNA–histone interactions to the view of higher-order arrangements of the packed DNA into supercoiled loops of DNA attached to a residual nuclear skeleton termed the *nuclear matrix* (36,37). As the concept of DNA rearrangement develops, it is important to understand these events at the molecular level, and this will undoubtedly involve nuclear structural elements. At present there is limited information regarding the role of nuclear structural elements in forming specific chromosomal bands and in DNA replication and transcription. However, there is increasing evidence that the nuclear matrix serves an important function in these processes. In this section, we will review these nuclear structural concepts with regard to their potential role in the process of DNA rearrangement.

For many years, it was believed that the nuclear membrane maintained the structural integrity of the eukaryotic nucleus. High molarity salt solutions (1 M NaCl) were used first by Zbarsky and Georgiev (38–41) and later by Busch and colleagues (42,43) to extract the majority of DNA and chromatin from the nucleus. This treatment left a relatively intact lipid-containing nuclear membrane that encompassed the nucleus, which still contained residual internal structures. The later development of mild nonionic detergents (Triton X-100) made it possible to extract the nuclear membrane lipid components. Thus, a series of extractions of nuclei with detergent, hypotonic, and hypertonic solutions could remove over 98% of the total nuclear phospholipid DNA and RNA and 90% of the nuclear proteins (44,45); these residual chromatin-depleted nuclear ghosts still maintained their spherical and structural integrity. Electron microscopy of this insoluble skeleton structure revealed a matrix network or scaffolding system that was composed on its periphery of residual elements of the nuclear membrane and nuclear pore complexes, surrounding an internal fibrillar network that extended from the periphery to residual elements of the nucleolus. This residual nuclear framework structure was first isolated from rat liver nuclei and was characterized and termed the nuclear matrix (44,45). The nuclear matrix represented only 5–10% of the total nuclear protein and yielded a few major polypeptide bands on sodium dodecyl sulfate polyacrylamide gel electrophoresis (45). The prominent protein fractions are free of histones and have apparent molecular weights of 60,000–70,000 and, by partial tryptic digest fingerprinting techniques, appear to be related structurally (45,46). In addition, there are several clusters of minor polypeptides of approximately 50,000 and 100,000–200,000 (45).

The nuclear matrix appears to play a central role in the structural organization and function of both DNA and nuclear RNA. DNA appears to be attached to the matrix and is organized in supercoiled loops, each containing 30,000–100,000 base pairs (bp), and each loop is anchored at its base to the matrix. During replication, these loops of DNA appear to be reeled through fixed sites for DNA synthesis that are attached to the matrix (36,47). Actively transcribed genes are enriched on the matrix (48–50), and it has been suggested that RNA may also be synthesized at fixed sites on the nuclear matrix (51). RNP particles are part of the matrix, and hnRNA and snRNA are associated almost exclusively with the matrix (24a,52–55). Thus, the nuclear matrix appears to play a central role in DNA replication and transcription and may be involved in DNA processing and transport. The nuclear matrix structure also contains specific binding sites for steroid hormones (56). For a recent review of nuclear matrix functions, see ref. 57. DNA arrangement with respect to the nuclear matrix may be important in many nuclear functions, which may be mediated by DNA loop topology (37).

One of the hallmarks of cancer pathology is the concomitant development of tumor agressiveness and changes in cell morphology. The nuclei become enlarged and invaginated and have a pleiomorphic structure that is reflected in both the nucleolus and in changes in the packing of the chromatin. Since the matrix determines the shape of the nucleus, such gross physical changes must involve the nuclear matrix in at least a passive way.

When gross chromosomal rearrangements of DNA occur in cancer, they would appear to require concomitant involvement of the nuclear matrix component to which the loops are attached (Fig. 2), since the higher-order topology of bands appears to be preserved when several

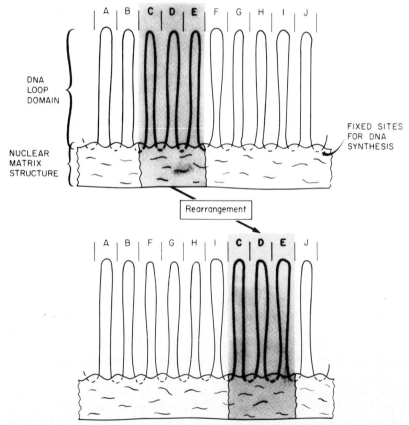

Fig. 2. DNA rearrangement may involve the concomitant movement of the nuclear matrix components to which supercoiled DNA loops are attached. Each DNA loop contains 50–100 kbp. The sites of DNA synthesis are fixed to the nuclear matrix structure at the base of each loop.

bands are translocated as a group from one chromosome to another. At present, it is not known what constitutes the chromosomal core structure and bands. However, some nuclear matrix proteins from the interphase nucleus do appear in the condensed metaphase chromosome (58,59). It has been reported recently that carcinogens bind to the nuclear matrix (60–62), and the nuclear matrix also appears to be the site of attachment of the polyoma viral transformation (T) antigen (63). Zbarsky (64) has reported changes in the nuclear matrix in tumor cells. The nuclear matrix contains the fixed sites for DNA replication (36), and the site of DNA replication may be the locus of sister chromatid exchange (65), a form of rearrangement induced by many carcinogens. It is interesting that cells from patients with Bloom's syndrome, which causes a predisposition to both chromosomal rearrangements and cancer, show slow "movement" of the replicating fork (66) that appears to be located on the nuclear matrix. We do not know if the nuclear matrix is involved actively in carcinogenesis. However, matrix-associated loops of DNA could resemble transposable elements (37).

In summary, we propose that the nuclear matrix is an important site for the control of DNA structure and function and that the matrix and the attachment of loop domains may be important elements in the process of DNA rearrangement.

REFERENCES

1. Feinberg, A. P., and Coffey, D. S. (1982) *In* "Tumor Cell Heterogeneity" (A. H. Owens, D. S. Coffey, and S. B. Baylin, eds.), pp. 469–494. Academic Press, New York.
2. Sandberg, A. A. (1980) "The Chromosomes in Human Cancer and Leukemia." Elsevier/North-Holland, Amsterdam.
3. Rowley, J. (1983) "Chromosomes and Cancer: From Molecules to Man." Academic Press, New York.
4. Temin, H. M. (1980) *Perspect. Biol. Med.* **14**, 11–26.
5. Blair, D. G., Oskarsson, M., Woud, T. G., McClements, W. L., Fischinger, P. J., and Vande Woude, G. G. (1981) *Science* **212**, 941–943.
6. Hayward, W. S., Neel, B. G., and Astrin, S. M. (1981) *Nature (London)* **290**, 475–480.
7. Neel, B. G., Hayward, W. S., Robinson, H. L. Fang, J., and Astrin, S. M. (1981) *Cell* **23**, 323–334.
8. Mitelman, F., and Levan, G. (1978) *Hereditas* **89**, 207–232.
9. Rowley, J. D. (1977) *Chromosomes Today* **6**, 345–355.
10. Sager, R. (1979) *Nature (London)* **282**, 447–448.
11. Klein, G. (1981) *Nature (London)* **294**, 313–318.
12. Ishidate, M., Jr., and Odashima, S. (1977) *Mutat. Res.* **48**, 337–354.
13. Latt, S. A., Schreck, R. R., Loveday, K. S., Dougherty, C. P., and Shuler, C. F. (1980) *Adv. Hum. Genet.* **10**, 267–331.

14. Perry, P., and Evans, H. J. (1975) *Nature (London)* **258**, 121–125.
15. Friedberg, E. C., Ehmann, U. K., and Williams, J. I. (1979) *Adv. Radiat. Biol.* **8**, 85–174.
16. German, J. (1972) *Prog. Med. Genet.* **8**, 61–101.
17. Cairns, J. (1981) *Nature (London)* **289**, 353–357.
18. Hicks, J. B., Strathern, J. N., and Klar, A. J. S. (1979) *Nature (London)* **282**, 478–483.
19. Hicks, J. B., and Herskowitz, I. (1976)*Genetics* **83**, 245–258.
20. Straus, D. S. (1981) *JNCI, J. Natl. Cancer Inst.* **67**, 233–241.
21. Nowell, P. C., and Hungerfore, D. A. (1960) *Science* **132**, 1497.
22. Rowley, J. D. (1980) *Annu. Rev. Genet.* **14**, 17–39.
23. Echols, H. (1981) *Cell* **25**, 1–2.
24. Sager, R. (1982) *In* "Tumor Cell Heterogeneity" (A. H. Owens, D. S. Coffey, and S. B. Baylin, eds.), pp. 411–423. Academic Press, New York.
24a. Herman, R., Weymouth, L., and Penman, S. (1978) *J. Cell Biol.* **78**, 663–674.
25. Morgan, T. H. (1911)) *Science* **34**, 384.
26. McClintock, B. (1950) *Proc. Natl. Acad. Sci. U.S.A.* **36**, 344–355.
27. Calos, M. P., and Miller, H. (1980) *Cell* **20**, 579–595.
28. Early, P., Huang, H., Davis, M., Calame, K., and Hood, L. (1980) *Cell* **19**, 981–992.
29. Max, E. E., Seidman, J. G., and Leder, P. (1979) *Proc. Natl. Acad. Sci. U.S.A.* **76**, 3450–3454.
30. Sakano, H., Huppi, K., Heinrich, G., and Tonegawa, S. (1979) *Nature (London)* **280**, 288–294.
31. Sakano, H., Maki, R., Kurosawa, Y., Roeder, W., and Tonegawa, S., (1980) *Nature (London)* **286**, 676–683.
32. Rovigatti, U. G., Rogler, C. E., Neel, B. G., Hayward, W. S., and Astrin, S. M. (1982) *In* "Tumor Cell Heterogeneity" (A. H. Owens, D. S. Coffey, and S. B. Kaylin, eds.), pp. 319–330. Academic Press, New York.
33. Tabin, C. T., Bradley, S. M., Bargmann, C. I., Weinberg, R. A., Papagorge, A. G., Scolnick, E. M., Dahr, R., Lowy, D. R., and Chang, E. H. (1982) *Nature (London)* **300**, 143–149.
34. Reddy, E. P., Reynolds, R. K., Santos, E., and Barbacid, M. (1982) *Nature (London)* **300**, 149–152.
35. Feinberg, A. P., and Vogelstein, B. (1983) *Nature* **301**, 89–92.
36. Pardoll, D. M., Vogelstein, B., and Coffey, D. S. (1980) *Cell* **19**, 527–536.
37. Feinberg, A. P., and Coffey, D. S. (1982) *Proc. Wistar Symp. Nucl. Envelope Nucl. Matrix, 2nd, 1981*, pp. 293–305.
38. Zbarsky, I. B., and Debov, S. S. (1948) *Dokl. Akad. Nauk SSSR* **63**, 795–798.
39. Zbarsky, I. B., and Georgiev, G. P. (1959) *Biochim. Biophys. Acta* **32**, 301–302.
40. Georgiev, G. P., and Chentsov, J. S. (1962) *Exp. Cell Res.* **27**, 573–576.
41. Zbarsky, I. B., Dmitrieva, N. P., and Yermolayeva, L. P. (1962) *Exp. Cell Res.* **27**, 573–576.
42. Narayan, K. S., Steele, W. J., Smetana, K., and Busch, H. (1967) *Exp. Cell Res.* **46**, 65–77.
43. Steele, W. J., and Busch, H. (1963) *Biochim. Biophys. Acta* **129**, 54–67.
44. Berezney, R., and Coffey, D. S. (1974) *Biochem. Biophys. Res. Commun.* **60**, 1410–1417.
45. Berezney, R., and Coffey, D. S. (1977) *J. Cell Biol.* **73**, 616–637.
46. Shaper, J. H., Pardoll, D. M., Kaufmann, S. H., Barrack, E. R., Vogelstein, B., and Coffey, D. S. (1979) *Adv. Enzyme Regul.* **17**, 213–248.
47. Vogelstein, B., Pardoll, D. M., and Coffey, D. S. (1980) *Cell* **22**, 79–85.

48. Pardoll, D. M., and Vogelstein, B. (1980) *Exp. Cell Res.* **128**, 466–470.
49. Nelkin, B. D., Pardoll, D. M., and Vogelstein, B. (1980) *Nucleic Acids Res.* **8**, 5623–5633.
50. Robinson, S., Nelkin, B. D., and Vogelstein, B. (1982) *Cell* **28**, 99–106.
51. Jackson, D. A., McCready, S. J., and Cook, P. R. (1981) *Nature (London)* **292**, 552–555.
52. vanEekelen, C. A. G., and vanVenrooij, W. J. (1981) *J. Cell Biol.* **88**, 554–563.
53. Berezney, R., and Coffey, D. S. (1977) *J. Cell Biol.* **73**, 616–637.
54. Miller, T. E., Huang, C.-Y., and Pogo, A. O. (1978) *J. Cell Biol.* **76**, 675–691.
55. Miller, T. E., Huang, C.-Y., and Pogo, A. O. (1978) *J. Cell Biol.* **76**, 692–704.
56. Barrack, E. R., and Coffey, D. S. (1980) *J. Biol. Chem.* **255**, 7265–7275.
57. Barrack, E. R., and Coffey, D. S. (1982) *Recent Prog. Horm. Res.* **38**, 133–195.
58. Matsui, S., Antoniades, G., Basler, J., Berezney, R., and Sandberg, A. A. (1981) *J. Cell Biol.* **91**, 60a.
59. Peters, K. E., Okada, T. A., and Comings, D. E. (1981) *J. Cell Biol.* **91**, 72a.
60. Blazsek, I., Vaukhonen, M., and Hemminski, K. (1979) *Res. Commun. Chem. Pathol. Pharmacol.* **23**, 611–626.
61. Hemminski, K., and Vainio, H. (1979) *Cancer Lett.* **6**, 167–173.
62. Ueyama, H., Matsuura, T., Numi, S., Nakayasu, H., and Ueda, K. (1981) *Life Sci.* **29**, 655–661.
63. Buckler-White, S. J., Humphrey, G. W., and Pigiet, V. (1980) *Cell* **22**, 37–46.
64. Zbarsky, I. B. (1981) *Mol. Biol. Rep.* **7**, 159–148.
65. Kato, H. (1980) *Cancer Genet. Cytogenet.* **2**, 69–78.
66. Hand, R., and German, J. (1975) *Proc. Natl. Acad. Sci. U.S.A.* **72**, 758–762.

A Point Mutation in β-Actin Gene and Neoplastic Transformation

TAKEO KAKUNAGA, JOHN LEAVITT,* HIROSHI HAMADA, AND TADASHI HIRAKAWA

Cell Genetics Section
National Cancer Institute
Bethesda, Maryland
and
*Linus Pauling Institute
Palo Alto, California

INTRODUCTION

It is obvious that neoplastic cell transformation consists of multistep and complex processes (1–5). Transformed phenotypes are qualitatively and quantitatively diverse, depending on the types of cells and tissues, stage of progression (time), and even the individual transformed cell. Numerous factors are involved in the transformation process. Transformed genotypes appear to be different between transformed cells. In the past, however, much research has been focused on the identification of changes in the proteins that are common and specific to all the transformed or tumor cells. This trend is rather strange if we consider the possible diversity of the process of transformation. It is well known that even in the case of simple mutation in bacteria, the changes in the protein molecules responsible for expression of mutated phenotypes can differ depending on the mutants. For example, arginine biosynthesis from glutamine in *Escherichia coli* involves eight enzymes. Synthesis of argine is affected by the mutation in *any one* of the genes encoding eight enzymes. Enzymatic inactivation of each enzyme can result from the mutation of the *different* mutable sites in *one* gene. Furthermore, different inactive enzyme molecules can result from *different* mutations of the *same* mutable site.

Thus, the possibility that there is one type of critical change in the particular protein that is responsible for expression of the transformed phenotype in all the transformed cells is extremely low, although it is not totally excluded.

It is our approach to investigate, thoroughly, the molecular nature of the changes in the proteins associated with neoplastic transformation of human fibroblasts regardless of whether the changes are commonly observed in many transformed cells or not. A universal picture of the mechanisms of neoplastic transformation may be developed from particulars. Here we describe the mutation in β-actin found in one chemically transformed human fibroblast cell line and propose the hypothesis on the role of this mutation in actin molecules in the neoplastic transformation.

A VARIANT FORM OF ACTIN IS PRESENT IN A CHEMICALLY TRANSFORMED HUMAN CELL LINE

Malignant human fibroblasts, transformed *in vitro* by single treatment with 4-nitroquinoline-1-oxide (6), and the untransformed parental cells, KD cells, were compared by two-dimensional electrophoresis of their proteins labeled with [^{35}S]methionine (Fig. 1). Approximatley 2% of protein species were lost or gained in the transformed cells among more than 1000 electrophoretically distinguishable protein species on the autoradiographs. Figure 1 shows autoradiographs of radioactive proteins extracted with Triton X-100 from normal human fibroblasts, KD cells, and the transformed line HuT-14, which were labeled with [^{35}S]methionine for 4 hours. Proteins A^x and a2 through all, are protein species present in only HuT-14 cells. Proteins a12 through a18 are present only in KD cells. The largest and most intensely labeled spot, designated A, in each gel consists of the normal β- and γ-actin (pI 5.3; molecular weight 42,000). A new protein designated A^x (pI 5.2, molecular weight 43,000), migrating very close to β- and γ-actin, was found in the soluble fraction of HuT-14 cells. A^x protein was also found in the Triton-insoluble fraction (7). The Triton-insoluble fraction of the cells was left on the culture dish after extraction of Triton-soluble components and consisted of the cytoskeleton and nuclear matrix (8).

The presence of A^x in the Triton-insoluble fraction and the close migration to β- and γ-actin suggest that A^x may be related to actin. Thus, the [^{35}S]methionine-labeled soluble proteins extracted from HuT-14 cells were incubated with purified anti-actin antibody. The resulting complexes were isolated by affinity binding of the antigen–antibody complex to *Staphylococcus aureus* and analyzed by one-

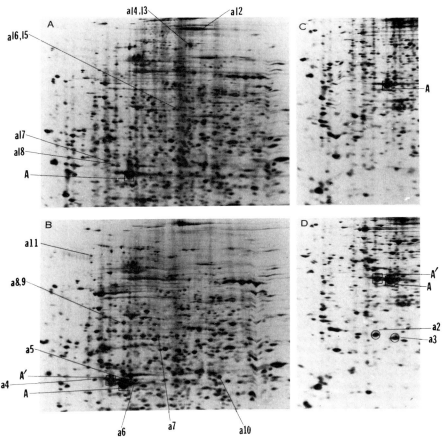

Fig. 1. Autoradiographs of two-dimensional gels containing [^{35}S]methionine-labeled Triton-soluble polypeptides. Polypeptides were separated in the first dimension by isoelectric focusing between pH range from 4 to 7 (right to left). A and C, polypeptides of KD cells; B and D, polypeptides of HuT-14 cells. A and B, 9% acrylamide gels; C and D, 12.5% acrylamide gels. From Leavitt and Kakunaga (7).

dimensional SDS gel-electrophoresis. A^x protein as well as β- and γ-actin were absorbed in the antibody–bacteria complex. Formation of a direct complex between A^x protein and anti-actin antibody was also demonstrated by incubation of A^x protein purified by two-dimensional electrophoresis with anti-actin antibody, indicating that A^x protein shares antigenic determinants with actin proteins (7).

To further examine the relation of A^x protein to actin, the tryptic peptide patterns of A^x protein and β- and γ-actin were compared

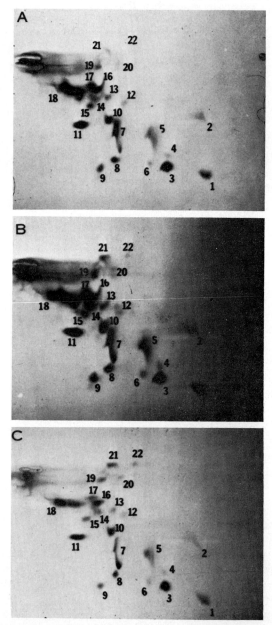

Fig. 2. Tryptic [^{35}S]methionine peptide patterns of β- and γ-actin and A^x polypeptides. (A) β- and γ-actin of KD cells; (B) β- and γ-actin of HuT-14 cells; (C) A^x of HuT-14 cells. Chromatography in the first dimension is from right to left. The numbered spots indicate positions of common tryptic peptides in each pattern. From Leavitt and Kakunaga (7).

(Fig. 2). The patterns exhibited by [^{35}S]methionine-labeled β- and γ-actin from KD cells, β- and γ-actin from HuT-14 cells, and A^x protein from HuT-14 cells were virtually identical. Thus it was concluded that A^x protein is a variant form of actin.

A VARIANT FORM OF ACTIN ENCODED BY THE β-ACTIN GENE THAT HAS A POINT MUTATION IN ITS STRUCTURAL GENE

The [^{35}S]methionine-labeled proteins were examined by autoradiography of two-dimensional gels with other normal and transformed human cells. A^x actin or any other new protein that migrates near actin were not detected in the cells freshly cultured from human embryo, human foreskin diploid fibroblasts, and human embryo lung diploid strain WI-38. Neither A^x-actin or similar protein were found in any of the transformed or tumor cells of human origin including HeLa, Widr (colon carcinoma cell line), HT1080 (fibrosarcoma cell line), and CCL-95.1 (SV40-transformed WI26 fibroblast line), nor has such an A^x-actin-like protein been found in a variety of normal and transformed cell lines from Syrian and Chinese hamster, monkey, mouse, and rat. These results suggest that the expression of A^x-actin originated from a genetic event that rarely occurs in nature.

Expression of A^x-actin may have originated from (1) altered post-translational processing or modification, (2) altered processing of actin mRNA, (3) derepression of a silent actin gene, and (4) mutation in actin genes. The first possibility was examined by analyzing the *in vitro* translation products synthesized in the presence of poly(A)$^+$ RNA extracted from HuT-14 cells (Fig. 3). Two-dimensional gel electrophoresis of the translation product showed that A^x-actin as well as β- and γ-actin were abundant translation products with either rabbit reticulocyte lysate or wheat germ lysate (Fig. 3A and B). On the other hand, A^x-actin was not detected among translation products of poly(A)$^+$RNA from another transformed human fibroblast line, HuT-11, which directed the synthesis of both β- and γ-actin (Fig. 3C). These results clearly indicate the presence of a mRNA species that codes for A^x-actin in the HuT-14 cells and exclude the first possibility that A^x-actin originated from altered post-translational processing or modification.

The relatedness of the coding sequences of the A^x, β-, and γ-actin gene were examined by assessing the complementarity of pcDd actin

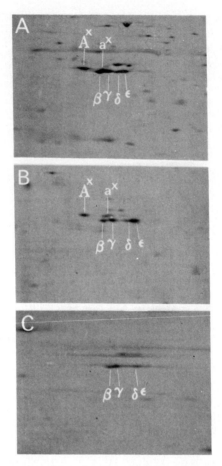

Fig. 3. Autoradiographs of two-dimensional gels containing *in vitro* translation products in the presence of poly(A)⁺ RNA from HuT-14 cells (A and B) or HuT-11 cells. (C). A and C, reticulocyte lysate system; B, wheat germ extract system. From Hamada et al. (9).

DNA, the DNA sequence complementary to *Dictyostelium* actin mRNA to the A^x-, β-, and γ-actin mRNA. Poly (A)⁺RNA was isolated from HuT-14 cells and hybridized to pcDd insert DNA, and the hybridized RNA was assayed for its ability to direct the synthesis of actin proteins in the reticulocyte lysate system. Two-dimensional gels of the translation products showed that the HuT-14 mRNA that hybridized to pcDd directed the synthesis of A^x- and β-actin and their respective unacetylated precursors; however, the mRNA that hybri-

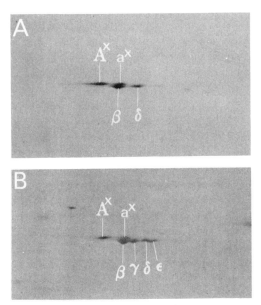

Fig. 4. Autoradiographs of two-dimensional gels containing *in vitro* translation products in the presence of HuT-14 poly(A)⁺RNA that hybridize to pcDd DNA (A) or pcHa-1 DNA (B). Rabbit reticulocyte lysate was used. From Hamada *et al.* (9).

dized to pcDd was incapable of directing synthesis of γ-actin and its precursor (Fig. 4A). In contrast, mRNA that hybridized to pcHa-1 DNA, the DNA sequence complementary to actin mRNA from HuT-14 cells (9), directed the synthesis of A^x-, β-, and γ-actin and their respective precursors (Fig. 4B). These results indicate that the pcDd-inserted DNA sequence is similar to the human β-actin gene sequence in that it hybridizes to β-actin mRNA and dissimilar to the human γ-actin mRNA to the degree in that it fails to hybridize to γ-actin mRNA. It is also indicated that A^x-actin mRNA has sequences homologous to β-actin mRNA. Thus, A^x-actin is likely a variant form of β-actin.

The sizes of A^x- and β-actin mRNA were compared by two different methods. First, HuT-14 poly(A)⁺RNA was fractionated on a sucrose gradient under undenaturing conditions, and then each fraction was tested for its ability to direct the synthesis of actin proteins in the reticulocyte system (9). Two-dimensional electrophoresis of the *in vitro* translates showed that A^x-actin as well as β- and γ-actin mRNAs are slightly larger than 18 S ribosomal RNA. Second, total RNA from

```
                                                        20                                            30
  2   Asp-Asp-Asp-Ile-Ala-Ala-Leu-Val-Val-Asp-Asn-Gly-Ser-Gly-Met-Cys-Lys-Ala-Gly-Phe-Ala-Gly-Asp-Asp-Ala-Pro-Arg-Ala-Val-
                        40                                             50                                            60
  Phe-Pro-Ser-Ile-Val-Gly-Arg-Pro-Arg-His-Gln-Gly-Val-Met-Val-Gly-Met-Gly-Gln-Lys-Asp-Ser-Tyr-Val-Gly-Asp-Glu-Ala-Gln-Ser-
                       70                    3Me                       80                                            90
  Lys-Arg-Gly-Ile-Leu-Thr-Leu-Lys-Tyr-Pro-Ile-Glu-His-Gly-Ile-Val-Thr-Asn-Trp-Asp-Asp-Met-Glu-Lys-Ile-Trp-His-His-Thr-Phe
                       100                                             110                                           120
  Tyr-Asn-Glu-Leu-Arg-Val-Ala-Pro-Glu-Glu-His-Pro-Val-Leu-Leu-Thr-Glu-Ala-Pro-Leu-Asn-Pro-Lys-Ala-Asn-Arg-Glu-Lys-Met-Thr-
                       130                                             140                                           150
  Gln-Ile-Met-Phe-Glu-Thr-Phe-Asn-Thr-Pro-Ala-Met-Tyr-Val-Ala-Ile-Gln-Ala-Val-Leu-Ser-Leu-Tyr-Ala-Ser-Gly-Arg-Thr-Thr-Gly-
                       160                                             170                                           180
  Ile-Val-Met-Asp-Ser-Gly-Asp-Gly-Val-Thr-His-Thr-Val-Pro-Ile-Tyr-Glu-Gly-Tyr-Ala-Leu-Pro-His-Ala-Ile-Leu-Arg-Leu-Asp-Leu-
                       190                                             200                                           210
  Ala-Gly-Arg-Asp-Leu-Thr-Asp-Tyr-Leu-Met-Lys-Ile-Leu-Thr-Glu-Arg-Gly-Tyr-Ser-Phe-Thr-Thr-Thr-Ala-Glu-Arg-Glu-Ile-Val-Arg-
                       220                                             230    234a
  Asp-Ile-Lys-Glu-Lys-Leu-Cys-Tyr-Val-Ala-Leu-Asp-Phe-Glu-Gln-Glu-Met-Ala-Thr-Ala-Ala-Ser-Ser-Ser-Leu-Glu-Lys-Ser-Tyr-
              240        Gly                                250                                         260
  Glu-Leu-Pro-Asp-Asp-Gln-Val-Ile-Thr-Ile-Gly-Asn-Glu-Arg-Phe-Arg-Cys-Pro-Glu-Ala-Leu-Phe-Gln-Pro-Ser-Phe-Leu-Gly-Met-Glu-
                       270                                             280                                           290
  Ser-Cys-Gly-Ile-His-Glu-Thr-Thr-Phe-Asn-Ser-Ile-Met-Lys-Cys-Asp-Val-Asp-Ile-Arg-Lys-Asp-Leu-Tyr-Ala-Asn-Thr-Val-Leu-Ser-
                       300                                             310                                           320
  Gly-Gly-Thr-Thr-Met-Tyr-Pro-Gly-Ile-Ala-Asp-Arg-Met-Gln-Lys-Glu-Ile-Thr-Ala-Leu-Ala-Pro-Ser-Thr-Met-Lys-Ile-Lys-Ile-Ile-
                       330                                             340                                           350
  Ala-Pro-Pro-Glu-Arg-Lys-Tyr-Ser-Val-Trp-Ile-Gly-Gly-Ser-Ile-Leu-Ala-Ser-Leu-Ser-Thr-Phe-Gln-Gln-Met-Trp-Ile-Ser-Lys-Gln-
                       360                                             370    374
  Glu-Tyr-Asp-Glu-Ser-Gly-Pro-Ser-Ile-Val-His-Arg-Lys-Cys-Phe
```

Fig. 5. Amino acid sequence of human cytoplasmic β- and A^x-actin isolated from HuT-14 cells determined by Vanderkerckhove *et al.* (10). The only amino acid difference between β-actin (Gly) and A^x-actin (Asp) in position **244** is enclosed in the box.

HuT-14 cells that synthesize A^x-actin in cells were fractionated on a agarose gel, transferred to diazobenzyloxymethyl paper, and hybridized to nick-translated pcDd insert [^{32}P]DNA that hybridizes with A^x- and β-actin mRNA. Only one band was observed in gels containing either HuT-14 or HuT-11 RNA. Thus, the size of A^x- and β-actin mRNA seems to be similar.

The entire amino acid sequences of A^x- and β-actin that were isolated from HuT-14 cells were determined in collaboration with Vandekerckhove and Weber (Fig. 5). A^x-actin had the N-terminal amino acid sequence B_1-Asp-Asp-Asp . . . , which is unique to β-actin, but differed from β-actin by only a single amino acid substitution. Glycine at 244 position in normal β-actin was substituted to aspartic acid in A^x-actin. This substitution corresponds to a GC → AT transition point mutation in the coding sequence. It is interesting to note that such a single amino acid substitution contributes to its significantly lower electrophoretic rate in the SDS dimension of the two-dimensional gel. The number of β-actin gene seems to be one copy per haploid human genome (8, and our unpublished data). Since HuT-14 cells express both A^x- and β-actin, there may be only one A^x-actin gene and one β-actin gene in the HuT-14 cells and a diploid β-actin gene set in the other human cells. Considering all of these findings together, it seems reasonable that A^x-actin is a product of a mutated β-actin gene in HuT-14 cells, and hereinafter A^x-actin is designated mutated β-actin (β^x-actin).

ADDITIONAL CHANGE IN THE β-ACTIN MOLECULE IN A VARIANT OF THE TRANSFORMED CELLS

In order to know further the role of a mutation of β-actin in the expression of the transformed phenotypes in HuT-14 cells, we have isolated cell variants from HuT-14 cells. First, 18 subclones were isolated from the mass population of HuT-14 cells by simple subcloning without treatment with mutagens or carcinogens. All of these subclones synthesized β^x-actin as well as β- and γ-actin, indicating that expression of β^x-actin is stable through cell generations. They also showed no change in their potentials to grow in soft agar and to form tumors in nude mice (Table I). Next, human cells were isolated and recultured from the tumors formed by subcutaneous injection of 2×10^6 HuT-14 cells per animal in nude mice. They also showed no change in their abilities to grow in soft agar and to form tumors in

TABLE I
Transformed Phenotypes of Subclones and Variants of HuT-14 Cells

Cells	Cloning efficiency in soft agar (%)	Tumorigenicity[a] (TD$_{50}$)
Normal diploid fibroblasts (KD)	<0.0001	>1 × 10^7
HuT-14	1.1	3 × 10^5
HuT-14-11[b]	0.9	3 × 10^5
HuT-14-12[b]	0.9	NT[f]
HuT-14-13[b]	1.0	NT
HuT-14-T[c]	0.8	3 × 10^5
HuT-14-3[d]	3.5	NT
HuT-14-3-1[d]	3.8	3 × 10^4
HuT-14-3-1-T[e]	8.5	<1 × 10^4

[a] Cell doses required for inducing tumors in 50% nude mice injected.
[b] Subclones isolated from HuT-14 cell population.
[c] Cells derived from the tumor, which was produced by subcutaneous injection of 2 × 10^6 HuT cells in nude mice.
[d] Subclones isolated from UV-irradiated HuT cell population.
[e] Cells derived from the tumor which was produced by subcutaneous injection of 2 × 10^6 HuT-14-3-1 cells in nude mice.
[f] Not tested.

nude mice: they also did not show alterations of expression of β^x-actin.

Thus, to induce the mutation or alteration of transformed phenotypes, HuT-14 cells were exposed to ultraviolet light (UV) and surviving cells were subcloned twice.

Although most of subclones obtained after exposure to UV did not show changes in their abilities to form colonies in soft agar or tumors in nude mice, one subclone, HuT-14-3-1, showed a significantly increased anchorage-independent cell growth and tumorigenicity (Table I). The cells were recovered from one of the tumors that arose during a period of 7 weeks after injection of 2 × 10^6 HuT-14-3-1 cells, established as a tumor cell line, and designated HuT-14-3-1 T. When the proteins synthesized in the HuT-14-3-1 T cells were examined using two-dimensional gel electrophoresis, the mutant β-actin in HuT-14-3-1 cells was shifted one additional unit negative charge as compared to mutant β-actin in HuT-14 cells (Fig. 6). The trace amounts of unacetylated precursor of this new mutant β-actin migrated in the electrophoretic position of the mutant β-actin of HuT-14

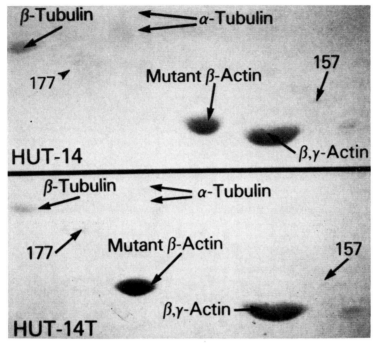

Fig. 6. Autoradiographs of two-dimensional gels containing [^{35}S]methionine-labeled (for 2 hr) actins and other neighboring proteins in Triton-soluble fraction of HuT-14 and HuT-14-3-1 T cells. From Leavitt et al. (11).

cells. The electrophoretic mobility of the new mutant β-actin in the SDS dimension was reduced like the original mutant β-actin of HuT-14 cells. This new species protein was identified as a variant form of actin by tryptic peptide fingerprint pattern, and tentatively designated new mutant β-actin or β^{xx-} actin. Examination of actin proteins of intermediate subclone HuT-14-3, and other subclones such as HuT-14-3-2 and HuT-14-3-3 revealed that the additional change in isoelectric point for the mutant β-actin occurred at the first step of isolation of HuT-14-3 line from HuT-14 cells with UV irradiation. The new mutant β-actin was synthesized by in vitro translation of poly(A)$^+$RNA extracted from either HuT-14-3-1 or HuT-14-3-1 T cells using reticulocyte lysate (Fig. 7). This result indicates that new mutant β-actin in the HuT-14-3 subclones is encoded within the messenger RNA.

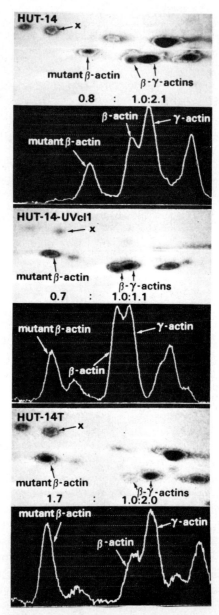

Fig. 7. Autoradiographs and densitometer tracing of two-dimensional gels containing [^{35}S]methionine-labeled proteins synthesized from poly(A)$^+$ RNA isolated from HuT-14, HuT-14-3-1, and HuT-14-3-1 T cells by *in vitro* translation using a rabbit reticulocyte lysate. From Leavitt *et al.* (11).

INCREMENTAL INCREASES IN THE MALIGNANCY ASSOCIATED WITH THE INCREASED ABNORMALITY OF MUTATED β-ACTIN MOLECULES

The density of each actin species on the autoradiographs of two dimensional gel containing [^{35}S]methionine-labeled proteins was quantified by computerized microdensitometry. The relative densities of mutant β-actin, normal β-actin, and γ-actin were compared to KD, HuT-14, HuT-14-3-1, and HuT-14-3-1 T with cellular soluble fraction, cellular insoluble fraction, and *in vitro* translation products, respectively (Table II). The ratio of mutant β-actin to normal β-plus

TABLE II
The Stepwise Changes in the Cytoplasmic Actins and Transformed Phenotypes in the Variants of HuT-14 Cells

		Cells		
	K_D	HuT-14	HuT-14-3-1	HuT-14-3-1T
Plating efficiency in soft agar (%)	<0.0001	1.1	3.8	8.5
Tumorigenicity[a] (TD_{50} in nude mice)	>10^7	3×10^5	3×10^4	<10^4
Relative ratio of actin[b] in cellular soluble fraction				
β^x or β^{xx}	0	1.2	NT[c]	1.7
β	1.0	1.0	NT	1.0
γ	1.2	1.2	NT	1.2
$\beta^x + \beta^{xx}/\beta + \gamma$	0	0.53	—	0.7
In insoluble fraction (cytoskeleton)				
β^x or β^{xx}	0	0.6	NT	0.04
β	1.0	1.0	NT	1.0
γ	1.2	1.2	NT	1.2
$\beta^x + \beta^{xx}/\beta + \gamma$	0	0.27	—	0.02
In *in vitro* translation				
β^x or β^{xx}	0	0.8	0.7	1.7
β	1.0	1.0	1.0	1.0
γ	1.0	2.0	1.1	2.0
$\beta^x + \beta^{xx}/\beta + \gamma$	0	0.26	0.36	0.57

[a] See footnotes to Table I.

[b] Relative density ratio of the actin species to the β-actin (1.0) on the autoradiograph of two dimensional gels containing [^{35}S]methionine-labeled proteins. β^x, mutant β-actin; β^{xx}, new mutant β-actin.

[c] Not tested.

γ-actin was increased in soluble fraction of HuT-14-3-1 T cells, and in the *in vitro* translation products synthesized in the presence of poly(A)⁺RNA of Hut-14-3-1 and HuT-14-3-1 T cells compared to HuT-14 cells. The ratio of mutant β-actin to normal β- plus γ-actin of cytoskeletal fraction was (1) always lower than that of soluble fraction or *in vitro* translation products and (2) significantly reduced in HuT-14-3-1 T cells compared to that in HuT-14 cells. There was no change in the ratio of normal β-actin to γ-actin between soluble and cytoskeletal fractions in all the cells tested. These results indicate that mutant β-actin is defective in incorporation into cytoskeleton.

On the other hand, the actual amount of the two mutant β-actin proteins determined by the densitometric measurement performed on Coomassie Blue-stained two-dimensional gels indicated that the quantity of mutant β-actin relative to normal β- and γ-actin was reduced by approximately 95% in HuT-14-3-1 T cells compared to that in HuT-14 cells (11). These findings suggest the increased rate of synthesis and short life of new mutant β-actin. In fact, the autoradiographic density of the new mutant β-actin in HuT-14-3-1 T cells was greatly enhanced when the cells were labeled with [^{35}S]methionine for a short time, either 8 or 15 min. With such a short labeling incubation, the new mutant β-actin in HuT-14-3-1 T cells accumulated at a 50% faster rate than its counterpart in HuT-14 cells. During continued incubation in the absence of label, but in the presence of excess nonradioactive methionine, the amount of label in the new mutant β-actin in HuT-14-3-1 T cells decreased rapidly, with a half-life of approximately 2 hr. In contrast, the mutant β-actin of HuT-14 cells was almost as stable as the normal β- and γ-actins, which do not appear to be degraded during chasing incubation time. It is noticeable that the amounts or efficiency of mRNA encoding the mutant β-actin in the HuT-14-3-1 T is significantly higher than those in other cells.

The comparison of these changes in the cytoplasmic actins with the changes in the transformed phenotypes, i.e., the potentials to grow in soft agar gel and to form tumors in nude mice, as summarized in Table II, indicate that incremental increases in malignancy of the transformed KD cell line and its subclones correlate with the increased abnormality of β-actin.

A HYPOTHESIS ON THE ROLE OF ACTIN MUTATION IN THE NEOPLASTIC TRANSFORMATION

Alteration of cell shape, motility, and membrane fluidity, which result in the disorganized cell arrangement and orientation, uncontrolled cell growth, and abnormal response to the environment, have been considered one of the most striking and universal phenotypes characteristics of neoplastic cells (12–19). Cytoplasmic actins, i.e., β- and γ-actin, are the major molecules of the cytoskeleton of which function is involved in maintaining and controlling cell morphology, motility, and membrane fluidity. Thus, it is conceivable that a mutation in β-actin leads the cells to express the transformed phenotypes by disrupting cytoskeletal functions. Changes in the organization of the cytoskeletal elements have been observed in many transformed cells (18,19). Indirect immunofluorescence staining with monoclonal, fluorescein-conjugated anti-actin antibody showed that organized, long linear cytoplasmic structures, referred to as *stress fibers* or *actin cables*, were observed in KD cells. In HuT-14 and its subclones, these well organized cytoskeletal structures were replaced by unorganized and short microfilamentous fibers and by a diffuse, nonstructural cytoplasmic staining. This disorganization of actin structures was more pronounced from HuT-14 to HuT-14-3-1 to HuT-14-3-1 T cells. In addition to the *in vivo* evidence as shown in Table II, suggesting defectiveness of the mutant β-actin in their ability to be incorporated into cytoskeletal structures, the mutated β-actin molecule has demonstrated deficiency in the *in vitro* polymerization of G-form (G-actin) to F-form (F-actin) (S. Taniguchi and T. Kakunaga, unpublished data).

Considering all the results described here and information accumulated on the cytoskeleton, we propose the hypothesis of the role of abnormal cytoskeletal function in the neoplastic transformation as shown in Fig. 8. Mutant β-actins are defective in polymerization, stability, or interaction with actin associated proteins. These deficiencies of β-actin will result in defects of structural stability or function of microfilaments or microfilament bundles.

Extraordinary conservation of cytoplasmic actins through evolutions and absence of growing cells or tissues lacking cytoplasmic actins strongly suggest the critical role of cytoplasmic actins in the survival or growth of cells. Thus, the defects produced by mutant β-actins may be most significant in the functional response to the regulatory signals or modifications. Similar abnormality of cytoskeletal function will also be induced by the mutations in actin associated proteins or the alteration of modification, such as phosphorylation, of actin-associated proteins that regulate the condition of microfilaments or microfilament bundles. The increased level of tyrosine phosphorylation of one of the actin associated proteins, vinculin, has been reported in cells transformed by Rous sarcoma virus (20). Alteration of tropomyosin has been observed in various transformed cells (F. Matsumura, personal communication). Mimicry of transformation by tumor-pro-

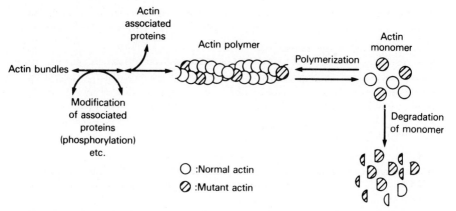

Fig. 8. Scheme for the disruption of cytoskeletal functions induced by a point mutation in β-actin in a chemically transformed human fibroblast line, HuT-14. Mutant β-actins are defective in polymeriization, stability, and interaction with actin associated proteins.

moting agents is preceeded by the disorganization of the actin cable (our unpublished data). Further studies are in progress to understand the exact role of a point mutation of β-actin in the neoplastic transformation.

ACKNOWLEDGMENT

We thank Dr. R. Firtel for providing us with pcDd actin ITL-1, Mr. A. Leavitt, Mrs. M. Petrino, C. Augl and G. Bushar for experimental help, and L. Nischan for secretarial assistance.

REFERENCES

1. Foulds, L. (1969) "Neoplastic Development," Vol. 1. Academic Press, New York.
2. Foulds, L. (1975) "Neoplastic Development," Vol. 2. Academic Press, New York.
3. Kakunaga, T., and Kamahora, J. (1969) *Symp. Cell. Chem.* **20**, 135–148.
4. Barrett, J. C., and Ts'o, P. O. P. (1978) *Proc. Natl. Acad. Sci. U.S.A.* **75**, 3761–3765.
5. Kakunaga, T. (1980) *Adv. Mod. Environ. Toxicol.* **1**, 355–382.
6. Kakunaga, T. (1978) *Proc. Natl. Acad. Sci. U.S.A.* **75**, 1334–1338.
7. Leavitt, J., and Kakunaga, T. (1980) *J. Biol. Chem.* **255**, 1650–1661.
8. Brown, S., Levinson, W., and Spudich, J. A. (1976) *J. Supramol. Struct.* **5**, 119–130.
9. Hamada, H., Leavitt, J., and Kakunaga, T. (1981) *Proc. Natl. Acad. Sci. U.S.A.* **78**, 3634–3638.

10. Vandekerckhove, J., Leavitt, J., Kakunaga, T., and Wever, K. (1980) *Cell* **22**, 893–899.
11. Leavitt, J., Bushar, G., Kakunaga, T., Hamada, H., Hirakawa, H., Goldman, D., and Merril, C. (1982) *Cell* **28**, 259–268.
12. Dulbecco, R. (1970) *Nature (London)* **227**, 802–806.
13. Abercrombie, M., and Heaysman, J. E. M. (1976) *JNCL, J. Natl. Cancer Inst.* **56**, 561–570.
14. Abercombie, M., Dunn, G. A., and Heath, J. P. (1977) *In* "Cell and Tissue Interaction" (J. W. Lash and M. M. Burger, eds.), pp. 57–70. Raven Press, New York.
15. Edelman, G. M. (1976) *Science* **192**, 218–226.
16. Nicolson, G. L. (1976) *Biochim. Biophys. Acta* **458**, 1–72.
17. Fox, C. H., Kakunaga, T., and Auer, G. (1978) *Acta Cytol.* **22**, 417–424.
18. Pollack, R., Osborn, M., and Weber, K. (1975) *Proc. Natl. Acad. Sci. U.S.A.* **72**, 994–998.
19. Pollack, R. (1980) *In* "Cancer: Achievements, Challenges and Prospects for the 1980's" (J. H. Burchenal and H. F. Oettgen, eds.), Vol. 1, pp. 501–515. Grune & Stratton, New York.
20. Sefton, B. M., Hunter, T., Ball, E. H., and Singer, S. J. (1981) *Cell* **24**, 165–174.

Studies of Mitochondria in Carcinoma Cells with Rhodamine-123

LAN BO CHEN, THEODORE J. LAMPIDIS,
SAMUEL D. BERNAL, KAREN K. NADAKAVUKAREN,
AND IAN C. SUMMERHAYES

Dana-Farber Cancer Institute
Harvard Medical School
Boston, Massachusetts

INTRODUCTION

We have discovered the use of rhodamine and cyanine dyes to localize mitochondria in living cells by epifluorescence microscopy (2,12–14). The remarkable specificity appears to result from the high membrane potential (negative inside) across mitochondria on the one hand, and the net positive charge carried by these dyes at physiological pH on the other. These mitochondrial supravital dyes have been employed in a number of studies. They include monitoring mitochondrial membrane potential in living cells (13), comparing normal and transformed fibroblasts for dye accumulation (12), assessing the efficacy of chemotherapy (1), detecting mitochondria fragmentation (2), monitoring mitochondrial proliferation during the cell cycle (11) and lymphoblast transformation (6), measuring differences in dye uptake between growing and resting fibroblasts (5) and between young and aged fibroblasts (9), following fusion between karyoplasts and cytoplasts (10), selection of cells with mutation in mitochondria (20), probing the role of mitochondria in erythroleukemia cell differentiation (15), following mitochondrial transformation of mammalian cells (3,4) and several other studies (7,16–19). The following is a summary of the recent findings on mitochondria in carcinoma cells.

PROLONGED RETENTION OF RHODAMINE-123 BY CARCINOGEN-TRANSFORMED EPITHELIAL CELLS AND CARCINOMA-DERIVED CELLS

We have previously shown that a confluent monolayer of mouse bladder epithelial cells accumulates small amounts of rhodamine-123 (Rh-123) (13). Cells at the periphery of bladder epithelial outgrowths and along the edge of a wound within an epithelial sheet accumulate Rh-123 more than cells in other regions. However, even those brightly stained cells release Rh-123 within 2 hr. In contrast, several 7,12-dimethylbenz(a)anthracene (DMBA)-transformed mouse bladder epithelial cells, benzo[a]pyrene-transformed rabbit bladder epithelial cells, and butyl(4-hydroxybutyl)nitrosamine (BBN)-induced mouse bladder transitional cells carcinoma retained a significant level of Rh-123 in dye-free medium even after 24 hr. Some of these tumorigenic cell lines—MB48, MB49 and RBC-1—still retain brightly stained mitochondria 4 days after incubation in dye-free medium. Most tumorigenic bladder epithelial cell lines tested retained mitochondrial fluorescence in dye-free medium for a prolonged period of time as assessed by fluorescent microscopy (Table I). One BBN-induced mouse bladder carcinoma line SH257 is a notable exception of being a short retainer. To examine whether extraordinary

TABLE I
Rhodamine-123 Retention of Normal and Transformed Bladder Epithelial Cells after 24 hr in Dye-Free Medium

Cell type of cell lines		Fluorescence
Primary bladder epithelial cells		
Mouse		0
Rabbit		0
Carcinogen-transformed		
MB48	DMBA-mouse	90
MB49	DMBA-mouse	90
BBN-6	BBN-mouse	70
SH264	BBN-mouse	62
SH257	BBN-mouse	0
RBC-1	Benzo[a]pyrene-rabbit	84
Human bladder tumor lines		
EJ	Transitional cell carcinoma	90
J82	Transitional cell carcinoma	15
RT4	Transitional cell carcinoma	70
RT112	Transitional cell carcinoma	75
HT1376	Transitional cell carcinoma	93

Rh-123 retention by mitochondria is also a characteristic of human bladder carcinoma cells, five cell lines—EJ, J82, RT4, RT112 and HT1376—were analyzed for dye retention. Four of them, with the exception of J82, retained mitochondrial-associated Rh-123 for a pro-

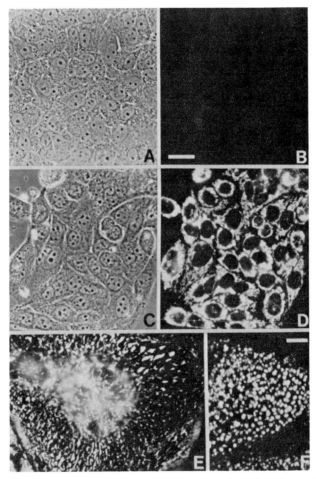

Fig. 1. Fluorescent micrographs (exposure time 2 sec) of Rh-123 retention by normal and carcinoma cells. Cells were stained with Rh-123 (10 μg/ml) for 10 min, washed, and kept in dye-free medium for 2 days. (B) Tertiary human breast epithelial cells; (D) human breast carcinoma, MCF-7; (E) human ovarian carcinoma, OVCA 433; (F) human cervical carcinoma, CaSki; (A) and (C) are phase contrast of (B) and (D), respectively. Bar represents 80 μm ub (B,D) and 30 μm in (E,F).

longed period, similar to that observed with carcinogen-transformed mouse bladder epithelial cells (Table I). We are now in the process of growing primary culture derived from normal human bladder for comparison.

In view of the difference in Rh-123 retention between normal mouse bladder epithelial cells and their tumorigenic counterparts, a

TABLE II
Rhodamine-123 Retention of Untransformed and Transformed Established Cell Lines after 24 hr in Dye-Free Medium

Cell lines		Fluorescence
Untransformed epithelial cell lines		
PtK-1	Kangaroo-rat	0
PtK-2	Kangaroo-rat	0
CV-1	Monkey kidney	0
BSC-1	Monkey kidney	0
Vero	Monkey kidney	0
MDCK	Canine kidney	5
Untransformed fibroblast lines		
3T3	Mouse embryo	18
Balb-3T3	Mouse embryo	5
NIH-3T3	Mouse embryo	10
BHK	Hamster kidney	15
Rat-1	Rat embryo	15
CCL 149	Rat lung	0
CCL 64	Mink embryo	40
C3H/10T½	Mouse embryo	12
Transformed epithelial cell lines		
THE	SV40-hamster	80
Ehrlich ascites	Mouse	60
CCL 51	Mouse, mammary	60
Lewis lung	Mouse	0
Transformed fibroblast lines		
SV3T3	SV40-3T3	10
Py3T3	Polyoma-3T3	10
NIH3T3-SRD	RSV-3T3	3
AnAn	RSV-rat	5
64F3	FeSV-CCL64	0
T8	Adenovirus II-rat	0
KiSV-3T3	KiSV-3T3	10
333-8-9	Herpes-hamster	15
Ann-1	Abelson MuLV-3T3	10
DMBA-Balb-3T3	DMBA transformed	12
X-Ray-C3H/10T½	X-Ray transformed	18
CCL 146	Gerbil fibroma	0
SV-BHK	SV40-hamster	15

TABLE III
Rhodamine-123 Retention of Human Tumor Lines after 24 hr in Dye-Free Medium

Cell lines		Fluorescence
Carcinoma		
MCF-7	Breast, adeno-	95
T47D	Breast, adeno-	95
ZR-75-1	Breast, adeno-	80
BT20	Breast, adeno-	80
HBL-100	Breast, adeno-	80
SW480	Colon, adeno-	80
HeLa	Cervix, adeno-	70
CaSki	Cervix, epidermoid	90
A431	Vulva, epidermoid	85
PaCa-2	Pancreas	90
PC-3	Prostate	90
SW13	Adrenal cortex, adeno-	90
OD562	Ovary	80
OVCA 433	Ovary	70
SCC-4	Tongue, squamous	60
SCC-9	Tongue, squamous	65
SCC-12	Skin, squamous	60
SCC-13	Skin, squamous	50
SCC-15	Tongue, squamous	40
SCC-25	Tongue, squamous	20
SCC-27	Skin, squamous	50
HUT 23	Lung, adeno-	75
HUT 125	Lung, adeno-	60
A549	Lung, adeno-	60
U1752	Lung, squamous	60
HUT 157	Lung, large cell	0
LX-1	Lung, poorly differentiated	0
OH-1	Lung, oat cell	0
HUT 60	Lung, oat cell	0
HUT 128	Lung, oat cell	0
HUT 231	Lung, oat cell	0
HUT 64	Lung, oat cell	0
Osteosarcoma		
N377/H135	Lung metastasis	10
Neuroblastoma		
CHP-100		0
CHP-134		0
CHP-212		0
Leukemia and lymphoma		
HL60		0
U937		0
K562		0
KG1		0
MOLT4		0
LA2-221		0
CEM		0
LA2-156		0

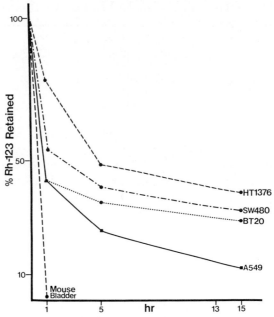

Fig. 2. Rh-123 retention (percent of initial uptake) measured by butanol extraction and fluorescent spectrophotometer. Cells were stained with Rh-123 (10 μg/ml) for 10 min, washed, and kept in dye-free medium. At various time points, Rh-123 was extracted with butanol and quantitated. Cells tested include mouse bladder epithelial cells and human carcinoma lines (HT1376, bladder; SW480, colon; BT20, breast; and A549, lung).

variety of established epithelial cell lines derived from kidney, ovary, pancreas, lung, adrenal cortex, skin, tongue, breast, prostate, cervix, vulva, and colon were examined for Rh-123 retention. All of these cell lines, verified to be of epithelial origin by keratin staining, were assayed for Rh-123 retention by fluorescent microscopy (Fig. 1; Table II and III), and a few of them were further analyzed for dye retention by butanol extraction (Fig. 2). With several exceptions the overall results show a remarkable pattern of short Rh-123 retention in nontumorigenic epithelial cell lines and cell lines established from normal epithelium, whereas many tumorigenic carcinoma-derived cells were characteristically long Rh-123 retainers. The most significant exceptions have been oat cell (five human lines and one mouse line, Lewis lung) and large cell carcinomas which display a short Rh-123 retention time. Unusual dye retention shown by mitochondria of carcinoma cells has thus far been rarely detected in established fibroblastic lines, transformed fibroblasts, sarcomas, leukemias, lymphomas, neuroblastomas, or osteosarcomas (Table II and III). The notable exception has been the mink fibroblastic line, CCL 64 (Table II).

SELECTIVE TOXICITY OF RHODAMINE-123 IN CARCINOMA CELLS *IN VITRO*

Figure 3 illustrates the difference in the effect of Rh-123 on the growth of transformed and normal epithelial cell lines. Section A and C illustrate the marked difference between treated and untreated cultures in two tumor-derived epithelial cell lines, human pancreatic carcinoma line, CRL 1420, and human breast carcinoma line, MCF-7. At 2 days of treatment, a marked cytotoxic effect is seen in cell line CRL 1420, while MCF-7 cells are inhibited in their growth. With further incubation, 3 days of CRL 1420 and 7 days of MCF-7, the cell number is reduced to less than 10^3/plate in each of the rhodamine-treated cultures. In marked contrast to these results, CV-1 and Ptk-1, derived from normal African green monkey kidney (epithelial) and normal marsupial kidney (epithelial), respectively, show no cytotoxicity

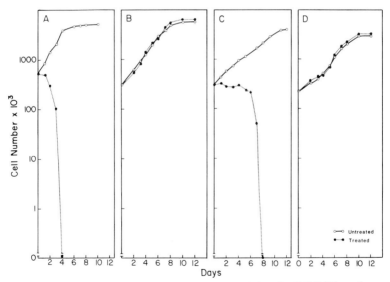

Fig. 3. Cell counts of carcinoma lines, CRL 1420 (A) and MCF-7 (C), and untransformed epithelial cell lines, CV-1 (B) and PtK-1 (D), continuously exposed to Rh-123 at 10 µg/ml. Each point represents the average count of triplicate cultures.

TABLE IV
Selective Killing by Rhodamine-123 in Carcinoma versus Normal Epithelial Cells

		Days of treatment required to produce ≥ 50% cell death rhodamine-123 (µg/ml)		Conversion of Rhodamine-123 staining from mitochondrial-specific to cytoplasmic
	Type nonspecific	10^a	50^a	
Carcinoma				
CRL 1420	Human pancreatic	3	2	Yes
EJ	Human bladder	3	2	Yes
CCL 77	Mouse Ehrlich-Lettre ascites	3	2	Yes
CCL 105	Human adrenal cortex adeno-	3	3	Yes
CCL 51	Mouse mammary	5	5	Yes
MCF-7	Human breast	7	3	Yes
CCL 15	Syrian hamster kidney	7	4	Yes
CCL 185	Human lung	7	6	Yes
CRL 1550	Human cervix	b	6	Yes
Normal epithelial				
CV-1	Monkey kidney	No effect^c (7)	(7)	No
BSC-1	Monkey kidney	No effect (7)	(7)	No
CCL 34	Canine kidney	No effect (7)	(7)	No
PtK-1	Marsupial kidney	No effect (7)	(7)	No
Tertiary	Human breast epithelial	No effect (7)	(7)	No
Primary	Mouse bladder epithelial	No effect (7)	(7)	NT^d

^{a} Applied continuously.
^{b} Growth inhibited.
^{c} Number of days of treatment at which no cytotoxic effects could be observed are indicated in parentheses. Cell viability was monitored by trypan blue exclusion.
^{d} NT Not tested.

when exposed to the same dose of Rh-123 (Fig. 3B and D). With even longer exposure (2 weeks at 10 µg/ml) or at higher doses (50 µg/ml for 7 days), cell growth of CV-1 and Ptk-1 remain relatively unaffected. In the course of these experiments, it was noted that a significant number of cells had detached from the culture dish surface after drug treatment (3 days for CRL 1420 and 7 days for MCF-7). We followed the fate of these cells in the presence of drug with increasing time and found that they gradually converted from trypan blue negative to trypan blue positive. In addition we have tested the ability of these detached cells, while still trypan blue negative, to proliferate in drug-free medium and found that they are no longer able to reattach to the culture dish surface nor divide in the suspension.

In order to determine whether this selective toxicity is a general phenomenon shared by other tumorigenic epithelial cells, a number of carcinoma lines were studied. Table IV is a composite of 15 different cell lines tested for sensitivity to Rh-123 when exposed continuously to various doses. It can be seen that all nine tumorigenic carcinoma-derived epithelial cell lines used in this study were susceptible to the cytotoxic effects of Rh-123 at 10, 25, or 50 μg/ml. The nontumorigenic epithelial cell lines derived from normal tissues were unaffected by treatment at these doses and continued to grow normally as untreated cells. In conjunction with these experiments, each cell line was tested for its sensitivity to the potent, broad spectrum antitumor agent Adriamycin. Although some of the carcinoma cell lines did appear to be slightly more sensitive to Adriamycin than the untransformed lines, the complete insensitivity of normal epithelial cells to Rh-123 at high doses, or for prolonged exposure times, suggests a stronger selectivity for Rh-123 than for Adriamycin in the cell lines tested in this study.

RHODAMINE-123 SELECTIVITY REDUCES CLONAL GROWTH OF CARCINOMA CELLS *IN VITRO*

MB49, a 7,12-dimethylbenz[a]anthracene (DMBA)-transformed mouse bladder epithelial line which is highly tumorigenic, retains a significant amount of Rh-123 in its mitochondria for 4 days after incubation in dye-free medium whereas primary cultures of normal bladder epithelial cells lose Rh-123 fluorescence within 2 hr. To determine whether this difference in retention results in greater inhibition of clonal growth in MB49 cells than in normal mouse bladder epithelial cells, we treated these cells grown *in vitro* with Rh-123 and assayed their colony forming ability. A 24-h exposure to 10 μg/ml of Rh-123 had a minimal effect on the colony forming units (CFU) of normal mouse bladder epithelial cells (92% of control) (Fig. 4A). In contrast, the CFU pf MB49 cells were markedly reduced to 4% of control by similar treatment with Rh-123. The effect of Rh-123 on MB49 cells depended upon the concentration and duration of exposure. Even a 6 hr

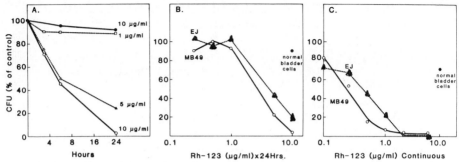

Fig. 4. Effects of Rh-123 on colong forming units (CFU) of normal mouse bladder cells (●), MB49 (○, □, △), and EJ (▲) bladder carcinoma cells.

exposure to 10 μg/ml of Rh-123 reduced CFU to 45% of control, whereas such treatment had no significant effect on the CFU of normal bladder epithelial cells.

We then compared the reduction of CFU of EJ (human bladder carcinoma line), MB49, and normal mouse bladder epithelial cells after 24 hr or continuous exposures to different concentrations of Rh-123 during the 2-week period of clonal cell growth (Fig. 4B, 4C). Continuous exposures of normal mouse bladder epithelial cells to 10 μg/ml of Rh-123 had only a small effect on CFU. However, both EJ and MB49 cells were susceptible to the inhibitory effects of Rh-123. Colony formation was reduced to 50% of control by 24 hr treatment with 2–5 μg/ml of Rh-123 (Fig. 4B) and by continuous treatment with 0.2– 0.5 μg/ml of Rh-123 (Fig. 4C).

We also compared the effects of Rh-123 on the CFU of other carcinoma and nontumorigenic cell lines, all of which have been confirmed to be of epithelial origin by antikeratin immunofluorescence. The carcinoma cells retained Rh-123 longer than nontumorigenic epithelial cells. BSC-1, a nontumorigenic monkey kidney epithelial line (Fig. 5A) and CCL 34, nontumorigenic dog kidney epithelial line (Fig. 5B) were found to be relatively insensitive to Rh-123. On the other hand, CCL 51, mouse breast carcinoma line (Fig. 5A), and HUT 23, human lung adenocarcinoma line (Fig. 5B), were very sensitive to the inhibitory effects of Rh-123. Similar to the bladder carcinoma, colony formation was reduced to 50% of control by continuous treatment with 0.2–0.5 μg/ml of Rh-123. Clonal growth of MCF-7 (human breast carcinoma line) and Ehrlich ascites (mouse carcinoma line) were also reduced to 50% of control by continuous treatment with 0.5 μg/ml of Rh-123, whereas PtK-1 (nontumorigenic marsupial kidney line) and CRL

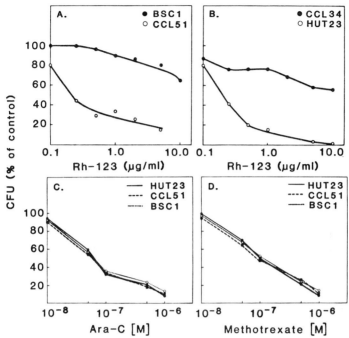

Fig. 5. Effect of Rh-123 on CFU of (A) BSC-1, a nontumorigenic monkey kidney epithelial line, and CCL 51, a mouse breast carcinoma line. (B) CCL 34, a nontumorigenic dog kidney epithelial line, and HUT 23, a human lung adenocarcinoma line. Effect of (C) Ara-C and (D) methotrexate on CFU on BSC-1, CCL 51, and HUT 23 cell lines.

1521 (normal human skin fibroblast line) had greater than 80% clonal growth under similar conditions. However, clonal growth of all the cell lines we tested were not reduced by 10 min exposure to 10 μg/ml of Rh-123. Thus, the conditions required for specific staining of mitochondria, for measurements of Rh-123 retention and for cell viability assays were not inhibitory to cells.

Unlike Rh-123, Ara-C and methotrexate (cell cycle-specific anticancer drugs) were not selectively inhibitory for carcinoma cells *in vitro* (Fig. 5C-5D). Since BSC-1, CCL 34, CCL 51, and HUT 23 had similar doubling times, it appeared that Ara-C and methotrexate inhibited clonal growth of cycling tumorigenic epithelial cells. These results also suggest that the selective inhibition of CFU by Rh-123 was not due to differences in cell-cycle kinetics between carcinoma and nontumorigenic epithelial cells.

ANTICARCINOMA ACTIVITY *IN VIVO* OF RHODAMINE-123 AND THE COMBINATION THERAPY WITH 2-DEOXYGLUCOSE

Control mice implanted intraperitoneally with 5×10^5 Ehrlich carcinoma cells had a median survival of 19 days (range 18–22 days) after tumor implantation in five experiments, reflecting consistency in mortality pattern (Fig. 6). Carcinoma-bearing mice treated with a maximum nonlethal dose of Rh-123 (15 mg/kg days 1,3,5) had a median survival of 50 days, T/C 260%. Lower doses of Rh-123 (5 mg/kg or 10 mg/kg days 1,3,5,7,9) were much less effective in prolonging survival (Fig. 6). When other maximum nonlethal doses, such as Rh-123 10 mg/kg (days 1–10) or 5 mg/kg (days 1–12) were used, T/C was 120% and 110%, respectively (Table V).

Treatment with 2-deoxyglucose alone (0.5 gm/kg days 1,3,5) did not prolong survival. However, combined treatment with 2-deoxyglucose (0.5 gm/kg days 1,3,5) and Rh-123 (15 mg/kg days 1,3,5) markedly pro-

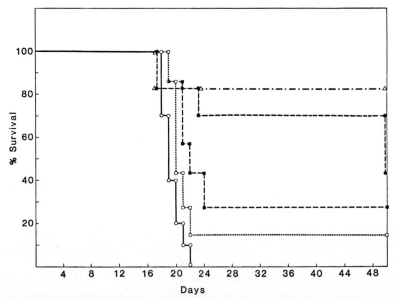

Fig. 6. Effect of Rhodamine-123 on survival of mice implanted with Ehrlich ascites carcinoma. The curves shown are representative of five separate experiments including 10 mice for controls of 7 mice in treatment groups for each experiment. ○, controls; □, Rh-123 (5 mg/kg days 1,3,5,7,9), T/C 105%; ■, Rh-123 (10 mg/kg days 1,3,5,7,9), T/C 115%; ●, Rh-123 (15 mg/kg days 1,3,5), T/C 260%; △, Rh-123 + 2-DG, T/C 420%.

TABLE V
Anticancer Activity of Rhodamine-123 on Long-Retaining Cells *in Vivo*

Type	T/C (treated/control median survival) (%)
Ehrlich ascites tumor	
Rh-123 (15 mg/kg days 1, 3, 5)	260
Rh-123 (10 mg/kg days 1, 3, 5, 7, 9)	115
Rh-123 (5 mg/kg days 1, 3, 5, 7, 9)	105
Rh-123 (10 mg/kg days 1–10)	120
Rh-123 (5 mg/kg days 1–12)	110
Rh-123 (15 mg/days 1, 3, 5) + 2-DG (0.5 g/kg days 1, 3, 5)	420
2-DG (0.5 mg/kg days 1, 3, 5)	100
MB49 bladder carcinoma	
Rh-123 (15 mg/kg days 1, 3, 5)	180
Rh-123 (10 mg/kg days 1, 3, 5)	110
Rh-123 (15 mg/kg days 1, 3, 5) + 2-DG (0.5 g/kg days 1, 3, 5)	250
2-DG (0.5 g/kg days 1, 3, 5)	100

longed survival, T/C 420% (Table V). Approximately 40% of the mice were cured (no evidence of tumor at 90 days).

We selected MB49 (mouse bladder carcinoma) for further experiments on the anticarcinoma activity of Rh-123 because we observed that Rh-123 markedly reduced clonogenic ability of MB49 and other bladder carcinoma cells without affecting clonogenicity of normal bladder epithelial cells *in vitro*. Control mice implanted intraperitoneally with 1×10^6 MB49 cells had a narrow distribution of deaths, with a median survival of 19 days (17–21 days). MB49-bearing mice treated with Rh-123 (15 mg/kg days 1,3,5) had a median survival of 27 days, T/C 180% (Table V). A lower dose of Rh-123 (10 mg/kg days 1,3,5) resulted in a T/C of 110%. Treatment with 2-deoxyglucose (0.5 gm/kg days 1,3,5) alone did not prolong survival of MB49-bearing mice. The combination of 2-deoxyglucose (0.5 gm/kg days 1,3,5) and Rh-123 (15 mg/kg days 1,3,5), however, markedly prolonged survival with a T/C of 250%. Mice treated with Ara-C (200 mg/kg days 1–5), used as a positive control, resulted in a T/C of 150%.

We have described that leukemia and melanoma cell lines, unlike carcinomas, release Rh-123 within 16–24 hours in rhodamine-free medium. Lewis lung carcinoma cells, like human oat cell carcinoma, contain keratin but are not long retainers of Rh-123 *in vitro*. Since L1210 leukemia, P388 leukemia, B16 melanoma, and Lewis lung carcinoma are commonly used to screen potential anticancer drugs, we

TABLE VI
Lack of Anticancer Activity of Rhodamine-123 on Short-Retaining Cells *in Vivo*

Type	T/C (treated/control median survival) (%)
L1210 Leukemia (IP)	
Rh-123 (10 mg/kg days 1–10)	110
Rh-123 (15 mg/kg days 1, 3, 5, 7, 9, 12)	110
Rh-123 (20 mg/kg days 1–3)	110
Rh-123 (20 mg/kg days 1, 3, 5, 7)	110
P388 Leukemia (IP)	
Rh-123 (10 mg/kg days 1–10)	110
Rh-123 (15 mg/kg days 1, 3, 5, 7, 9, 11)	110
Rh-123 (20 mg/kg days 1–4)	110
Rh-123 (20 mg/kg days 1, 3, 5, 7)	110
B16 Melanoma (IP)	
Rh-123 (10 mg/kg days 1–10)	115
Rh-123 (15 mg/kg days 1, 3, 5, 7, 9)	110
Rh-123 (20 mg/kg days 1–3)	100
Rh-123 (20 mg/kg days 1, 3, 5, 7)	100
Lewis lung carcinoma (IM)	
Rh-123 (5 mg/kg days 1–10)	110
Rh-123 (10 mg/kg days 1–10)	100
Rh-123 (20 mg/kg days 1–3)	100

tested the antitumor activity of Rh-123 in these various tumor models. As shown in Table VI, Rh-123 did not significantly prolong survival of mice implanted with these tumors.

MEASUREMENT OF RHODAMINE-123 UPTAKE BY CARCINOMA CELLS

Figure 7 illustrates the uptake over a 6 hour period of continuous incubation with 1 µg/ml of Rh-123, for a number of cell lines. The nonspecific contribution to the uptake measurements was subtracted before the data were plotted on a per cell basis. The lines that exhibited minimal uptake were the normal epithelial-derived nontumorigenic cell lines CV-1 (African green monkey kidney) and MDCK (canine kidney), and the feline sarcoma virus transformed mink fibroblast line, 64F3. Even after 6 hr of continuous incubation with 1 µg/ml Rh-123, the average uptake of Rh-123 not exceed 2×10^{-5} ng/cell.

In contrast to the above, there are cell lines characterized by a sub-

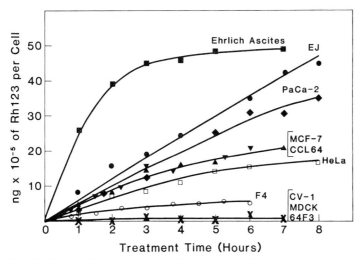

Fig. 7. Rhodamine-123 uptake per cell by various cell lines. Rh-123 (1 µg/ml) was present continuously. The background has been subtracted. EJ, human bladder carcinoma; PaCa-2, human pancreatic carcinoma; MCF-7, human breast carcinoma; HeLa, human cervix carcinoma; CCL 64, mink fibroblast; 64F3, FeSV-transformed CCL 64; CV-1, normal African Green monkey kidney epithelial; MDCK, normal dog kidney epithelial.

stantial uptake of Rh-123. Among those which we have studied, the cell line with the highest uptake was the Ehrlich ascites tumor (Fig. 7). This cell line was unique among those studied in that the uptake of Rh-123 was initially very rapid, but approached a plateau after only 3 hr at an average value of 5×10^{-4} ng/cell. This level of dye uptake is more than 20 times that of the CV-1, MDCK, or 64F3 lines. Two other lines that exhibited high levels of uptake, but did not appear to reach plateaus within 3–4 hr, were the human bladder carcinoma derived line EJ and the human pancreatic carcinoma derived line PaCa-2. Moderate but significant increases in Rh-123 uptake with time were seen in the normal mink fibroblast line CCL 64, the human breast carcinoma line MCF-7, and HeLa cells. The human adenovirus (type 2) transformed rat fibroblast lines, F4 and F18, showed lesser but detectable increases in Rh-123 uptake with time.

It has been shown that the accumulation of Rh-123 in living cells as observed by fluorescence microscopy is affected by mitochondrial membrane potential (13). Therefore, we examined the uptake of Rh-123 in the presence of and in the absence of the respiratory uncoupler FCCP, which is known to dissipate mitochondrial membrane poten-

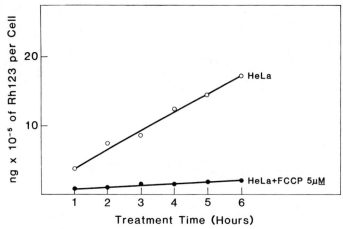

Fig. 8. Effect of FCCP (a proton ionophore that dissipates mitochondrial membrane potential) on Rh-123 uptake by HeLa cells. Rh-123 (1 µg/ml) or FCCP (5 µm) was present continuously in culture medium.

tial. Figure 8 illustrates the effect of FCCP (5 µm) on the uptake of Rh-123 by HeLa cells. In the absence of FCCP, HeLa cells exhibited a nearly linear uptake to the dye over the 6 hr time period shown. However, addition of 5 µm FCCP to the Rh-123-containing medium greatly reduced the progressive accumulation of the dye. At any given point, uptake in the presence of FCCP was a factor of 6–8 less than that in the absence of FCCP.

BASIS FOR RHODAMINE-123 TOXICITY

Why can Rh-123 kill carcinoma cells described above? In our earlier works in which most of the cell types used were normal fibroblasts or epithelial cells, Rh-123 was not found to be toxic. Thus, we have not explored possible inhibitory activities of Rh-123. Since Rh-123 is cytotoxic to Rh-123 long-retaining carcinoma, it is essential to investigate the basis for this toxicity. Rhodamine-6G has previously been shown to inhibit oxidative phosphorylation in isolated mitochondria (8). We perform similar experiments with rhodamine-123 (in collaboration with Drs. C. Salet and G. Moreno at Muséum National d'Histoire Naturelle, Paris, France). Figure 9 shows that Rh-123 at 5 µg/ml reduced the respiratory control ratio (RCR) of isolated liver mitochondria from 4.0 to 1.0, whereas rhodamine-B and rhodamine-116, which are un-

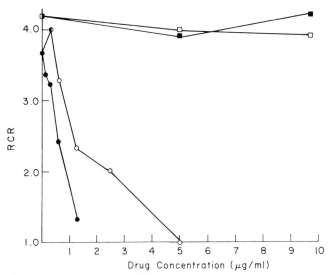

Fig. 9. Respiratory control ratios (RCR) (the rate of O_2 consumption in the presence of ADP divided by the rate of O_2 consumption without ADP) were determined in isolated rat liver mitochondria for each of the four rhodamines tested at the indicated drug concentrations. Mitochondria were isolated in a medium consisting to 0.25 M sucrose, 1 mM EDTA, and 1 mM Tris at pH 7.4. The last centrifugation was done in the sucrose medium without EDTA, and the final pellet was suspended in 0.25M sucrose in order to achieve a concentration of 100 mg/ml of proteins. In a thermostated (25°C) water-jacketed glass vessel, 20 μl of the mitochondrial suspension were added to 1.4 ml of respiratory medium (24 mM glycylglycine, 10 mM $MgCl_2$, 60 mM KCl, 7 mM KH_2PO_4, 87 mM sucrose). Oxygen consumption was measured polargraphically with a Clark oxygen electrode (Yellow Springs Instrument) connected to a dual-channel chart recorder. Succinate (14 mM) was used as the substrate. Experiments were conducted on different days and control RCRs are indicated for each drug tested: ●, 6G; ○, 123; ■, 116; □, B.

charged at physiological pH and unable to stain mitochondria in living cells, do not have any effect on RCR in the same experiments. Rhodamine-6G, which has positive charges at physiological pH and able to stain mitochondria, is inhibitory for RCR as reported previously and confirmed here. It is likely that the basis for cytotoxic activity of Rh-123 on carcinoma cells is the inhibition of mitochondrial oxidative phosphorylation. It should be noted, however, that there may be additional targets for Rh-123. Before cell death we detected that Rh-123 was released from mitochondria and stayed in the cytoplasm for a significant period of time (12–24 hr). Rh-123 may damage cytoplasm further, thereby causing cell death. Potential cytoplasmic targets for Rh-123 are currently under investigation.

SUMMARY

In spite of modern therapies, cancers of the lung, breast, and colon are still the major causes of deaths due to cancer in the United States. Therefore, there is a particular need to develop new and more selective drugs against these epithelial tumors. Rhodamine-123 (Rh-123), a supravital dye for mitochondria, is used to compare the mitochondria of normal epithelial cells with carcinoma cells. Whereas normal or untransformed cells release the dye within 2-16 hr, many carcinoma cell lines retain a significant level of dye even after 2-4 days. This phenotype is exploited as a target for chemotherapy of carcinoma cells. Successful treatment of Ehrlich ascites carcinoma in a mouse with Rh-123 and 2-deoxyglucose is demonstrated. It appears that the selective killing of carcinoma cells is achieved by longer retaining of Rh-123, a drug appeared to be a mild inhibitor for mitochondrial oxidative phosphorylation.

ACKNOWLEDGMENTS

This work has been supported by grants from the National Cancer Institute (CA 22427, CA 22659, CA 29793) and American Cancer Society to L. B. Chen. T. J. Lampidis is supported by NCI New Investigator Award. S. D. Bernal is supported by an NCI Postdoctoral Fellowship. K. K. Nadakavakaren is supported by an American Cancer Society Postdoctoral Fellowship. L. B. Chen is supported by the American Cancer Society Faculty Research Award.

REFERENCES

1. Bernal, S. D., Shapiro, H. M., and Chen, L. B. (1982) *Int. J. Cancer* **30**, 219-224.
2. Chen, L. B., Summerhayes, I. C., Johnson, L. V., Walsh, M. L. Bernal, S. D., and Lampidis, T. J. (1982) *Cold Spring Harbor Symp. Quant. Biol.* **46**, 141-155.
3. Clark, M. A., and Shay, J. W. (1982) *Nature (London)* **195**, 605.
4. Clark, M. A., and Shay, J. W. (1982) *Proc. Natl. Acad. Sci. U.S.A.* **79**, 1144.
5. Cohen, R. L., Muirhead, K. A., Gill, J. E., Waggoner, A. S., and Horan, P. K. (1981) *Nature (London)* **290**, 593-595.
6. Darzynkiewicz, A., Staiano-Coico, L., and Melamed, M. R. (1981) *Proc. Natl. Acad. Sci. U.S.A.* **78**, 2383-2387.
7. Darzynkiewicz, Z., Traganos, F., Staiano-Coico, L., Kapuscinski, J., and Melamed, M. R. (1982) *Cancer Res.* **42**, 799-806.
8. Gear, A. R. L. (1974) *J. Biol. Chem.* **249**, 3628.
9. Goldstein, S., and Korczack, L. B. (1981) *J. Cell Biol.* **91**, 392-398.

10. Hightower, M. J., Fairfield, R., and Lucas, J. J. (1981) *Somatic Cell Genet.* **7**, 321–329.
11. James, T. W., and Bohman, R. (1981) *J. Cell Biol.* **89**, 256–263.
12. Johnson, L. V., Summerhayes, I. C., and Chen, L. B. (1982) *Cell* **28**, 7–14.
13. Johnson, L. V., Walsh, M. L., Bockus, B. J., and Chen, L. B. (1981) *J. Cell Biol.* **88**, 526–535.
14. Johnson, L. V., Walsh, M. L., and Chen, L. B. (1980) *Proc. Natl. Acad. Sci. U.S.A.* **77**, 990–994.
15. Levenson, R., Macara, I. G., Smith R. L., Cantley, L., and Housman, D. (1982) *Cell* **28**, 855–863.
16. Maro, B., and Bornens, M. (1982) *Nature (London)* **295**, 334.
17. Pagano, R. E., Longmuir, K. J., Martin, O. C., and Struck, D. K. (1981) *J. Cell Biol.* **91**, 872.
18. Siemens, A., Walter, R., Liaw, L. H., and Berns, M. W. (1982) *Proc. Natl. Acad. Sci. U.S.A.* **79**, 466–470.
19. Steinkamp, J. A., and Hieber, R. D. (1982) *Cytometry,* **2**, 232.
20. Ziegler, M. L., and Davidson, R. L. (1981) *Somatic Cell Genet.* **7**, 73–88.

Index

A

Acetate, inhibition of tumor promotion and, 128
N-Acetoxy-N-2-acetylaminofluorene
  mechanism of mutagenesis
    discussion, 76–80
    material and methods, 70–71
      DNA sequence analysis of mutants, 71
      E. coli transformation and selection of mutants, 71
      irradiation of cells before transformation, 71
    results, 72–73
      hot spot sequence 1, 76
      hot spot sequence 2, 76
      mutation frequency and restriction enzyme analysis of mutants, 73–75
      sequence analysis of class II mutants, 75–76
Acetylaminofluorene
  development of resistance to MTX and, 272, 277, 280
  modification of (polydG-polydC)-(polydG-polydC) by, 25–31
  orientation to axis of DNA, 17
  reactive metabolite, 14
  tumor induction by, thyroidectomy and, 198
Actin(s)
  cytoplasmic, function of, 365
  mutation, hypothesis on role in neoplastic transformation, 364–366
  variant form
    in a chemically transformed human cell line, 352–355
    encoded by the β-actin gene having a point mutation, 355–359

β-Actin
  additional changes in a variant of the transformed cell, 359–362
  increased abnormality of mutated molecules, incremental increases of malignancy associated with, 363–364
  mutant
    half-life of, 364
    incorporation into cytoskeleton, 364
    rate of synthesis, 364
    ratio to normal actins, 363–364
    stability through cell generations, 359
  point mutation in a transformed cell, 359
Adenosine triphosphatase, Na, K-dependent, thyroid hormone and, 203
Adenosine triphosphate, endocytosis and, 189
Adenovirus
  entry into cells, 193
  transformation of cells in culture, thyroid hormone modulation of, 204–206
Adriamycin, sensitivity of carcinoma cell lines to, 377
Adult T-cell leukemia virus
  properties of
    associated antigens, 215–216
    biochemical properties, 215
    biologic activities, 217–218
    cell lines carrying virus, 218–219
    monoclonal antibodies to, 216–217
    morphological observations, 214–215
  relation to ATL
    detection of virus in fresh leukemic cells of patients, 219
    healthy carriers, 219–220

natural human antibody, 220-221
African green monkey, kidney cells, phases of SV40 infection in, 296-297
Afterburners, elimination of nitroarenes from diesel engine emissions, 92
Age, natural antibody to ATL virus and, 220
Ah locus
  role in determining histological response to TCDD, 149-153
  TCDD receptor and, 148
Amino acid(s)
  pyrolysates, new mutagens from, 112-114
  sequences
    of A^x- and β-actin, 358-359
    oncogenes and, 248
    of rat and human TGFs, 175-176
2-Amino-3,4-dimethylimidazo[4,5-f]quinoline, from broiled sardines, 113
2-Amino-3,8-dimethylimidazo[4,5-f]quinoxaline, from broiled beef, 113
3-Amino-1,4-dimethyl-5 H-pyrido[4,3-b]indole, pyrolysis of tryptophan and, 112
2-Amino-dipyrido[1,2-a:3′,2′-d]imidazole, pyrolysis of glutamic acid and, 112-113
2-Aminofluorene, tumor induction by, thyroidectomy and, 198
2-Amino-6-methyldipyrido[1,2-a:3′,2′-d]imidazole, pyrolysis of glutamic acid and, 112
2-Amino-3-methylimidazo[4,5-f]quinoline, broiled sardines and, 113
2-Amino-3-methyl-9H-pyrido[2,3-b]indole, from pyrolysis of soy bean globulin, 113
3-Amino-1-methyl-5H-pyrido[4,3-b]indole
  activation of, 114
  pyrolysis of tryptophan and, 112
2-Amino-9H-pyrido[2,3-b]indole, from pyrolysis of soy bean globulin, 113
Ampicillin, selection of mutants resistant to, 71
Anchorage-independent growth, of adenovirus transformed cells, thyroid hormone and, 206

Anchorage transformation, portion of SV40 DNA and, 300-301
Anthralin, skin tumor promotion and, 126
Antibodies
  to cytidine
    AAF-modified polymers of dG and dC, 30
    AF-modified DNA and, 21-22
  diversity, DNA deletions and, 342
  monoclonal, to ATL virus, 216-217
  natural human, to ATL virus, 220-221
  against pp 60^{src} using antigen produced in prokaryotes, 312-313
  to v-ras gene product, detection of human transforming gene product and, 288-290
  to Z-DNA, Drosophila melanogaster polytene chromosomes and, 23-24
Anticarcinoma activity in vivo, of rhodamine-123 and combination therapy with 2-deoxyglucose, 380-382
Antigens
  ATL virus-associated, 215-216
  coded for by SV40 DNA, 295
Anti-immunoglobulin M, uptake by cells, 186
Aplysiatoxin
  biological effects of, 117, 119-120
  identification of, 119
  structure, comparison to TPA, 102-104
  tumor promotion and, 100, 101-103
Arachidonate
  inhibition of tumor promotion and, 129
  release from lymphocytes by PMA, 63-64
Arachidonate cyclooxygenase, inhibition of, 58, 62
Arachidonate lipoxygenase, inhibitors of, 58, 62
Arochlor 1254, tumor promotion and, 157, 158
Aromatic hydrocarbons, halogenated, epidermal changes produced by, 146-159
Aryl hydrocarbon hydroxylase, induction of, 144, 146, 152
Arylhydroxylamines
  esterification of, 89-90
  formation from nitroarenes, 88, 90-91
  reaction with DNA, 89

INDEX

Asialoglycoproteins, uptake by cells, 186
Autoimmune disease, Z-DNA antibodies and, 24
Avian leukosis virus
   activation of c-*myc* gene by, 228–229
   MC29 virus and, 226
   activation of c-*myc* and, 256
   mechanism of tumor formation, 242,
Avian sarcoma virus
   phosphorylation induced by, comparison to EGF-stimulated phosphorylation, 314–319
   protein kinases produced by, 309–310
5-Azacytidine, expression of poly(A)$^+$ RNAs and, 257

B

Bacillus Calmette-Guerin, vaccination, inhibition of tumor promotion and, 129
Bacteriophage λ, use in cloning human transforming genes, 287–288
BBN, *see* Butyl (4-hydroxybutyl) nitrosamine
B-cells, differentiation, c-*myc* expression in, 227
Beef, broiled, new mutagen from, 113
Benzo[a]pyrenes
   metabolites, activity of, 104
   promoterlike activity of, 104
   transformed bladder cells, retention of rhodamine-123 by, 370
Benzo(*e*)pyrene, skin tumor promotion and, 126
Benzoyl peroxide, skin tumor promotion and, 126
Bgl I restriction site, linkage to human c-*mos*, 237
Bladder carcinoma cells, human, retention of rhodamine-123 by, 371–372
Bladder epithelial cells, rhodamine-123 accumulation by, 370
Bloom's syndrome, nuclear matrix and, 347
Breast cancer, thyroid hormone and, 198
Bromodeoxyuridine, expression of poly (A)$^+$ RNAs and, 257
Butylated hydroxyanisole, inhibition of tumor promotion and, 129

Butylated hydroxytoluene, inhibition of tumor promotion and, 129
Butyl (4-hydroxybutyl) nitrosamine, bladder transitional cell carcinoma, retention of rhodamine-123 by, 370
Butyrate, inhibition of tumor promotion and, 128

C

Calcium ionophore A23187, as first-stage promoter, 135
Cancer, *see also* Tumors
   DNA rearrangement in, 339–341
   inherited disorders and, 341
   initiation of, 41
   predisposition to, genetic disorders and, 340–341
   reversibility of, 341
Carcinogen(s), *see also specific compounds; viruses*
   amplification of integrated SV40 sequences and, 282
   chemical, activation of, 13
   nuclear matrix and, 347
   other than retroviruses, 10–11
Carcinogenesis
   DNA methylation and, 343–344
   multistage, 10, 97–98
   steps in, 98
Carcinogenicity
   of new mutagen produced by pyrolysis, 114
   of nitroarenes, 91, 92
   in vivo, of dihydroteleocidin B, teleocidin and lyngbyatoxin, 118–119
Carcinoma cells
   clonal growth *in vitro*, selective reduction by rhodamine-123, 377–379
   measurement of rhodamine-123 uptake by, 382–384
   selective toxicity of rhodamine-123 *in vitro*, 375–377
   sensitivity to adriamycin, 377
   transforming genes in, 285
Carcinoma-derived cells,
   prolonged retention of rhodamine-123 by, 370–375
Carriers, healthy, of ATL virus, 219–220

Cell(s)
  giant multinuclear, ATL virus and, 216
  normal and transformed, study of onc genes in, 254–255
  phenotype, transformation and, 249
  retrovirus-transformed, EGF binding by, 166–168
  transformation, nitroarenes and, 91
  transformd by carcinogens or radiation, poly(A)$^+$ RNA fraction, LTR probe and, 256–260
  type, development of MTX resistance in, 274, 276
Cell culture
  modulation of neoplastic transformation in, by thyroid hormone, 199–204
  transformation by adenovirus, thyroid hormone and, 204–206
Cell cycle, regulation of DHFR during, 277
Cell division, oncogenes and, 248–249
Cell lines
  carrying ATL virus, 218–219
  established, rhodamine-123 retention by, 372, 374
  normal, transformed or tumor, absence of A^x-actin in, 355
Cell surface, see also Membrane
  morphological structures and endocytosis
    coated pit mediated, 185–186
    electron microscope experiments, 188–189
    fluid phase, 183–185
    fluorescence experiments, 186–187
    formation and structure of receptosomes, 189–191
Cellular transcripts
  regulation in cells transformed by a variety of agents, 328–329
  regulation in SV40 transformed cells, 324–328
  regulation of levels after serum stimulation of nontransformed cells, 331–333
c-erb gene, activation by ALV-induced erythroleukemias, 229
Chromatin, structure, DNA function and, 344

Chromosomes, tumor, DNA rearrangement in, 340
Chromosome effects, tumor promoters and, 105–106
Clastogenic factor
  production by PMA in lymphocytes, 60–61
  release in certain diseases, 59
Clathrin
  antibody to, coated pits and, 190
  coated pits and, 186
c-mos gene
  in cells transformed by chemicals or radiation, methylation of, 255
  human, activation of mouse cells, 234–240
  human and mouse, direct DNA sequence comparison, 234, 235
c-myc gene
  activation by ALV, 228–229
  activation, avian leukosis virus and, 228–229, 256
  amplification in human promyelocytic leukemic cell line HL-60, 246–248
  coding sequences of, 228
  expression in human leukemia and lymphoma, 244–246
  insertion of viral promoter adjacent to, 243
  occurrence of, 227
Coated pits
  endocytosis and, 185–186
    physiological significance of, 193–194
  entry of toxins and viruses into cells and, 193
  structure of, 186
Coding sequences, of A^x, β- and γ-actin gene, relatedness of, 355–357
Codons, differences between humos and mumos, 239
Comformation, of AF modified DNA, 20–21
c-ras gene, in cells transformed by carcinogens or radiation, 255
c-rel gene, composition of, 5
Croton oil, skin tumor promotion and, 126
c-src gene, discovery and occurrence of, 241

Cutaneous T-cell lymphoma, antibody to ATL virus and, 220–221
Cyanine dyes, uses of, 369
Cyclic nucleotides, inhibition of tumor promotion and, 128, 129
Cycloheximide, x-ray induced transformation and, 203, 204
Cystine residues, of TGFs, 175–176
Cytoplasm, damage by rhodamine-123, 385
Cytosine arabinoside, toxicity for carcinoma cells, 379, 381
Cytoskeleton
 incorporation of mutant β-actin into, 364
 of transformed fibroblasts, 365

D

Dark basal keratinocytes, skin tumor promotion and, 127, 128, 135, 136–137
d(CpGpCpGpCpG) hexamer, conformation of, 23
Debromoaplysiatoxin
 biological effects of, 117, 119–120
 identification of, 119
Decanoyl peroxide, skin tumor production and, 126
Density gradient centrifugation, of ATL virus, 215
2-Deoxyglucose, combination therapy with rhodamine-123, 380–382
3-(Deoxyguanosin-N^2-yl)AAF, formation of, 14, 70
N-(Deoxyguanosin-8-yl)AAF, formation of, 14, 70
Deoxyribonucleic acid
 alpha sequence
  preparation of, 43
  location of pyrimidine dimers in, 43–45
 ATL proviral, detection of, 219
 carcinogenic electrophiles and, 13
 of class II mutants, sequence analysis of, 75–76
 copy
  nature of, 4
  RNAs in SV40 transformed cells, 324–325
 RNAs in cells transformed by a variety of agents, 328–329
 serum stimulation of nontransformed cells and, 331–333
 direct sequence comparison, of human and mouse c-mos genes, 234–235
 function, chromatin structure and, 344
 N-hydroxyacetylaminofluorene adducts, 14, 20–22
 inhibition of replication, gene amplification and, 276, 277, 280
 mutant, sequence analysis of, 71
 proviral, hypermethylation of, 167
 reaction with metabolites of new mutagen obtained by pyrolysis, 114
 reaction with N-acetoxy AAF, 19
 reaction with arylhydroxylamines, 89
 rearrangement in cancer, 339–341
 rearrangement in normal development and expression of oncogene, 342–344
 repair disorders of, cancer and, 340
 saltatory replication, gene amplification and, 272
 strand breakage, TPA and, 106
 synthesis, addition of fresh fetal calf serum to cultures and, 332
 UV-irradiated, 42
 alkali labile sites in, 45
 sites of pyrimidine-pyrimidone (6-4) photoproducts in, 50–52
Dexamethasone, antipromotional effect of, 58
2,7-Dichlorodibenzo-p-dioxin, tumor promotion and, 157, 158
Diesel engine, nitroarene emissions and, 92
Diethylnitrosamine, carcinogenesis and, TCDD and, 145
Differentiation, oncogenes and, 249–250
α-Difluoromethylornithine, inhibition of tumor promotion and, 129, 136–138
Dihydrofolate reductase
 gene amplification
  hydroxyurea and, 277
  under nonselecting conditions, 277–279
  TPA and, 282
 regulation during cell cycle, 277

Dihydroteleocidin B
  biological effects of, 116–118, 120
  preparation of, 116
  skin tumor promotion and, 126
  in vivo carcinogenicity tests with, 118–119
Dimethylbenzanthracene
  transformed bladder cells, retention of rhodamine-123 by, 370
  TCDD and, 146
  tumor initiation by, 377
    tumor promoters and, 118–119, 155–159
Dimethylsulfoxide, inhibition of tumor promotion and, 128
Diphenylhydramine, inhibition of tumor promotion and, 129
Disulfiram, inhibition of tumor promotion and, 129
Double minute chromosomes
  formation of, 272
  in nuclei of solid tumors, 281
  redistribution of, 279
Drosophila melanogaster, polytene chromosomes, Z-DNA and, 24
Drug metabolizing enzymes, polycyclic aromatic hydrocarbons and, 105
dTpdC, UV photoproduct, properties of, 48–50

E

Eco RI restriction site, of Mo-MSV, linkage to human *c-mos*, 236–237
Ehrlich ascites tumor, uptake of rhodamine-123 by, 383
5,8,11,14-Eicosatetraynoic acid, clastogenicity of PMA and, 62
Electron microscopic experiments, endocytosis and, 188–189
Embryogenesis, phorboid receptors and, 101
Endocytosis
  coated pit-mediated, physiological significance of, 193–194
  morphological structures on cell surface
    coated pit-mediated, 185–186
    electron microscope studies, 188–189
  fluid phase, 183–185
  fluorescence studies, 186–187
  formation and structure of receptosomes, 189–191
  pathological processes, toxins and viruses, 193
Endonuclease, repair of pyrimidine dimers and, 43
env gene,
  probes for, in carcinogen transformed cell, 259
Enzymes, lysosomal, uptake by cells, 186
Epidermal growth factor
  binding
    polycyclic aromatic hydrocarbons and, 104
    by retrovirus-transformed and human tumor cells, 166–168
  cell transformation and, 99
  conjugated to horseradish peroxidase, labeling of receptosomes by, 191
  phosphorylation stimulated by, comparison to ASV-induced phosphorylation, 314–319
  purified transforming growth factor and, 170–175
  receptor, tyrosine phosphorylation of, 176–180
  uptake by cells, 185, 189
Epidermis
  changes produced by halogenated aromatic hydrocarbons in, 146–159
  newborn, proteins of, 130
Epithelial cells, carcinogen-transformed, rhodamine-123 retention by, 370–375
Erythroblasts, failure to be transformed by acute leukemia virus, 250
Escherichia coli
  expression of RSV *src* gene in, 306–310, 312–313
  irradiation prior to transformation, 71
  *lac*I gene, pyrimidine dimers, alkali-labile regions and mutations in after UV irradiation, 45–48
  mutation frequency and restriction enzyme analysis of mutants, 73–75
  transformation and selection of mutants, 71

Ethanol, Z-DNA spectrum and, 26–27
Ethylphenylpropiolate
 hyperplasia and, 128
 tumor promotion and, 130–133, 135
Exocytosis, model for, 192

F

F-actin, formation, mutated β-actin and, 365
FCCP, see para-Trifluromethoxyphenylhydrazone
Feline sarcoma virus, EGF binding and, 166–167
Fetal bovine serum, removal of thyroid hormones from, 202
Fibroblasts
 malignant, variant form of actin in, 352–355
 transformed, exposure to UV light, 360–362
Firemaster-FF-1, tumor promotion by, 157, 158
Fluid phase, endocytosis, morphological structures of cell surface and, 183–185
Fluocinolone acetonide
 antipromotional effect of, 58
 inhibition of tumor promotion and, 128, 136–138
Fluorescence Activated Cell Sorter, determination of number of DHFR gene copies per cell and, 278
Fluorescence experiments, endocytosis and, 186–187
Foods, proteinaceous, new mutagens from pyrolysates of, 112–114
Free radicals, tumor promotion and, 130

G

gag gene, probes for, in carcinogen transformed cells, 259
β-Galactosidase, uptake by cells, 186, 192
Gene(s)
 for actins, number per genome, 359
 of normal cells, activation into transforming genes, 290
 SV40, transformation and, 297–299
 transposition, LTR sequences and, 256
Gene amplification
 development of resistance to cancer chemotherapeutic agents and, 280
 DHFR gene
  hydroxyurea and, 277
  under nonselecting conditions, 277–279
  TPA and, 282
 examples of, 270
 malignancy and, 281
 mechanism of, 270–272
 retroviruses and, 281
Genetic disorders, predisposition to cancer and, 340–341
Genetic elements, movable, nature of, 3
Genomic changes, carcinogenesis and, 98
Genotoxicity, of nitroarenes, 87
α-Globin gene, recombination with retrovirus, subsequent loss of sequences, 9
Globulin, soybean, new mutagens by pyrolysis of, 113
Glucocorticoid hormones, tumor promotion and, 105
Glutamic acid, pyrolysis, new mutagens from, 112–113
Gnidilatin, phorbol ester binding and, 100
Gnilatimacrin, phorbol ester binding and, 100
Golgi system
 coated pits and, 186
 compartmentalization within, 192–193
 exocytosis and processing of VSV G protein, 192
 receptosomes and, 187, 190, 191–192
Growth state, synthesis of RNA complementary to SV40 cDNA and, 332–333
Guanine, reaction with nitroarene metabolites, 86
Guanosine
 deletion
  in mutants treated with N-AcO-AAF, 76
  probability of same residue being affected in a number of mutants, 78

modified at C-8 with AAF, conformation of, 14–18
modified at C-8 with AF, 20–22
modified at N^2 with AAF, 18–20

H

Hairless mice, epidermal changes in, TCDD and, 146–149
Hamburger, cooked, mutagens in, 112
Hamster embryo cells, neoplastic transformation of, 199
Hematopoietic tissues, c-*myc* gene expression and, 227
Hepatomas
  growth, thyroid ablation and, 199
  induction of, thyroidectomy and, 198
2,4,5,2',4',5'-Hexabromobiphenyl, tumor promotion and, 157, 158
3,3',4,4',5,5'-Hexabromobiphenyl
  epidermal response in hairless mice, 149
  tumor promotion and, 157, 158
Histamine, inhibition of tumor promotion and, 129
Homogeneously staining regions
  gene amplification and, 272
  in nuclei of solid tumors, 281
*Hpa*I restriction site, linkage to *Sma*I or *Kpn*I restriction sites, 236
hr locus
  Ah locus and, 146–149
  interaction with *Ah* locus, 149, 152–153
HRS/J mice
  response to TCDD, 149
  tumor promotion by TCDD in, 153–159
Hydrogen peroxide, as first-stage promoter, 135
Hydroperoxy arachidonate, as clastogen, 62
N-Hydroxyacetylaminofluorene, DNA adducts, 14, 20
Hydroxylaminoarenes, as ultimate mutagens, 88
4-Hydroxy-2,3-*trans*-nonenal, as clastogen, 62
Hydroxyurea, development of resistance to MTX and, 272, 275, 276–277

Hyperplasia, epidermal, tumor promoters and, 127–128

I

Image intensification, fluorescence experiments on endocytosis and, 187
Imidazol, clastogenicity of PMA and, 62
Immunofluorescence staining, of transformed fibroblasts, cytoskeleton and, 365
Immunoglobulins, maternal, uptake by cells, 186
Introns, cellular oncogenes and, 255
5'-Iodo-2-deoxyuridine, induction of ATL virus and, 215, 216
Indomethacin
  as antipromotor, 58
  clastogenicity of PMA and, 62
Inherited disorders, cancer and, 341
Insulin, uptake by cells, 185
Interferon, uptake by cells, 186, 194
Ionophores, vesicle formation and, 191
Isobutylmethylxanthine, inhibition of tumor promotion and, 128–129

K

Keratin
  changes in papillomas and squamous cell carcinomas, 138–140
  modifications, tumor promoters and, 130
*Kpn*I restriction site
  linkage to human c-*mos*, 236
  linkage to *Sma*I or *Kpn*I restriction site, 236

L

Lauroyl peroxide, skin tumor promotion and, 126
Lectins, uptake by cells, 186
Leukemia, treatment with rhodamine-123, 381–382
Leukemic cells
  human oncogene expression in, 244–246
  detection of ATL virus in, 219
Leukemic cell line HL-60, amplification of c-*myc* DNA sequences in, 246–248

Lipid-hydroperoxides, damaging effects of, 56–58
Long terminal repeats
    cellular oncogenes and, 255
    retrovirus, function of, 256, 343
    U3 region, in carcinogen transformed cells, 259–260
    viral promoter region and, 242, 253
Long terminal repeat-like sequences, poly(A)$^+$-RNA containing, studies on expression of, 255–260
Low density lipoproteins, uptake by cells, 185, 189
Lung carcinoma, treatment with rhodamine-123, 381–382
Lymphocytes
    human, transformation by ATL virus, 217
    production of clastogenic factor in by PMA, 60–61
    release of arachidonate from by PMA, 63–64
Lymphomas
    B-cell, integration of provirus in, 228
    levels of c-*myc* mRNAs in, 228
    transforming genes and, 285
Lyngbyatoxin A
    biological effects of, 116–118
    isolation of, 116
    in vivo carcinogenicity tests with, 118–119
Lysine, pyrolysis, new mutagens from, 112
Lysosomes
    pinocytic vesicles and, 185
    receptosomes and, 187, 191–192

M

α_2-Macroglobulin, uptake by cells, 185, 189
Macromolecules, uptake, bristle-coated pits and, 185–186
Magnesium ions, Z-type transformation and, 33–34
Maize, gene rearrangement in, 342
Malignancy
    gene amplification and, 281
    incremental increases in associated with increased abnormality of mutated β-actin, 363–364

Malondialdehyde, as clastogen, 62
Mammalian cells
    cultured, mutagenic nitroarenes in, 86
    metabolism of nitroarenes in, 90–91
MC 29 virus
    derivation of, 226
    neoplasms induced by, 226
Melanoma, treatment with rhodamine-123, 381–382
Membrane, *see also* Plasma membrane
    ATL antigen and, 216
    phorboid receptors in, 99–100
    receptors for PMA in, 59–60
Membrane transport, defects, drug resistance and, 280
Methotrexate
    development of resistance to, agents enhancing, 272
    development of resistance to MTX and, 272, 275, 277
    fluorescent analog, determination of number of DHFR gene copies per cell and, 278
    inhibition of carcinoma cells by, 379
    mechanisms for resistance to, 274
Methotrexate resistance
    pretreatment regimens and enhanced frequencies of, 276–277
    single step selection for, 272–276
N-Methyl-N'-nitro-N-nitrosoguanidine, tumor initiation by, tumor promoters and, 156–157
Mezerein, tumor promotion and, 128, 130
    multistage, 133–138
Microfilaments, mutated β-actin and, 365
Microsomal enzymes, activation of new mutagens obtained by pyrolysis by, 113–114
Microtubules, movement of intracellular organelles and, 185
Mitochondria
    membrane potential, rhodamine-123 uptake and, 383–384
    respiratory control ratio, rhodamine-123 and, 384–385
Moloney murine leukemia virus, helper virus, transduction with, 262–264

Moloney murine sarcoma virus
  development of expression and transduction vector from, 260–261
  portions, recombination with human c-*mos*, 234, 236–237
Mouse ascites tumor, double minute chromosomes in, 281
Mouse embryo cells, neoplastic transformation of, 199
Mutations
  frequency at TC and CT sequences, 52
  hot spots for, 76, 79–80
  multiple possibilities for, 351
Mycosis fungoides, ATL virus and, 220
Myelocytomatosis, isolation of MC29 virus and, 226

N

Neoplastic transformation, *see also* Transformation
  hypothesis on role of actin mutation in, 364–366
  modulation by thyroid hormone in cell culture, 199–204
  thyroid dependence of, historical background, 198–199
Neuroblastoma, transforming genes in, 285
Nitroarenes
  activity, metabolism to ultimate mutagens, 86
  carcinogenicity of, 91, 92
  cell transformation and, 91
  diesel engines and, 92
  genotoxic effects of, 87
  metabolism of, 87–88, 90–91
  mutagenicity
    in cultured mammalian cells, 86
    in *Salmonella typhimurium*, 85
6-Nitrobenzo(a)pyrene, mutagenicity and metabolism of, 91
1-Nitropyrene, occurrence in atmosphere, 84
4-Nitroquinoline-1-oxide, carcinogenesis by, 352
Nitroreductases, metabolism of nitroarenes and, 87–88
Nitrosoarenes
  dismutation to nitrenium ion, 88
  as ultimate mutagens, 88

Nuclear matrix
  DNA replication and transcription and, 344–345, 347
  of tumors, 346–347
Nucleus
  receptors for T_3 in normal and transformed cells, 206–207
  structure and function, 344–347

O

Oncogene(s)
  cellular
    differences from viral homologs, 255
    in normal and transformed murine cells, 254–255
  evolution and, 248
  expression,
    in human leukemic cells, 244–246
    reactivation of, 343
  function of, 248–250
  presence in RSV, 242
  retroviral
    reduced EGF and, 167–168
    relationship to human transforming genes, 288–290
    tyrosine-specific protein kinase activity and, 176–177
  viral, derivation of, 5
Oncogenesis, promoter insertion mechanism of, 242–244
  amplification of c-*myc* DNA sequences in promyelocytic leukemic cell line HL-60, 246–248
  oncogene expression in human leukemic cells, 244–246
Ornithine, pyrolysis, new mutagens from, 112
Ornithine decarboxylase, induction by tumor promoters, 101, 115, 127, 128, 133, 137–138
Osteosarcoma, transforming genes in, 285
Oxygen, activated, tumor promoters and, 106

P

Papillomas
  changes in proteins of, 138–140
  promotion of, 155–157

Parahydroxyanisole, inhibition of tumor promotion and, 129
Pathobiology, human, active oxygen and, 58–59
pH, of receptosome, 190
Phenylalanine, pyrolysis, new mutagens from, 112
Phenylalanine chloromethylketone, inhibition of tumor promotion and, 128, 129, 136–138
Phorbol dibutyrate
  binding, inhibition by serum factor, 100–101
  membrane receptors for, 99–100
Phorbol esters, skin tumor promotion and, 126, 127, 130–133
Phorbol-myristate-acetate
  antipromotors and, 58
  clastogenic action, membrane mediation of, 59–64
  effects of, 55
  release of arachidonate from lymphocytes by, 63–64
Phospholipase A_2, inhibitors of, 58
Phosphoprotein phosphatases, tyrosine-specific, 319
Phosphorylation, EGF-stimulated, comparison to ASV-induced, 314–319
Photoreactivation, UV damage and, 42
Pinocytosis, cell surface structures and, 184
Piranhalysis, lysosomes and, 185
Plasma membrane, *see also* Membranes
  labeling with ruthenium red and concanavalin A-peroxidase, coated pits and, 190–191
  receptosome and, 189, 190
Point mutations, carcinogenic agents and, 340
pol gene, probes for, in carcinogen transformed cells, 259
Polyamines, tumor promotion and, 129, 137–138
Poly(A)⁺RNA-containing LTR-like sequences, studies on expression of, 255–260
Polycyclic aromatic hydrocarbons
  promoterlike activity of, 104–105
  structures of, 84
Poly(dG)-poly(dC), AAF and, 27

Poly(dG-dC)-poly(dG-dC), modified with AAF, Z-type conformation in, 25–31
Poly(dG-m⁵dC)-poly(dG-m⁵dC), AAF-modified, adoption of Z-type conformation in, 31–36
Polyoma virus
  transformation antigen, nuclear matrix and, 347
Polypeptides
  of ATL virus, 216
  monoclonal antibodies and, 217
  of nuclear matrix, 345
Polyriboinosinic:polyribocytidylic acid, inhibition of tumor promotion and, 129
Promoters, of carcinogenesis, types of agents causing, 56
Promoter sequence, transposable, gene activation and, 343
Promyelocytic leukemia cells
  c-myc expression in, 227–228
  transforming genes in, 285
  tumor promoters and, 115–116
Prostaglandin(s), inhibitors of synthesis, inhibition of tumor promotion and, 129
Prostaglandin G_{21} as clastogen, 62
Protease inhibitors, inhibition of tumor promotion and, 128, 129, 136–138
Protein(s)
  capsid, synthesis in SV40 infection, 297
  changes in papillomas and squamous cell carcinomas, 138–140
  critical changes in during skin tumor promotion, 130–133
  pyrolysates, new mutagens from, 112–114
  synthesis
    neoplastic transformation and, 203–204
    thyroid hormone and, 203
Protein A^x, isolation from malignant fibroblasts, 352–353
Protein kinase
  as expression of RSV *src* gene, 306–308
  search for other substrates, 320
  substrate, distribution in various tissues, 317–318

tyrosine-specific, transforming growth factor production and, 169
Proton pump, coated pit membrane and, 191
Proto-oncogenes, descendants of, 5
Proviruses, defective, c-*myc* activation and, 229
Pseudomonas toxin, uptake by cells, 186, 193
*PVU*II restriction site, linkage to *Sma*I or *Kpn*I restriction sites, 236
Pyrimidine dimers, ultraviolet light and, 42
Pyrimidine dimer *N*-glycosylase, 43
Pyrimidine-pyrimidone (6-4) photoproducts, formation at different sites in DNA, 50–52

R

Rat embryo cells, transformation by adenovirus, 204
Receptor
 for EGF, phosphorylation of, 176–180
 for halogenated aromatic hydrocarbons, 152–153
 phorboid, cell membranes and, 99–100
 receptosomes and, 189
 for TGF, 179
Receptosome
 formation and structure of, 189–191
 intracellular fate of, 191–192
 labeling of, 187
 movement of, 187
Recombinants, miniplasmid πmos-hu inserted in homologous v-*mos* region, 237–239
Respiratory control ratio, mitochondrial, rhodamine-123 and, 384–385
Restriction endonuclease, analysis of human transforming genes and, 285–287
Restriction sites, loss of, 73
Reticuloendotheliosis virus
 recombination with c-*rel*, 9
 strain T, cells transformed by, 4
Reticulosarcoma, thyroid dependence of, 199
Retinoic acid
 c-*myc* expression and, 228

 inhibition of tumor production and, 128, 136–137
Retrovirus
 cells transformed by, EGF binding and, 166–168
 characteristics of ATLV, 214–215
 gene amplification and, 281
 highly oncogenic, evolution of, 5–10
 LTRs and, 242–243, 253
 weakly oncogenic, 11
Reverse transcriptase, ATL virus and, 215
Rhodamine dyes, uses of, 369
Rhodamine-123
 anticarcinoma activity and combination therapy with 2-deoxyglucose, 380–382
 basis for toxicity, 384–386
 prolonged retention by carcingen-transformed epithelial cells and carcinoma-derived cells, 370–375
 selective reduction of clonal growth of carcinoma cells *in vitro*, 377–379
 selective toxicity in carcinoma cells *in vitro*, 375–377
 uptake, mitochondrial membrane potential and, 383–384
 uptake by carcinoma cells, measurement of, 382–384
Rhodamine-6G, oxidative phosphorylation and, 384
Ribonucleic acids
 association with nuclear matrix, 345
 of ATL virus, 215
 of cells transformed by a variety of agents, regulation of, 328–329
 c-*rel*, occurrence of, 9
 from leukemic white cells, hybridization with v-*myc* or c-*myc* specific sequences, 245–246
 messengers
  abundance in SV40 transformed cells, 324–325
  for A^x-actin, 355
  comparison of sizes of A^x- and β-actin, 357, 359
  for doubly mutated β-actin, 361
  stimulation of cellular levels after SV40 infection, 330–331
 messenger for c-*myc* gene, lymphomas and, 228

regulation of synthesis in SV40 infection, 297
Ribonucleotide reductase, hydroxyurea and, 276
Ribosomal protein S6, phosphorylation of, in presence of either heated or unheated pp60^{src}, 320
Rous sarcoma virus
cells transformed by, 4
genome of, 305–306
mechanism of tumor formation, 242
presence of oncogene in, 242
src gene, expression in E. coli, 306–310
src gene product, autophosphorylation of, 310–312
transformation by, vinculin phosphorylation and, 365

S

Salmonella typhimurium
deficient in nitroreductase, nitroarenes and, 87–88
activity of new mutagens obtained by pyrolysis in, 113
mutagenicity of nitroarenes for, 85
strain resistant to 1,8-dinitropyrene and 2-nitrofluorene, 89–90
Saltatory motion, of intracellular organelles, 185
Sardines, broiled, mutagens in, 112, 113
Serum factor, phorbol dibutyrate binding and, 100–101
Sézary syndrome, ATL virus and, 220
Simian virus 40
antigens produced by, 295
genes, transformation and, 297–299
infection, stimulation of cellular mRNA levels after, 330–331
integrated, amplification by carcinogens, 282
regulation of cellular transcripts in cells transformed by, 324–328
Sister chromatid exchange
carcinogenic agents and, 340, 347
tumor promoters and, 105
Skin
irritation, tumor promoters and, 115
morphological and biochemical responses to tumor promoters, 127

Skin tumors
initiation stage, 125
promotion, 126–128
critical protein changes during, 130–133
inhibition of, 128–130
multistage, 133–138
SmaI restriction site, linkage to human c-mos, 236
S, nuclease
AAF-modified DNA and, 17–18
C-8 position of guanosine and, 17–18
N^2 position of guanosine and, 19–20
AAF-modified poly(dG-dC)-poly(dG-dC) and, 28
AAF-modified poly(dG-m^{5}C)-poly(dG-m^{5}C) and, 34
AF modified DNA and, 21
Squamous cell carcinomas, protein changes in, 138–140
src gene, of RSV, product of, 306
Steric hindrance, of guanosine modified at C-8 with AAF, 15–16, 18
Steroid(s), anti-inflammatory, inhibition of skin tumor promotion and, 128, 129
Steroid hormones, nuclear matrix and, 345
Streptomyces mediocidicus, teleocidin from, 116
Superoxide dismutase
anticlastogenic activity of, 61–62
antipromotor activity of, 58, 59
Sup F gene, use as marker in isolation of human transforming genes, 287
Survival rate, of hr/hr mice, TCDD and, 159
Swimmer's itch, causative agent, 119

T

T antigen
of SV40, functions of, 297, 298–299
of SV40 mutant, with spliced hydrophobic C terminal, 299–300
Teleocidin
biological effects of, 116–118, 120
isolation of, 116
structure, comparison to TPA, 102–104
tumor promotion and, 100, 101–103

in vivo carcinogenicity tests with, 118–119
2,3,7,8-Tetrachlorodibenzo-*p*-dioxin
  binding of, 144, 146
  carcingenicity studies on, 145–146
  effects of lethal dose, 145
  local and systemic effects, 157
  role of *Ah* locus in determining histological response to, 149–153
  tumor multiplicity and, 159
  tumor promotion in HRS/J mice by, 153–159
2,3,7,8-Tetrachlorodibenzofuran, tumor promotion by, 157, 158
Tetracycline, selection of mutants resistant to, 71
  nature of mutants, 79
12-*O*-Tetradecanoyl phorbol-13-acetate
  cell culture effects, 99
  DHFR gene amplification and, 282
  local and systemic effects, 157
  protein synthesized in presence of, 120
  structural comparison to teleocidin and aplysiatoxin, 102–104
  tumor multiplicity and, 157, 159
  tumor promotion by, 130–140, 155–159
Tetramethylrhodamine, endocytosis studies and, 186
Thromboxane, inhibitors of synthesis, inhibition of tumor promotion and, 129
Thymidine, incorporation, hydroxyurea and, 276
Thymidine kinase, levels in cells, growth phase and, 331
Thy(6-4)Pyo, structure and postulated intermediates, 48, 49
Thyroid hormone
  adenovirus transformation of cells in culture and, 204–206
  modulation of neoplastic transformation of cells in culture and, 199–204
  neoplastic transformation and, historical background, 198–199
  removal from fetal bovine serum, 202
Transduction, with Mo-MuLV helper virus, 262–264

Transferrin, uptake by cells, 186
Transformation, *see also* Neoplastic transformation
  cell phenotypes and, 249
  expression of viral oncogenes and, 250
  SV40 genes and, 297–299
Transforming genes
  human
    detection of, 285–287
    relationship to known retroviral oncogenes, 288–290
    isolation from human sources, 287–288
Transforming growth factor
  production of, 168–169
  properties of, 169
  rat and human, amino acid sequences of, 175–176
  rat, mouse and human, purification of, 170–175
Transmission, of ATL virus, 219–220
p-Trifluromethoxyphenylhydrazone, 383–384
Triiodothyronine
  concentration in culture medium, x-ray induced transformation and, 203
  entry into cells, 194
  nuclear receptors in normal and transformed cells, 206–207
Tropomyosin, alterations in transformed cells, 365
Tryptic peptide patterns, of β- and γ-actin and protein A^x from malignant fibroblasts, 353–355
Tryptophan, mutagens formed from by pyrolysis, 112
Tumors, *see also* Cancer
  multiplicity, promoters and, 157, 159
  nuclei of, 346–347
  solid, evidence of gene amplification in, 281
Tumor cells
  human, EGF binding by, 166–168
  transforming growth factor and, 169
Tumor initiators,
  characteristics of, 98
Tumor lines, human, rhodamine-123 retention by, 373
Tumor promoters
  characteristics of, 98

chromosomal effects and activated oxygen, 105–106
mechanism of action, 99–104
new, naturally occurring potent compounds, 114–116
screening for, 115
Tyrosine
of EGF receptor, phosphorylation of, 176–180
RSV protein kinase and, 308

U

Ultraviolet light
development of resistance to MTX and, 272, 275, 277, 280
DNA and, 42
Uridine, cellular RNA of SV40 infected cells labeled with, hybridization to cDNA, 330–331

V

Vector, of expression and transduction, development of, 260–261
Vesicular stomatitis virus, G protein, exocytosis and processing of, 192
Vincristine, development of resistance to, 280
Vinculin, tyrosine phosphorylation in, RSV transformation and, 365
Viral promoter, long terminal repeat and, 242
Viral transforming gene products, normal cellular homologues of, 313–314
Viruses
tumor, DNA rearrangement and, 340
uptake by cells, 186, 193
Vitamin A, derivatives, inhibition of tumor promotion and, 128, 136–137

v-*mos* gene, as probe of cells transformed by chemicals or radiation, 254–255
v-*myc* gene, identification of, 226
v-*ras* gene
origin of, 288
product, monoclonal antibodies and, 288–290
relationship to human transforming genes, 288
v-*rel* gene, composition of, 5

X

Xanthine guanine phosphoribosyl transferase, cloned, development of expression and transduction vector and, 260–261
Xanthine oxidase, 1-nitropyrene and, 90
X-rays
induction of neoplastic transformation and, 199
therapeutic use of, 281

Y

Yeast, change in phenotype, 341
Yolk proteins, uptake by cells, 186, 189

Z

Z-DNA, conformation
in alternating purine-pyrimidine polymers, 23–25
induction of Z-type conformation in poly(dG-dC)-poly(dG-dC) modified with AAF, 25–31
Z-type conformation in AAF modified poly(dG-m^{5}dC)-poly(dG-m^{5}dC), 31–36